T0325110

Philosophy of Science for Biologists

Biologists rely on theories, apply models, and construct explanations, but rarely reflect on their nature and structure. This book introduces key topics in philosophy of science to provide the required philosophical background for this kind of reflection, which is an important part of all aspects of research and communication in biology. It concisely and accessibly addresses fundamental questions such as the following: Why should biologists care about philosophy of science? How do concepts contribute to scientific advancement? What is the nature of scientific controversies in the biological sciences? Chapters draw on contemporary examples and case studies from across biology, making the discussion relevant and insightful. Written for researchers and advanced undergraduate and graduate students across the life sciences, its aim is to encourage readers to become more philosophically minded and informed to enable better scientific practice. It is also an interesting and pertinent read for philosophers of science.

KOSTAS KAMPOURAKIS is the author and editor of books about evolution, genetics, philosophy, and history of science, and the editor of the Cambridge University Press book series *Understanding Life*. He is a former editor-in-chief of the journal *Science & Education*, and the book series *Science: Philosophy, History, and Education*. He is currently a researcher at the University of Geneva, where he also teaches at the Section of Biology and the University Institute for Teacher Education (http://kampourakis.com).

TOBIAS ULLER is Professor of Evolutionary Biology at Lund University, Sweden. He works on the relationships between development, heredity, and evolution, using an integrative approach guided by mathematical modeling and conceptual analysis. He has held fellowships in the United Kingdom, United States, and Sweden, and was the 2018 recipient of the Tage Erlander Prize, awarded by the Royal Swedish Academy of Sciences for research in Natural Sciences and Technology.

Philosophy of Science for Biologists

Edited by

KOSTAS KAMPOURAKIS
University of Geneva

TOBIAS ULLER
Lund University

CAMBRIDGE
UNIVERSITY PRESS

CAMBRIDGE
UNIVERSITY PRESS

Shaftesbury Road, Cambridge CB2 8EA, United Kingdom

One Liberty Plaza, 20th Floor, New York, NY 10006, USA

477 Williamstown Road, Port Melbourne, VIC 3207, Australia

314–321, 3rd Floor, Plot 3, Splendor Forum, Jasola District Centre, New Delhi – 110025, India

103 Penang Road, #05–06/07, Visioncrest Commercial, Singapore 238467

Cambridge University Press is part of Cambridge University Press & Assessment, a department of the University of Cambridge.

We share the University's mission to contribute to society through the pursuit of education, learning and research at the highest international levels of excellence.

www.cambridge.org
Information on this title: www.cambridge.org/9781108491839

DOI: 10.1017/9781108648981

First published 2020

A catalogue record for this publication is available from the British Library

ISBN   978-1-108-49183-9   Hardback
ISBN   978-1-108-74070-8   Paperback

Contents

List of Contributors | *page* vii
Preface | ix

1 Why Should Biologists Care about the Philosophy
of Science? | 1
Tobias Uller and Kostas Kampourakis

2 What Constitutes an Explanation in Biology? | 21
Angela Potochnik

3 What Is Biological Knowledge? | 36
Kevin McCain

4 What Is the Nature of Theories and Models in Biology? | 55
Emily C. Parke and Anya Plutynski

5 How Are Biology Concepts Used and Transformed? | 79
Ingo Brigandt

6 Why Does It Matter That Many Biology Concepts Are
Metaphors? | 102
Kostas Kampourakis

7 How Do Concepts Contribute to Scientific Advancement?
Evolutionary Biology As a Case Study | 123
David J. Depew

8 How Can Conceptual Analysis Contribute to Scientific
Practice? The Case of Cultural Evolution | 146
Tim Lewens

9 What Methods Do Life Scientists Use? A Brief History
with Philosophical Implications | 168
Erik L. Peterson

10 Is It Possible to Scientifically Reconstruct the History of
Life on Earth? The Biological Sciences and Deep Time | 193
Carol E. Cleland

11   What Is the Basis of Biological Classification? The Search
     for Natural Systems   216
     Thomas A. C. Reydon

12   What Is the Nature of Scientific Controversies in the
     Biological Sciences?   235
     Michael R. Dietrich

13   What Is the Relation between Facts and Values in
     Biological Science? Biology *in* Society   255
     Carrie Friese and Barbara Prainsack

14   A Philosopher in the Age of Creationism: What Have
     I Learned after Fifty Years Doing Philosophy of Biology
     That I Want to Pass On to Biologists   275
     Michael Ruse

15   How Can We Teach Philosophy of Science to Biologists?   299
     Kostas Kampourakis and Tobias Uller

     Further Reading   313
     Index   319

Contributors

Ingo Brigandt
University of Alberta

Carol E. Cleland
*University of Colorado
Boulder*

David J. Depew
University of Iowa

Michael R. Dietrich
University of Pittsburgh

Carrie Friese
*London School of Economics and
Political Science*

Kostas Kampourakis
University of Geneva

Tim Lewens
University of Cambridge

Kevin McCain
*University of Alabama at
Birmingham*

Emily C. Parke
University of Auckland

Erik L. Peterson
*University of Alabama at
Tuscaloosa*

Anya Plutynski
*Washington University in
St. Louis*

Angela Potochnik
University of Cincinnati

Barbara Prainsack
*University of Vienna, and King's
College London*

Thomas A. C. Reydon
Leibniz University Hannover

Michael Ruse
Florida State University

Tobias Uller
Lund University

Preface

Is philosophy of science of any use to biologists? A well-known response is that philosophy of science is as helpful to science as ornithology is to birds. Whether or not it was Richard Feynman who actually said this does not affect the fact that many biologists that we have met, especially those older than us, would readily agree. Among these biologists one can find top researchers, with prestigious grants and publications, who think that any philosophical discussion is a waste of time. The experienced researcher, they would say, knows what has to be done; the inexperienced has to learn from the experienced ones in the lab or in the field. Whatever Kuhn or Popper said (they have not heard about Lakatos, or any philosopher after him) is irrelevant to the actual practice of science. Philosophy of science is, at best, a nice endeavor for retired scientists, if they decide to reflect upon their own career and work. Or so the story goes.

This response is a caricature, of course, and many biologists do not think like this. But even those who are not in principle opposed to philosophical reflection and discussion usually do little to promote it. They have data to analyze, papers to write, and grant proposals to submit. Science is a full-time job, and there is little time left for philosophizing, which thus becomes a luxury. However, our aim with the present book is to show that it is not a luxury but a necessity. Philosophical reflection is inherent in any scientific activity, and what is necessary is to guide the experienced researchers to make it explicit, and the inexperienced ones to understand it. We hope that the chapters in the present book show how important philosophy of science is for biology, and how much biologists will benefit from thinking and reflecting in a philosophical manner.

We must note that the chapters in this book cover some philosophical aspects only, focusing on those that we considered the most important for biologists, especially the younger ones, to understand. We begin with a chapter we wrote that sets the context by explaining in some detail why biologists should care about the philosophy of science. The next three chapters discuss some very fundamental issues: what constitutes an explanation in biology (by Angela Potochnik); what biological knowledge is (by Kevin McCain); and what the nature of theories and models in biology is (by Emily Parke and Anya Plutynski). Then we focus on concepts, devoting four chapters on the nature and role of concepts, discussing how biology concepts are used and transformed (by Ingo Brigandt); why it matters that many biology concepts are metaphors (by Kostas Kampourakis); how concepts contribute to scientific advancement (by David Depew); and how conceptual analysis can contribute to scientific practice (by Tim Lewens).

The subsequent chapters discuss the methods used in the life sciences (by Erik L. Peterson); how biologists study the past and why this kind of work can be as solid as experimental science (by Carol Cleland); what the basis of biological classification is (by Thomas Reydon); what the nature of scientific controversies in the biological sciences is (by Michael R. Dietrich); and what the relation is between facts and values in biological science (by Carrie Friese and Barbara Prainsack). Last, but not least, Michael Ruse, one of the founders of the field we call philosophy of biology, shares his fifty-year-long experience of doing philosophy of biology. We conclude with some practical suggestions of our own about how to teach philosophy of science to biologists.

We are of course indebted to the contributors to this volume for their high-quality chapters and their excellent collaboration. We are also indebted to Katrina Halliday, publisher for life sciences at Cambridge University Press, who supported this – rather unusual for the life sciences series – book project right from the start and toward publication. We are also very grateful to Olivia Boult and Sam Fearnley for their work and collaboration during production, as well as to Chris Bond for his meticulous copyediting of the book. Finally, we are indebted to each other for an excellent collaboration, the outcome of which you are now holding in your hands. We hope that you will enjoy reading it as much as we did.

Kostas Kampourakis and Tobias Uller

1 Why Should Biologists Care about the Philosophy of Science?

TOBIAS ULLER AND KOSTAS KAMPOURAKIS

1.1 Introduction

To many biologists, science and philosophy may appear an odd couple without much in common. Perhaps the word "philosophy" will even bring to mind endless arguments and speculation about whether the chicken or the egg came first, without ever getting anywhere. After all, are philosophers not still arguing over the same things as Aristotle and his fellow Greeks? Well, yes. But biologists too are concerned with the questions that occupied Aristotle: what living beings are and where they come from; how they develop, function, and interact with one another; and why there are so many forms and how those forms should be classified. There has been tremendous progress in biology, of course. But it does not appear that biologists will ever run out of questions. This is because good science does not only reveal new things about the world; it also reveals that there are things we did not even know we could know. So we want to know more.

Not all research is equally effective in promoting the advancement of science, and it is therefore useful to reflect on what works and what does not. In fact, while biologists may think of themselves as busy enough just doing science, they have been and still are preoccupied with metascientific questions and issues, which is what philosophy of science is about: how to think about genes or ecosystems, the nature of species, the causes of evolution, the value of experimentation versus observation, and if molecular or evolutionary biology is the most fundamental of the biological sciences, to take just some examples. Perhaps even more significantly, twentieth-century genetics and molecular and evolutionary

We are grateful to Nathalie Feiner for thoughtful comments on this chapter.

biology were all shaped by attempts to ensure that the biological sciences meet the criteria of a "hard science," mapped on ideals in physics.

This ideal no longer seems as appealing as it once did. Over the last decades, studies in philosophy of science have revealed that scientific aims, methods, models, and concepts are much more diverse than previously envisaged by either scientists or philosophers. The biological sciences have been important for reaching this conclusion because biologists turn out to be very flexible in their scientific exemplars. What goes on in ecology and molecular or evolutionary biology can at times bear limited resemblance to each other, let alone to traditional exemplars in physics. Yet, the biological sciences are hugely successful and influential. This success is difficult to account for if the standards of mid-twentieth-century physics were the only way to ensure knowledge.

The present volume was put together because we believe that being aware of the rich conceptual and methodological issues of science can make biologists better at what they are doing – as students, teachers, scientists, and professionals. Philosophy of science is not the preoccupation of armchair philosophers or retired scientists, but it is rather central to any scientific endeavor. The problem is that, more often than not, biologists are not educated and encouraged to pursue this kind of reflection. The present volume is an attempt to support them in doing this.

1.2 Science and the Philosophy of Science

While biologists study life, philosophers of science study science (see Lewens 2015 for an accessible introduction to philosophy of science). Not unlike biologists, some philosophers are motivated by the big picture, whereas others are obsessed with a particular problem (Box 1.1). Many philosophers of science work on particular "study systems," such as molecular biology or evolutionary theory. This requires familiarity with the aims, methods, and knowledge of each particular area, which is one reason why philosophers need to engage with scientists. Another reason is that science is shaped by human abilities, interests, and values. Sometimes it is not possible to understand science without understanding the scientists themselves. This work often draws on the history of science to reveal how events actually played out or how facts and values

BOX 1.1 **What Kinds of Questions Do Philosophers of Science Ask?**

Philosophy of science is concerned with what science is, how it works, and what it can tell us. Some of the most fundamental questions concern very general features of science:

- What makes science different from non-science?
- How are scientific knowledge and understanding generated?
- How is science organized?
- What are the limits of science?

Major topics in philosophy of science are those that analyze and clarify the main components of scientific investigation, including

- What is a scientific explanation?
- How are scientific concepts used and transformed?
- What is the role of idealization in science?
- What is the relationship between theory and data?

Answers to these questions often require careful study of more narrowly defined questions, and most philosophers of science are therefore working on particular problem agendas that can range from quite general to very specific:

- What is the difference between reductionist and holistic approaches to the study of life?
- What is a biological mechanism?
- What do biologists mean when they refer to genes?
- What is the utility of Hamilton's rule in evolutionary theory?

Philosophers of biology are those philosophers of science who are particularly concerned with the biological sciences. Philosophy of biology did not begin in earnest until the 1970s, and earlier philosophy of science was largely concerned with physics or chemistry. However, philosophy of biology is now one of the main areas of inquiry and philosophers of biology have made important contributions to philosophy of science as well as biological theory. For examples, see the Further Reading section at the end of this book.

influence each other over time (see Chapter 13). Being scrutinized can feel uncomfortable for scientists, in particular if they believe that science is – or should be – free of such biases. As illustrated by several of the chapters in this book, this belief is not only mistaken but can actually be detrimental to science itself. The questions, methods, and models that a scientific community considers exemplar are shaped in part by shared attitudes and beliefs. Ignoring these social aspects of science can make biologists less well equipped to identify and solve scientific problems, and it can make them struggle to handle controversies between scientists and between science and other parts of society (see e.g., Chapters 7, 12, and 14)

For philosophy of science to become useful to biologists, scientists and philosophers need to find ways to communicate, share ideas and results, and perhaps occasionally work together. As biologists who have attempted to work with people from other disciplines will testify, collaboration is easier said than done. One main hurdle is simply ignorance about each other's work. Another is to become familiar with terminology and habits of mind that are often specific to particular disciplines. These hurdles can be overcome. Nevertheless, just as it will take time and effort for a cancer researcher to figure out if insights from evolutionary theory will be useful to her, it will take time and effort to figure out if philosophy of science will be useful to you. We hope that this collection of chapters will be helpful for those who are willing to dedicate their time. At the end of the book, we provide suggestions for further reading on both general topics in philosophy of science, as well as topics that are likely to be of particular interest to biologists. In the last chapter we also make concrete suggestions about how topics from philosophy of science could be taught to biology students.

1.3 Kuhn and Popper As Caricatures

Perhaps no philosopher of science is more familiar to scientists than Karl Popper. Popper was concerned with the big questions in philosophy of science, and his work has had a long and lasting intellectual impact, not the least on scientists (Lewens 2015). His idea that scientific hypotheses can never be proven, only falsified, is commonly introduced to beginners in the natural sciences as the fundamental feature of science. Falsification not only separates science from non-science but Popper also meant that the repeated failure to falsify hypotheses can account for the growth of scientific knowledge. Another philosopher of science who is likely to be mentioned in introductory science classes is Thomas Kuhn, famous for introducing the idea of a paradigm shift (Kuhn 1996). Scientists tend to be more ambivalent toward Kuhn since he emphasized that science is a collective, social endeavor where scientists sometimes appear irrational. But Kuhn's concept of paradigm shifts can be interpreted as a radical theory change introduced in the face of repeated falsification of established theories. For many biologists, this view of how science works – steadily securing knowledge through hypothesis testing and rarely interrupted by radical theory change when major hypotheses are disproven – may be the entire philosophy of science they are exposed to during their studies, perhaps even for their entire academic career.

No shame on Popper and Kuhn, and scientists are often taught a caricature of their work (like we just did!), but this is not really enough to understand how science works. Believe it or not, philosophy of science has progressed! While falsifiability remains an important litmus test for a scientific hypothesis, it is now widely recognized that the building of knowledge through falsification of *a priori* hypotheses is a poor characterization of many successful sciences, including biology. Scientific knowledge and understanding is generated through much more diverse standards and activities than envisaged by these early philosophers of science. There are good reasons for this diversity. The world is immensely complex, and humans are limited beings. Thus, it is reasonable that different scientific questions demand different approaches or methods. However, a diversity of scientific standards does not imply an absence of standards. It is important to understand what works and why.

In practice, biologists tend to pick up most ideas of what science is and how it works from fellow biologists, typically those who work on similar problems using similar methods. But if there is no universal standard of science, this can make it difficult to recognize or understand the importance of research that uses different standards or, for that matter, the limitations of one's own approach. Such failure can lead to inefficient science, missed opportunities for scientific breakthroughs, or even long and fruitless controversy. In what follows, we reflect on three features of science – its aims, methods, and concepts – to make a case for why biologists can benefit from insights gathered from the philosophy of science.

1.4 Scientific Aims

What are the aims of science? A short list would likely include description, classification, prediction, and explanation. Biologists *describe* and *classify* new species, molecules, and biological processes; they *predict* the effects of human activities on biodiversity or the spread of disease; and they *explain* how cells work and why populations evolve. A main reason for these activities is that many biologists ultimately strive to *understand* living systems, such as cells, organisms, and ecosystems. This understanding has practical consequences for technology, medicine, and many other features that make up societies, and it is therefore important far outside academic circles.

A phenomenon can be said to be understood when one can give it a satisfactory explanation (see Chapter 2).[1] Given that we explain phenomena all the time, it will perhaps come as a surprise to learn that it is neither obvious what it means to explain something, nor what, if anything, that makes scientific explanations different from everyday explanations. The traditional point of view on behalf of philosophers of science is that scientific explanations consist of statements that demonstrate that the phenomenon to be explained follows from natural law (Woodward 2017). This account of explanation is heavily influenced by physics, and biologists hardly find it very appealing since there is a widespread skepticism toward the existence of biological laws.

A more promising idea is that explanation is linked to causality, manipulability, and control (e.g., Woodward 2003).[2] It will feel natural to biologists to think of causes as difference-makers (Illari & Russo 2014). Rain causes seeds to germinate because if it had not rained the seeds would remain dormant. Loss of genetic variation causes population extinction because if it were not for the loss of genetic variation the population might have adapted to the environment. One view of scientific explanation is that it is achieved when the information provided by the explanation allows one to answer a range of such what-if-things-had-been-different questions (e.g., Woodward et al. 2003; Strevens 2008; Potochnik 2017). For example, an explanation for how ATP is generated

---

[1] Philosophers speak of the phenomenon to be explained as the *explanandum* and the sentences that do the explaining as the *explanans*.

[2] There are various versions of this theory of causal explanation, Woodward 2003 and Strevens 2008 are useful starting points.

may refer to biochemical features of glycolysis. This explanation reveals something about the causal tapestry of the world; the molecular detail makes it possible to grasp the consequences of a change in the concentration of pyruvate or the chemical structure of the reacting molecules. According to some philosophers, this is what it means to understand how ATP is generated, and the more what-if-things-had-been-different questions about ATP production we can answer the better we understand it.

Not all explanations in biology are mechanistic like this, however, and many explanations in biology look more like historical explanations (see Chapter 10). An explanation for the extinction of dinosaurs may refer to a meteorite that struck the earth and caused long-term changes in the earth's climate. Nevertheless, the reason why this explanation generates understanding is similar to the case of ATP; reference to the meteorite and its effect on climate makes it possible to grasp what would have happened to the dinosaurs if the meteor had not have struck the earth, or if it had been smaller, or if there had been no competing mammals around. There may be other kinds of scientific explanations, but being able to give answers to what-if-things-had-been-different questions appears to at least be one important feature of many scientific explanations.

A good thing about this notion of explanation is that one need not take truth too seriously. What really is "out there" may forever be out of reach, but representations of the world can be sufficiently good approximations that enable one to foresee what would have happened if things had been different. It is not always possible to support the explanation through active intervention, of course (this is difficult for the dinosaur extinction, for example). But scientists can nevertheless ensure that their theories are empirically justified – or true enough – by imagining and studying a range of different situations. This is why it is important that scientific theories are falsifiable; if a theory makes no falsifiable claims, it also appears impossible to predict the consequences of an intervention.

Another helpful feature of the causal theory of explanation is that it brings attention to the fact that scientists need to manage causal complexity (Potochnik 2017). Biological systems are enormously complex, and any representation of a living system will only capture some of its actual causes. This is in itself not a problem. In fact, too much detail makes it harder to grasp what would have happened if things had been different. A diagram of all causal interactions in a cell would describe the cell but not explain how it works. To explain phenomena, scientists leave things out (abstraction) and make assumptions that are false (idealization). Abstraction and idealization play positive roles in explanation because they foreground the causal relations that are of interest – idealization makes the phenomenon appear as if it were produced by the focal causes alone (Potochnik 2017). As a result, how one thinks about biological processes influences which of the myriad of actual causes of a particular phenomenon that are picked out as being explanatory causes.

It may be helpful to illustrate this feature of idealization using a nonbiological example. Consider the frequent delays of trains arriving into Stockholm Central. One possible cause of these delays is that late departures of trains that are not headed toward Stockholm can propagate through a jammed train system – a kind of cascading effect. To see if this can explain the arrival delays into Stockholm C, transport planners may benefit from assuming that trains run at a constant speed unless they have to stop to let other trains pass. Plugging in real data on train speeds and how they vary seems unnecessarily complicated. Doing so might even make it harder to grasp how improving departure punctuality of trains throughout Sweden will affect arrival times of trains bound for Stockholm C.

The transport planners will feel satisfied if there is a good fit between their model and actual arrival times into Stockholm C. They could claim that they now understand why trains are delayed, because they can explain it in terms of cascading effects of delayed departures of trains bounded for other destinations. However, imagine that it turns out that, contrary to what the model predicts, the actual arrival times are unaffected by such departure delays. One possible explanation for this mismatch between model and reality is that train drivers adjust the speed to compensate and ensure that trains headed for Stockholm have a free pass. This may appear to imply that train speed is a cause of punctuality but not of delays. But this cannot be the case because slow-downed trains can also jam tracks and propagate delays.

What is happening here? Firstly, note that train speed initially appeared unable to account for the arrival delays because it was idealized away from the model. Secondly, it is only when the proposed model was unable to account for the phenomenon that we looked for another cause. This is why train speed appeared as a possible explanation for why trains were not delayed, but not an explanation for why they were delayed. But there is no fundamental causal asymmetry here. Interventions on either departure times or train speeds can cause arrival delays of trains bound for Stockholm C because both can result in interference between trains. As a result, a satisfactory explanation for the late arrivals may require the use of multiple different idealizations, each one suitable for picking out the contribution of a particular cause or set of causes. Trying things out and keeping what appears important may eventually allow more complex representations that have greater explanatory power.

The challenges that transport planners face are also faced by biologists. Biological phenomena are produced and sustained by many factors, and these are often causally intertwined. As a result, there can be several legitimate explanations of the same phenomenon, each drawing on only some of its causes. These explanations are often sufficiently different to happily coexist. One example is the distinction between what biologists commonly refer to as ultimate and proximate explanation. Roughly speaking, ultimate explanations are considered historical explanations that trace events that occur within a population or a lineage, whereas proximate explanations are considered mechanistic explanations at the level of the individual. An ultimate explanation for why mammals maintain a high body temperature may, for example, refer to its fitness benefits in cold climate, which implies that this trait became increasingly common and sophisticated as a result of natural selection. A proximate explanation for the same phenomenon may refer to the autonomic, neuronal, and molecular mechanisms that underlie the ontogenetic development of endothermy.

Following the highly influential work of Ernst Mayr (1961), it is customary in evolutionary biology to consider that causes that feature in proximate explanations should not be invoked to explain evolutionary adaptation (e.g., Dickins & Rahman 2012). However, a closer look at the rationale for this distinction reveals that it relies on an idealization of the evolutionary process that foregrounds fitness differences and screens off other putative causes of adaptive change (see Walsh 2015; Pocheville 2019; Uller & Helanterä 2019). This reflects that the main agenda for evolutionary biology has been to understand the role of natural selection in adaptive evolution, not the role of development, physiology, or behavior. The assumptions made in evolutionary theory tend to turn the latter into constraints; they can account for the absence of adaptive fit but not its presence.

This line of thought is so common to biologists that many take it for granted. However, a comparison to the explanations for the delayed train arrivals is a reason to treat this conclusion with caution. That is, that one particular idealization of evolution by natural selection privileges genes and natural selection does not imply that there is an inherent causal asymmetry in evolutionary processes (Laland et al. 2011). The role of proximate causes in adaptive evolution is in fact one of the most persistent controversies in biology (Amundsen 2005). Contemporary examples include the disagreement over the explanatory role of development, plasticity, extra-genetic inheritance, and niche construction in evolution (see Laland et al. 2014, 2015). One possible reason that these issues are difficult to resolve is that the genetic representation of evolution is commonly taken at face value, rather than being understood as an idealization designed to explain evolutionary phenomena in terms of natural selection. An increased awareness of the relationship between idealization and explanation may reduce the risk that causes that are idealized away become permanently neglected, facilitate capitalization of insights from other disciplines, and put a restraint on unproductive scientific controversy.

While there are good reasons why a biological phenomenon like adaptation can have several explanations, biologists may sometimes wish to determine which of a number of different explanations is the most satisfactory (see Chapter 3). Consider cichlid fish, famous for the ability to evolve very similar morphologies in different lakes (Seehausen 2006). Evolutionary biologists have demonstrated that this convergence happened because the local habitat and foods are often similar in different lakes, which favors a limited set of life styles such as bottom-dwelling grazers and open water predators (e.g., Muschick et al. 2012). Thus, natural selection explains the convergent evolution of cichlid fish. But biologists have also pointed out that some of the recurring features of these fish, such as the shapes of bodies and jaws, tend to be plastic (Schneider & Meyer 2017). That is, those characters respond to the habitat or diet that individual fish encounter during their lifetime. Some biologists believe that plasticity has contributed to the striking convergence between different species because plasticity can make some features more likely to become selected than other, perhaps equally fit, phenotypes (see West-Eberhard 2003). In terms of explanation, the question is whether or not an explanation for adaptive convergence in terms of plasticity *and* natural selection is better or more satisfactory than an explanation in terms of natural selection alone (Uller et al. 2019).

Picking the best of two explanations is easy if only one of them is backed up by empirical evidence (although negative evidence is not always enough to give up a hypothesis, as explained in the next section). Beyond this, there may be nothing inherent in these explanations for adaptive convergence that make one better than the other; perhaps picking the best explanation is simply a matter of one's interest (assuming both explanations are empirically justified). However, scientists sometimes prefer one explanation over others if it applies to many different phenomena rather than a few, if it is easy to understand, or based on its elegance or simplicity (Ylikoski & Kourikoski 2010). Explanatory standards such as these are important since the criteria for picking the best explanation influences what biologists consider knowledge (Chapter 3). Such explanatory standards can vary between biological disciplines (e.g., those concerned with mechanistic versus historical explanations), but these differences may not be well recognized by biologists. As a result, it can be difficult to assess the quality and scope of biological research that lies outside of one's immediate expertise.

These examples from evolutionary biology reveal exactly how difficult it is to do science. We have discussed scientific explanation at some length here because we believe that it is a good illustration of how philosophy of science can be helpful to biologists, but similar issues also come up for other scientific aims, such as classification (Chapter 11). In our opinion, the philosophy of science that is usually taught in introductory classes does not pay sufficient tribute to the challenges that scientists face when they attempt to produce knowledge and understanding of a world that is enormously complex. How biologists choose to represent the world influences the questions they consider worthwhile, how they organize their research, and the answers they look for. This means that scientific advancement requires the flexible use of methods and concepts.

1.5 Scientific Methods

It is probably obvious that biologists do not follow a single universal scientific recipe, but rather several more or less distinct approaches (see Chapter 9). Biological science certainly appears to frequently stray away from a strict method of falsification. A careful look behind the scenes of scientific papers that communicate the results of a test of a well-defined hypothesis will often reveal a process that looks very different to what we have been told science should look like. This is one reason why p-values are so problematic; scientists' formulation of hypotheses often develops in parallel with observation, practice, and data collection rather than in a strictly ordered fashion. It may be tempting to conclude that failure to adhere to the strict rules of hypothesis testing makes some biological research fundamentally flawed. A more optimistic view is that scientific practice simply reflects that there is a diversity of scientific methods, all of which may be appropriate. Regardless, it is important to examine how different scientific methods achieve scientific aims (e.g., understanding), what tools are available to meet these aims, and how well those tools actually work in practice. Biologists are highly engaged in these issues, in particular with respect to the appropriate use of statistics, lab- vs field-based methods, the use of model systems, and so on. These discussions may benefit from a greater attention to the literature on philosophy and history of science, which often has a fair bit to say on the matter (e.g., Chapter 10; Chapter 9).

Another peculiar break with falsification is that scientists only occasionally and reluctantly abandon a hypothesis when the data fail to support it. For example, a frequent failure to demonstrate that offspring of males with exaggerated sexual signals were fitter than other offspring did not generally make behavioral ecologists abandon the "good genes" hypothesis (Roughgarden 2009, pp. 213–224). This widespread practice is difficult to make sense of if biologists really believed in falsification as a method to scientific progress. Eventually a hypothesis may of course fail to be confirmed and become abandoned, but this will not result in a wholesale rejection of the theory. For example, the biologists that did conclude that there was little if any evidence for good genes rarely used this to argue against the theory of sexual selection (but see Roughgarden 2009). This is not irrational behavior on behalf of the scientists. Instead, examples such as these demonstrate that scientific theories are organized into more fundamental theoretical frameworks that are not revised on the same basis as a more specific hypothesis. Philosophers of science have introduced and use many different concepts to make sense of this feature of science, including thought styles (Fleck 1979), paradigms (Kuhn 1962), research programs (Lakatos 1978), and problem agendas (Love 2008; see Chapter 7).

An example from population biology can illustrate this point. During range expansions, populations on the front line are often small but rapidly growing. The combination of bottlenecks and rapid population growth can make particular genetic variants become very common even if they do not bring any fitness advantage (Excoffier et al. 2009). The formulation of this idea – sometimes referred to as allele surfing – relies on a set of population genetic principles (see Charlesworth & Charlesworth 2010). These principles derive from abstraction and idealization of complex biological processes, which are considered appropriate to solve a particular problem or kind of problem. Biologists may refer to both allele surfing and the set of population genetic principles as "theories," but it is only the former that can be falsified (theoretically by demonstrating that the conclusions do not follow from the premises, and empirically by demonstrating that allele surfing is not something that actually happens in real populations). Principles relevant to some discipline, methodology, or problem are typically not falsifiable. In fact, idealizations are commonly made despite full knowledge they are false (Potochnik 2017). Putting these "theories," such as population genetic theory, to the test means to assess how well those theories deliver satisfactory explanations, not to try to prove them wrong. In the next section, we argue that this is one reason why conceptual analysis can advance science.

Despite these reasons to doubt the classic view of the scientific method, many biologists adhere to its core features including, of course, the notion that experiments are the key to scientific knowledge and understanding. However, not even experiments are fundamental to all biologists. Many evolutionary biologists, for example, rely heavily on observation and "traces" of past events to reconstruct what happened and explain why it happened (Currie 2018; Chapter 10). Even molecular biology – perhaps the "ideal" reductionist science of biology – makes use of nonexperimental inference, for example, when species comparisons are used to substantiate claims that the activity of transposable elements ("jumping genes") causes genome evolution (Bourque et al. 2018). In her chapter, Cleland (Chapter 10) argues that experimental science is not as reliable as popularly thought, and historical science is not so unreliable. This may be welcome news to some biologists, and worrying to others. Regardless, it will be useful to become aware of the possible advantages and disadvantages of different scientific methods.

1.6 Scientific Concepts

Progress in the biological sciences is commonly driven by new discoveries and technological breakthroughs. The history of genetics provides many good examples, including PCR, high-throughput sequencing, and CRISPR-Cas. Nevertheless, data alone is often not enough, and conceptual analysis too can advance science (see Chapter 5, Chapter 8). This should perhaps not be surprising since concepts organize research agendas and feature in models, theories, and explanations. Since these are core features of science, understanding concepts may help to understand phenomena, or at least make science more effective at understanding phenomena. Concepts should not be confused with terminology, since a single concept can have several terms and the same term can be used for several concepts.

A good example of the latter is "gene," which in modern biology routinely refers to several different concepts (Griffiths & Stotz 2013; Kampourakis 2017). Biologists sometimes express the feeling that this multitude of meanings only creates problems and confusion. The solution, in their view, is a clear definition and a precise one-to-one map from concept to term. One illustrative example is many biologists' frustration over the term "epigenetics" (e.g., Deans & Maggert 2015; see Baedke 2018 for a conceptual analysis). For example, the journal *Cell Reports* requires authors to adhere to a strict definition of epigenetic (and epigenetics) and does not allow authors to refer to epigenetic as a stand-alone term.[3] This certainly removes ambiguity with respect to what biological feature that epigenetic refers to in papers published in this journal. This can be a good thing. Nevertheless, there are reasons to be skeptical toward this puritanism since concepts are more than

---

[3] This example is based on personal experience and email communication between authors and editors concerning a paper published in the journal (Tobi et al. 2018).

placeholders of specific biological entities or phenomena. For example, the concepts of "gene," "genetic," or "genetics" do not have a similar one-to-one identity as what *Cell Reports* demands for epigenetic. That biologists appear reasonably content with genetic, whereas the ambiguity in epigenetic is a major source of frustration and confusion, suggests that there is something to be learnt from understanding how biological concepts are formed, used, and transformed (see Chapter 5).

Conceptual analysis aims to clarify the use of concepts and the different roles that those concepts play in science. One role is to circumscribe a phenomenon or entity, which is often accompanied by a definition. This is how *Cell Reports* views the epigenetic concept. Another important role of concepts is to set a research or problem agenda (Love 2008). Arguably this too is a major role of the epigenetics concept, similar to one of the meanings of genetics. Since its inception, epigenetics has changed meaning several times, but it has consistently been used to refer to a set of biological problems, processes, and entities that are tied to a particular explanatory aim. To Waddington, who is often considered to have coined the term, this explanatory aim was to understand development (e.g., Waddington 1957). To molecular biologists today it may be to understand gene regulation (e.g., Allis et al. 2015). To epidemiologists, it may be to understand the molecular mechanisms that can cause the lifestyle of parents to influence health and disease of their offspring (e.g., Mill & Heijmans 2013). This use of a single concept to refer to a multitude of research agendas can obviously lead to confusion. But recognition that concepts such as epigenetics can be useful even without a consensus definition should make the situation easier to handle. Historical analysis of the gene concept actually suggests that the fuzziness of the concept sometimes has been a strength since it allowed formulation of a research agenda that was more inclusive than it otherwise might have been (Rheinberger 2000). Familiarity with such examples from the history of biology will make biologists better able to handle the conceptual complexity of contemporary biology and avoid some of the more common pitfalls.

One such pitfall arises from the fact that many scientific concepts are metaphors (see Chapter 6). Two familiar examples are membrane "pumps" and "channels." These concepts make use of familiar objects to make it easier to grasp an important biological difference in cell membranes; a pump requires ATP to move ions across the membrane, whereas channels allow ions to freely flow from one side to the other. But it would be a mistake to attach to the biological pumps and channels some other properties that are distinctive of water pumps and channels. For example, water pumps need to be constructed, controlled, or programmed, whereas water channels may passively form and function without external control. It is not obvious that these differences between water pumps and channels also apply to membrane pumps and channels; applying them may even be misleading. Human brains are, however, prone to make such associations. This is why it is often pointed out that the "cost of metaphor is eternal vigilance." One widely discussed example where vigilance may have slipped is the concept of a genetic program. As Susan Oyama (2000), Evelyn Fox Keller (2000), and others have demonstrated, the metaphorical use of "program" or "blueprint" exercised an enormous influence on research on genes, development, and evolution, and it continues to exercise an important hold over scientific and public understanding of biology. Some biologists shrug this off and claim that everyone always knew there was no "blueprint," perhaps offering "recipe" as a better metaphor. Familiarity with the history and philosophy of science will make one wary of such recollections of the past, and encourage vigilance when new metaphors replace the old ones.

As explained in the previous section, concepts also function to organize larger bodies of theory or research efforts. Such conceptual "frameworks" differ from scientific hypotheses or conjectures, even if both of them sometimes are referred to as "theory." Biologists usually keep track of this distinction, but confusion over the structure of scientific concepts can be one reason for scientific controversy (see Chapter 12). The controversy over the role of development and other proximate causes in evolutionary explanations already discussed may be one example. Some biologists appear to think that this controversy will be resolved by proving the gene-centric conceptual framework right or wrong (at least, this is one way to interpret Noble 2013 and Wray et al. 2014, to exemplify with two opposite conclusions). However, there are reasons to doubt that falsification is an appropriate criterion here. Population and quantitative genetic theory are not conjectures but collections of principles, adopted to address particular kinds of problems. Rather than empirical falsification, a more realistic aim is to clarify the explanatory limits of such conceptual frameworks and suggest alternative representations that may be better suited to move beyond those limits. The literatures on developmental plasticity, extra-genetic inheritance, and niche construction are good illustrations of this endeavor (Laland et al. 2015).

This endeavor is complicated by the fact that it is possible to explore the role of plasticity, non-genetic inheritance, and niche construction in evolution without giving up on idealizations that will grant a privileged role of natural selection and genes (see Section 1.4). One particularly good illustration is the large body of research on "plasticity-led evolution" that has followed since the publication of Mary-Jane West-Eberhard's book *Developmental Plasticity and Evolution* (West-Eberhard 2003). Much of this research represents plasticity as a property of genotypes (biologists call these reaction norms; Levis & Pfennig 2016). As a result, any contribution of plasticity to adaptive evolution can be considered a *consequence* of natural selection on genes, rather than as a primary cause of adaptive change (see Uller et al. 2019). Examples such as these illustrate that simply "extending" a theory to include new phenomena need not resolve contention. It is also important to be aware that biologists' interpretative understanding of these phenomena will be dictated by their conceptual framework, that is, how they think about living beings.

1.7 Concluding Remarks

As biologists we feel that we have benefited from reading philosophy of science and engaging with philosophers. Not all biologists will feel the same, of course. But we do not believe that we are particularly unusual. Any field probably benefits from a diversity of perspectives, and this diversity tends to grow from encouraging reflection and critical assessment, not the least from scientists within the field. A little bit of philosophy of science is one way to make this happen.

The chapters in this book can be read in any order and where one would like to start depends on one's interest. Some chapters are easier to digest than others, and not everything will be to everyone's liking. Our advice to biologists is to look for issues that feel most relevant to their own work and begin there. If you are engaged in a field that is controversial, consider if the controversy could partly be dissolved through conceptual analysis or a more explicit formulation of the idealizations that are used for the phenomena you study. If you struggle to see the value of someone's research, or even an entire field, consider if your opinion is a result of different aims, problem agendas, or methods, or if it is shaped by preferences, values, and beliefs. If you look for new and exciting ways to tackle your problem, identify causes that are currently screened off, and alternative concepts and metaphors that may prove fruitful. If you are engaged in public outreach, consider if your research can be communicated more effectively by emphasizing the process by which knowledge is generated. Above all, stay curious, not just about biology, but also about the nature of the biological sciences.

References

Allis, C. D., Caparros, M-L., Jenuwein, T., Reinberg, D., & Lachlan, M. (2015). Epigenetics, 2nd ed. Cold Spring Harbor, NY: Cold Spring Harbor Laboratory Press.

Amundsen, R. (2005). *The Changing Role of the Embryo in Evolutionary Thought: Roots of Evo-Devo.* Cambridge: Cambridge University Press.

Baedke, J. (2018). *Above the Gene, beyond Biology: Toward a Philosophy of Epigenetics.* Pittsburgh, PA: University of Pittsburgh Press.

Bourque, G., Burns, K. H., Gehring, M., Gorbunova, V., Seluanov, A., Hammell, M., et al. (2018). Ten Things You Should Know about Transposable Elements. *Genome Biology* 19(1): 199. doi:10.1186/s13059-018-1577-z

Charlesworth, B. & Charlesworth, D. (2010). *Elements of Evolutionary Genetics.* Greenwood Village, CO: Roberts & Company Publishers.

Currie, A. (2018). *Rock, Bone and Ruin: An Optimist's Guide to the Historical Sciences.* Cambridge, MA: MIT Press.

Deans, C. & Maggert, K. A. (2015). What Do You Mean, "Epigenetic"? *Genetics* 199(4): 887–896.

Dickins, T. E. & Rahman, Q. (2012). The Extended Evolutionary Synthesis and the Role of Soft Inheritance in Evolution. *Proceedings of the Royal Society B: Biological Sciences* 279(1740): 2913–2921.

Excoffier, L., Foll, M., & Petit, R. J. (2009). Genetic Consequences of Range Expansions. *Annual Review of Ecology Evolution and Systematics* 40: 481–501.

Fleck, L. (1979 [1935]). *Genesis and Development of a Scientific Fact.* Chicago: Chicago University Press.

Griffiths, P. E. & Stotz, K. (2013). *Genetics and Philosophy: An Introduction.* Cambridge: Cambridge University Press.

Illiari, P. & Russo, F. (2014). *Causality: Philosophical Theory Meets Scientific Practice.* New York: Oxford University Press.

Kampourakis, K. (2017). *Making Sense of Genes.* Cambridge: Cambridge University Press.

Keller, E .F. (2000). *The Century of the Gene.* Cambridge, MA: Harvard University Press.

Kuhn, T. S. (1996) [1962]. *The Structure of Scientific Revolutions (3rd edn.).* Chicago: University of Chicago Press.

Lakatos, I. (1978). *The Methodology of Scientific Research Programmes: Philosophical Papers Volume 1.* Cambridge: Cambridge University Press.

Laland, K., Uller, T., Feldman, M., Sterelny, K., Muller, G. B., Moczek, A., et al. (2014). Does Evolutionary Theory Need a Rethink? – POINT Yes, Urgently. *Nature* 514 (7521): 161–164.

Laland, K. N., Sterelny, K., Odling-Smee, J., Hoppitt, W., & Uller, T. (2011). Cause and Effect in Biology Revisited: Is Mayr's Proximate–Ultimate Dichotomy Still Useful? *Science,* 334(6062): 1512–1516.

Laland, K. N., Uller, T., Fellman, M. W., Sterelny, K., Muller, G. B., Moczek, A., et al. (2015). The Extended Evolutionary Synthesis: Its Structure, Assumptions and Predictions. *Proceedings of the Royal Society B: Biological Sciences* 282(1813). doi:10.1098/rspb.2015.1019

Levis, N. A. & Pfennig, D. W. (2016). Evaluating "Plasticity-First" Evolution in Nature: Key Criteria and Empirical Approaches. *Trends in Ecology & Evolution* 31(7): 563–574.

Lewens, T. (2015). *The Meaning of Science.* Milton Keynes, UK: Penguin Random House.

Love, Alan C. (2008). Explaining Evolutionary Innovations and Novelties: Criteria of Explanatory Adequacy and Epistemological Prerequisites. *Philosophy of Science* 75(5): 874–886.

Mayr, E. (1961). Cause and Effect in Biology – Kinds of Causes, Predictability, and Teleology Are Viewed by a Practicing Biologist. *Science* 134(348): 1501–1506.

Mill, J. & Heijmans, B. T. (2013). From Promises to Practical Strategies in Epigenetic Epidemiology. *Nature Reviews Genetics* 14(8): 585–594.

Muschick, M., Indermaur, A., & Salzburger, W. (2012). Convergent Evolution within an Adaptive Radiation of Cichlid Fishes. *Current Biology* 22(24): 2362–2368.

Noble, D. (2013). Physiology Is Rocking the Foundations of Evolutionary Biology. *Experimental Physiology* 98(8): 1235–1243.

Oyama, S. (2000). *The Onotgeny of Information.* Durham, NC: Duke University Press.

Pocheville, A. (2019). A Darwinian Dream: On Time, Levels, and Processes in Evolution. In T. Uller & K. N. Laland (eds.), *Evolutionary Causation: Biological and Philosophical Reflections,* pp. 265–298. Cambridge, MA: MIT Press.

Potochnik, A. (2017). *Idealization and the Aims of Science.* Chicago: Chicago University Press.

Rheinberger, H-J. (2000). Gene Concepts: Fragments from the Perspective of Molecular Biology. In P. Beurton, R. Falk, & H-J. Rhienberger (eds.), *The Concept of the Genes in Development and Evolution,* pp. 219–239. Cambridge: Cambridge University Press.

Roughgarden, J. (2009). *The Genial Gene: Deconstructing Darwinian Selfishness*. Los Angeles: University of California Press.

Schneider, R. F. & Meyer, A. (2017). How Plasticity, Genetic Assimilation and Cryptic Genetic Variation May Contribute to Adaptive Radiations. *Molecular Ecology* 26(1): 330–350.

Seehausen, O. (2006). African Cichlid Fish: A Model System in Adaptive Radiation Research. *Proceedings of the Royal Society B: Biological Sciences* 273(1597): 1987–1998.

Strevens, M. (2008). *Depth: An Account of Scientific Explanation*. Cambridge, MA: Harvard University Press.

Tobi, E. W., van den Heuvel, J., Zwaan, B. J., Lumey, L. H., Heijmans, B. T., & Uller, T. (2018). Selective Survival of Embryos Can Explain DNA Methylation Signatures of Adverse Prenatal Environments. *Cell Reports* 25(10): 2660–2667.

Uller, T., Feiner, N., Radersma, R., Jackson, I. S. C., & Rago, A. (2019). Developmental Plasticity and Evolutionary Explanations. *Evolution & Development*, 22(1–2): 47–55.

Uller, T. & Helanterä, H. (2019). Niche Construction and Conceptual Change in Evolutionary Biology. *The British Journal for the Philosophy of Science* 70(2): 351–375.

Waddington, C. H. 1957. *The Strategy of the Genes: A Discussion of Some Aspects of Theoretical Biology*. London: George Allen & Unwin Ltd.

Walsh, D. M. (2015). *Organism, Agency and Evolution*. Cambridge: Cambridge University Press.

West-Eberhard, M. J. (2003). *Developmental Plasticity and Evolution*. New York: Oxford University Press.

Woodward, J. (2003). *Making Things Happen: A Theory of Causal Explanation*. New York: Oxford University Press.

Woodward, James, "Scientific Explanation," *The Stanford Encyclopedia of Philosophy* (Fall 2017 Edition), Edward N. Zalta (ed.), https://plato.stanford.edu/archives/fall2017/entries/scientific-explanation/

Wray, G. A., Hoekstra, H. E., Futuyma, D. J., Lenski, R. E., Mackay, T. F. C., Schluter, D., & Strassmann, J. E. (2014). Does Evolutionary Theory Need a Rethink? COUNTERPOINT No, All Is Well. *Nature* 514(7521): 161–164.

Ylikoski, P. & Kuorikoski, J. (2010). Dissecting Explanatory Power. *Philosophical Studies: An International Journal for Philosophy in the Analytic Tradition* 148(2): 201–219.

2 What Constitutes an Explanation in Biology?

ANGELA POTOCHNIK

2.1 Introduction

"Explaining" and "explanation" are words that tend to feature prominently in even the most basic descriptions of science. This is a big part of what science is about: generating explanations of our world (see also Chapter 3). It seems to be a big part of what biology is all about as well. Research in biology undoubtedly leads to practical applications in pursuits from medicine to agriculture to conservation, but one of its fundamental aims is to generate *understanding* of the living world around – and within – us.

The centrality of explanation to the scientific enterprise is matched by philosophers' enthusiasm for debating the nature of explanation in science. Philosophers of science have been up to our elbows in debates about the nature of scientific explanations since at least the middle of the twentieth century. Pet theories abound, but some basic insights into the broad contours of scientific explanation have also emerged.

In this chapter, I aim to provide a relatively nonpartisan discussion of the nature of explanation in biology, grounded in widely shared philosophical views about scientific explanation. At the same time, this discussion reflects what I think is important for philosophers and biologists alike to appreciate about successful scientific explanations. So, some points will be controversial, at least among philosophers. Along the way, I indicate which ideas are controversial and say something about the nature of controversy. I make three main points: (1) causal relationships and broad patterns have often been granted importance to scientific explanations, and they are in fact both important; (2) some explanations in biology cite the components of or processes in systems that account for the systems' features, whereas other explanations feature large-scale or structural causes that influence a system; and (3) there can be multiple different explanations of a given biological phenomenon, explanations that respond to different research aims and can thus be compatible with one another even when they may seem to disagree.

2.2 Causes, Patterns, and Causal Patterns

Today, philosophers of science by and large see explanation as deeply related to causation. At its most basic, the idea is that to explain something is to show what is responsible for that thing – and whatever is responsible for something is its *cause*. You explain the suddenly dark room by appealing to the fact that a fuse was blown; you account for a polymorphism by noting there was frequency-dependent selection; and you account for a high concentration of female caribou antlers by pointing to the fact that this was a calving site (Miller et al. 2013). Philosophers sometimes disagree about what qualifies as a causal explanation or about the features that causal explanations must have, and some philosophers question whether *all* scientific explanations cite causes. But, to my knowledge, all contemporary philosophers accept that causation is central to at least most scientific explanations. The central role that philosophers accord causation in scientific explanation is akin to the call in biology for identifying the cause, process, or mechanism responsible for something. (Though, as I make clear in the next section, philosophers of science use the concept of mechanism in a more restricted way than this.)

It wasn't always so: Philosophers didn't always take causes to be at the heart of explanation. At the start of contemporary discussions of explanation, Carl Hempel famously posited that explanations were derivations. So-called *deductive-nomological* explanations involved explaining a phenomenon by deriving it logically from a scientific law and the initial conditions that, given the law, brought about the phenomenon in question (Hempel 1965). Thus, to explain something was taken to consist in showing how it resulted from laws of nature, thereby rendering it unsurprising. Later philosophers argued that Hempel had a variety of related ideas here, and though the official deductive-nomological account of explanation faced a number of significant difficulties, the "unofficial" account fared better. This was supposed to be the idea that explanations *unify* by showing how disparate phenomena turn out to arise for the same reason or to fit the same pattern (Friedman 1974; Kitcher 1981). Newton famously assimilated the movement of celestial bodies in orbit to the freefall of bodies here on Earth. Cooperative behaviors have emerged in many different species and biological communities because there is a set of circumstances in which cooperation can be advantageous to organisms. Such accounts seem to explain by showing how the phenomena in question fit into a pattern.

This has also been touted as an important philosophical insight about scientific explanation: Explanations should unify disparate phenomena, cite general patterns, or something along those lines. Some philosophers have endorsed this instead of the idea that explanations should cite causes. A growing number of philosophers argue that not all scientific explanations are causal, that some phenomena are explained by statistical or mathematical facts or other regularities that do not seem to be causal (see, e.g., Reutlinger & Saatsi 2018). One interpretation of this is advocacy of the explanatory importance of patterns and regularities, some of which are causal and others of which are not.

In my view, both of these ideas – that causal dependence is explanatory, and that patterns and regularities are explanatory – are important insights into the nature of scientific explanation. Some philosophers have combined both insights into a single account of explanation with the idea that scientific explanations should cite patterns in causal influence. On this kind of an approach, it's not just information about causal relationships that is explanatory, but also information about the *scope* of those causal relationships, or the range of circumstances in which a given causal relationship holds. This kind of view has been advocated more or less explicitly by James Woodward (2003), Michael Strevens (2004, 2008), and me (Potochnik 2015, 2017), among others. Thus, I think scientific explanations – at least by and large – depict *causal patterns*. The causal content of an explanation is a way to show a dependence: that the phenomenon to be explained came about due to the featured causes. That an explanation depicts a *pattern* in causal dependence means that it also shows the extent – or scope – of the causal dependence in question.

Consider, for example, explaining the bright coloration of the scarlet ibis (*Eudocimus ruber*). Pointing out the bird's ability to metabolize carotenoids in its diet to influence its pigmentation is explanatory causal information. Likening this to the source of the flamingo's pink, the cardinal's red, and the goldfinch's yellow indicates something about the scope of this causal pattern. The scarlet ibis's coloration is due to one main form of avian pigmentation, a process that can create red, pink, orange, and yellow coloration. It can also be enlightening to point out a difference between the scarlet ibis and white ibis: The former but not the latter has a substantial volume of a carotenoid carrier protein in its blood (Trams 1969). This indicates why the scarlet ibis is able to absorb and transport carotenoids in this way while the related white ibis is not – also indicating something about the scope of the explanatory causal pattern.

The idea that causal patterns explain is a causal approach to explanation, but it is not merely a causal approach. On this view, simply providing some causal information is not sufficient for explanation. An explanation also needs to indicate the scope of the explanatory causal dependence, the range of circumstances in which similar causal dependence obtains. The explanatory value of this relates to the insight that explanations should unify by showing how disparate phenomena arise for the same reason or fit the same pattern. With the explanation of scarlet ibis coloration mentioned previously, insight is provided into conditions in which similar coloration occurs. Such explanations show how (potentially disparate) phenomena fit the same pattern. One might even say that causal pattern explanations are akin to the original deductive-nomological view of explanation, for they show how phenomena result from regularities in our world, though the regularities are not universal laws but limited in scope and may have exceptions.

I have suggested that the idea that causal patterns explain relates comfortably to a range of other ideas philosophers have had about explanation. But why think it is true? That is, why think that causal patterns are the sort of things that help us explain our world? Grasping the nature of a causal dependence and the scope in which that dependence holds is key to determining the causal structure of the world, which Gopnik (1998) has influentially argued is the endpoint of explanation. Grasping what I call causal patterns is also, as Woodward (2003) and others have argued, key to effective action. This indicates how and in what conditions we can act to bring about or prevent the focal phenomenon. Additionally, research in cognitive science suggests that causal information and broad generalizations are both the kinds of information that strike our intellects as explanatory (Lombrozo & Carey 2006) and that we learn more through the act of explaining (Lombrozo 2011). Uncovering causal patterns thus seems to do exactly the tasks we expect from our scientific explanations, and grasping causal patterns seems to look a lot like explaining.

Perhaps in some explanations, the causal dependence does more of the explanatory lifting than information about the scope of dependence, and vice versa in some other explanations. Explaining the production of ATP in anaerobic respiration by detailing the steps of glycolysis seems to get more explanatory "oomph" from the detailed causal information about the chemical reactions involved than from indications of the scope of this pattern, that is, the conditions in which anaerobic respiration occurs. In contrast, explaining a trait as a product of natural selection assimilates it to a broad range of phenomena – all physical and behavioral traits that have been positively selected – while giving relatively little causal detail. Natural selection, after all, has myriad ecological sources and leads to an astonishing variety of outcomes. Kim Sterelny (1996) calls these two approaches "actual sequence" and "robust process" explanations, respectively. It may be that some explanatory dependence patterns aren't even causal in nature. The statistical pattern of regression to the mean has a broad scope of applicability; this pattern can explain a number of phenomena, such as why the outliers in some quantitative trait tend to have offspring with less extreme values for that trait. But this pattern does not seem to be causal. The insight that causal patterns are explanatory can accommodate this variety. I say more later about which causal patterns are explanatory in which circumstances, but I suspect most scientific explanations occur between these extremes. Science generates understanding by depicting causal dependence and the patterns of when that dependence holds. That is, scientists by and large explain phenomena by depicting causal patterns.

2.3 From Mechanisms to Large-Scale Causes

My first point about explanation was the idea that information about causal dependence and information about the scope of that dependence are both important to scientific explanations. Scientific explanations feature causal patterns. The second point is that not all explanations cite components and processes. There is sometimes a tendency among philosophers and biologists alike to expect that information about causal dependence boils down to information about the parts of an entity or system and the processes they collectively carry out. Attending to the role of not just causes but also patterns in causal action makes clear this is not so. Biological explanations vary from component- and process-based to those that feature large-scale or structural causes. By "large-scale causes," I mean influence from some spatial or temporal distance, and by "structural causes," I mean contextual influences that shape a phenomenon but that do not change to precipitate the phenomenon. These varieties of causal influence are just as important, just as causal, and just as explanatory as components of a system and the processes they carry out.

Let us begin this discussion by returning to the idea, mentioned at the beginning of the previous section, that there is regularly a call in biology to identify the cause or mechanism responsible for something. Some philosophers of science take very seriously such appeals to mechanisms. The so-called "new mechanists" have put a lot of work into defining exactly what a mechanism is, and they think explanation in biology and related disciplines consists in describing mechanisms. These philosophers disagree about some details regarding the nature of mechanisms and their role in explanation, but the general picture is that mechanisms are integrated networks of components that carry out certain activities, thereby bringing about predictable outcomes. Paradigmatic examples of mechanisms are the ATP cycle (Bechtel & Richardson 1993), protein synthesis (Darden 2006), and the action potential (Craver 2006).

This view of mechanisms emphasizes the explanatory value of identifying processes carried out by the components of an entity. This approach fits very well with some areas of research in biology, but advocates of mechanisms tend to go further, expecting processes carried out by the components of a given entity to be central to *any* explanation, at least in biology. For example, Connolly et al. (2017) suggest that ecology needs to focus exclusively on component- and process-based models, which depict the causal roles of components of a system or the processes that precipitate some outcome. These authors claim that the value of such models is that they can capture the causal structure of a system. Philosophers who emphasize the explanatory value of mechanisms tend to equate a lack of detail about causal processes with a failure to explain (see, e.g., Craver 2006). This is related to McGill & Nekola's (2010) characterization of ecologists sometimes justifying the value of their work by appealing to it being more mechanistic, resulting in "ideological squabbles" about what qualifies as a mechanism.

I urge a much broader interpretation of the call to identify causes. Not all explanatory causal information regards processes carried out by components. Some explanations feature causal patterns regarding the environmental context, such as optimal foraging models for evolved food preferences, which cite ecological factors such as patterns of food distribution that give rise to certain selection pressures. These cite structural causes, by which I mean contextual factors that may not have changed to precipitate the phenomenon. The distribution of food need not have changed to bring about selection for given food preferences, but had this distribution been different in certain respects, the evolved food preferences would have been as well. Other explanations cite large-scale causes that are distant in space or time from the phenomenon, such as evolutionary or phylogenetic explanations for traits (temporally distant causes) and some ecological explanations, such as wetlands suffering due to decreased snowpack in the mountains (spatially distant cause). Finally, still other explanations describe highly general causal patterns to which focal phenomena cohere, such as appealing to the second law of thermodynamics to explain ice melting, cooling a beverage in the process; indicating that some trait is a product of natural selection; or indicating how scarlet ibis coloration is an instance of carotenoid pigmentation in birds. All of these are causal pattern explanations – explanations that feature information about causal dependence and information about the scope of that dependence – but none are naturally described as processes carried out by the components of the system. (Of course, as I have already suggested, some other causal pattern explanations *are* naturally characterized as such.)

Philosophers who advocate mechanistic explanation disagree with me on this. For any of these cases of structural or large-scale causes or highly general causal patterns, most mechanists will either claim that the causal pattern is aptly characterized as a mechanism or call into question whether there is a genuine causal explanation. But I think causal patterns featuring contextual factors, large-scale causes, and highly general regularities significantly shape the phenomena of our world. This leads to their explanatory significance.

The expectation that all explanatory causes are local, component-based processes thus inhibits our recognition of a range of important causal patterns. These include, among many others, how ecological features shape selection pressures, phylogenetic influences on traits, and the highly general pattern of carotenoid pigmentation in birds. In my view, focusing on a narrow sense of mechanism both results from and contributes to an inaccurately reductionist view of the world, where causal significance is expected to be local and component-based. Such an expectation renders large-scale causal patterns less visible or even "spooky" seeming.

In this section, I have suggested that the call in biology for causes, process, or mechanism should be interpreted as a call for information about causal patterns, wherever they are found. This is a broader interpretation than that of philosophers of science who emphasize the significance of mechanisms, understood roughly as processes carried out by components. Yet this broader interpretation of the call for mechanism goes beyond establishing correlation, or the existence of a "mere" pattern. Causal patterns are more than just patterns: they are regularities in how causes exert their effects. This gives information about what to expect in different circumstances and about how to intervene on a phenomenon to bring about a desired effect. And this is so whether the causal pattern in question is local and component-based or large-scale. In the next section, I give reason to think that the very same phenomena will sometimes be explained by citing local, mechanistic causes and other times large-scale causes.

Much more argument would be needed for me to provide full support for this broad conception of explanatory causal patterns. But I hope this brief discussion is sufficient for two purposes. First, to provide some initial motivation for the idea that biologists should look beyond the local components of a system in their hunt for causes, explicitly including consideration of the significance of large-scale causal factors. And, second, simply to highlight that a philosophical question about scientific explanation is the degree to which explanations must be component- and process-based.

2.4 A Variety of Explanations Without Conflict

Scientists are regularly in the position of trying to discern whether a proffered explanation is right. This is a challenging task, and the details of how those decisions are made are, for the most part, beyond the scope of this chapter. (That said, philosophy of science does have significant resources to offer on this issue as well; see, for instance, Chapter 3.) But, the third point I want to make about scientific explanation regards a related issue. Alongside the need to discern whether a proffered explanation is right, scientists are often in the position of adjudicating between two different explanations of the same phenomenon. For instance, biologists may ask whether some trait is the product of natural selection or phenotypic plasticity, or whether an evolutionary game theory model or quantitative genetic model is more apt. If I am right that causal patterns explain, including broad causal patterns, this has an interesting consequence for the task of adjudicating between different potential explanations of the same phenomenon. At least sometimes, multiple explanations of a phenomenon *all* may be right.

The thought here is that the same phenomenon is influenced by lots of different causal factors, and thus may be explained by a number of different causal patterns. A good illustration of this is Mayr's (1961) proximate and ultimate causes of traits: physiological or developmental influences on the one hand and evolutionary influences on the other. Patterns exist in how proximate causes exert their influence, and patterns exist in how ultimate causes exert their influence. There's no immediate reason to think that either pattern impinges on the other. Rather, some trait – let's say, scarlet ibis feather color – has a physiological explanation about the production of feathers with carotenoid pigmentation, as well as an evolutionary explanation about why this species of birds evolved to have feathers of this color.

What is more, feather color (and other traits) don't just have a proximate and an ultimate explanation. For scarlet ibis plumage coloration, for example, there may be an explanation focused on the role of selection – such as this coloration's role in mate attraction, as well as an explanation focused on the steps by which gene complexes supporting the necessary enzymes and selective transport of carotenoids evolved in birds. Or an explanation may detail the specific carotenoid-carrier protein in the blood of scarlet ibises but lacking from white ibises. There may also be something to say about how parasite load influences carotenoid metabolism and transport. Each of these explanations addresses why the scarlet ibis has the feather color(s) it does, but each does so by depicting different causal patterns, all of which may play a role in scarlet ibis coloration.

In this way, I think scientific explanations proliferate. There can be multiple different explanations of a given phenomenon without conflict – Mayr's proximate and ultimate explanations are just the tip of the iceberg. And, in my view, some disputes in biology over which kind of explanation is more successful – evolutionary game theory or quantitative genetic, selective or evolutionary developmental, etc. – are pursued in error. The question is often (not always, but often) not which explanation is correct, but simply which kind of causal pattern is sought. (See Potochnik 2013 for a fuller development of this point.)

The list I sketched previously of different explanations for scarlet ibis coloration vary in terms of which causes of the coloration they feature. This is one way in which different causal pattern explanations, even of the same phenomenon, vary. Causal pattern explanations also may be more general, say, focusing on carotenoid metabolism in birds in general, or more nuanced, such as how several of these causal factors combine for the scarlet ibis in particular. Which of these causal patterns is explanatory depends on the research priorities. To generate explanatory knowledge, the question is not simply, "what caused this?" but instead, "which of the potential causes are we interested to explore?" and only then, "what role did those causes play?"

It is certainly possible for one explanation of a phenomenon to reveal that another proffered explanation for that phenomenon is wrong. If it is discovered that some population of flamingos is white due to a lack of beta-carotene in their diet, it would be mistaken to look for a natural selection explanation for why flamingos are white. White coloration in this population is not the product of natural selection, but was directly, and recently, caused by environmental change. Yet it is much more likely for different explanations of a phenomenon to be compatible than for them to conflict. Phenomena have many different causal influences, and there are many different patterns in how these influences play out.

In some circumstances, biologists may be interested to piece together as much as possible about all the causal influences on some phenomenon – say, all the phylogenetic, evolutionary, developmental, metabolic, and environmental influences on carotenoid pigmentation in the scarlet ibis, as well as the interplay among those influences. Other times – and much more often, I suspect – the investigation of a phenomenon occurs in the context of some broader research interests that occasion focus on a particular causal pattern to the exclusion of others. For the present example, such a focus may be the evolution of carotenoid metabolism and selective transport in birds, or whether and in what ways sexual selection is responsible for carotenoid coloration in the scarlet ibis, or how parasite load influences carotenoid metabolism and selective transport, etc.

The relationship among multiple explanations of a given phenomenon has come up in a few different contexts in philosophical discussions of scientific explanation. A handful of influential philosophers have suggested that any given explanation is partial, so there are inevitably multiple different explanations of any one phenomenon (Railton 1981; Lewis 1986). Philosophers have disagreed about whether and in what ways such multiple explanations should relate to one another, with some anticipating integration of these explanations (e.g., Mitchell 2003) and others arguing, as I have also suggested here, that multiple different explanations of a phenomenon remain independent from one another. From my perspective, this explanatory independence, as I have called it elsewhere (Potochnik 2010), arises due to varying research interests even among those investigating the same phenomenon and to different causal patterns being explanatory in light of those various interests.

In some cases, these different interests are obvious. Other times, biologists may take themselves to disagree about causal facts, but the disagreement is also motivated by different research priorities and thus different aims for the explanations they are developing. To illustrate this point, let's return to Mayr's proximate–ultimate distinction but this time for a different purpose. Laland et al. (2011) challenged an implication this distinction has often been taken to have, namely that developmental processes are evolutionarily unimportant (one might say, merely proximate). These researchers emphasize that, to the contrary, feedback loops exist by which developmental processes influence evolution. They conclude: "It is now vital to recognize that developmental processes frequently play some role in explaining why characters possess the properties that they do, as well as in accounts of the historical processes that explain their current state" (p. 1516). This is an important observation that has significant implications for evolutionary theory. But, I do not think the significance of developmental processes to evolution is a reason to replace selective explanations of traits with selective-developmental explanations. Rather, in my view, this is better interpreted as the identification of a neglected kind of causal pattern, namely patterns in how development influences potential evolutionary outcomes. These different causal patterns are explanatory in light of different research questions. Sometimes a classic evolutionary explanation suffices, and the influence of development on evolution can be ignored; other times the latter is central to what biologists aim to understand.

I have suggested that there can be multiple noncompeting causal pattern explanations for any given phenomenon. In light of this idea, I urge biologists to take seriously the possibility that the apparent conflict among different research programs arises not due to different competing explanations of the same phenomenon, but rather due to different research agendas that lead to emphasizing different causal patterns. Sometimes a breakthrough in understanding warrants revisiting what we thought we knew, including the nature of the causal patterns we had posited in our explanations. It may be that, for some traits, there is no evolutionary explanation – no causal patterns to be found – without taking both selection and development into account. Other times, breakthroughs in understanding bring to light new causal patterns but do not undermine our existing explanations.

2.5 Conclusion

Explanation is taken to be an important aim, if not the central aim, of science. In this chapter, I have motivated three ideas about the nature of scientific explanations. These ideas are grounded in philosophical debates about explanation, even as they also reflect my particular views. First, I have suggested that philosophical debates about the definitive features of explanation support the idea that both causal dependence and the scope of that dependence – that is, causal patterns – are important to explanation. If this is so, biologists might explicitly think about both the causal content and the generality or scope of the explanations they develop. It is not always more explanatory to build in more detail. When scarlet ibis coloration is investigated in the context of explaining carotenoid metabolism and selective transport in birds, any reference to, say, sexual selection for the ibis's coloration is mere distraction. Omissions and simplifying assumptions are ways of signaling that those details don't matter given the present research aims – that the pattern in question is independent of them.

Second, I have suggested that while some explanations focus on components and processes, others focus on large-scale causes, including contextual features, distant causes, and highly general patterns. I urge biologists to look beyond the local components of a system in their hunt for causes, and to explicitly include consideration of the significance of large-scale causal factors. I am inclined to think that all of us – scientists, philosophers, and the public alike – share certain reductionist tendencies. Among these is a tendency to consider the large-scale to be a fixed background and the local and tiny to be where the causal action is. Across science, again and again, this expectation has been revealed to be incorrect. And yet the tendency persists.

Third and finally, I have suggested that phenomena of interest in biology may have multiple explanations, each occasioned by different research agendas and featuring different causal patterns. In my view, some disputes about research strategies and methods are, at root, disagreements about which explanations are most interesting – which causal patterns enlightening – rather than disagreements about which explanations are accurate. This idea also relates to the idea I motivated about the explanatory value of large-scale causal patterns. One way in which large-scale causes have been rendered invisible is by pointing out that there is already an explanation in terms of components or other local factors. But if phenomena have multiple explanations, the recognition of local, small-scale influences shouldn't lead us to expect the absence of large-scale explanations. Cancer has genomic causes, but it also has developmental, environmental, and socioeconomic causes. And yet our research dollars seem to go disproportionately to studying the tiny molecular bits residing inside us.

This reveals the error in what I take to be another reductionist tendency: an implicit expectation that events have just one or a few causes. To the contrary, complex causal relations abound, with any event bearing the influence of many causes, and causal interaction and feedback common (Love 2017). Recognizing and emphasizing that biological phenomena embody multiple causal patterns, and that different causal patterns can figure into explanations tailored to different questions, is one step toward counteracting these reductionist tendencies.

References

Bechtel, W. & Richardson, R. (1993). *Discovering Complexity: Decomposition and Localization as Strategies in Scientific Research*. Cambridge, MA: MIT Press.

Connolly, S. R., Keith, S. A., Colwell, R. K., & Rahbek, C. (2017). Process, Mechanism, and Modeling in Macroecology. *Trends in Ecology & Evolution* 32(11): 835–844.

Craver, C. (2006). When Mechanistic Models Explain. *Synthese* 153(3): 355–376.

Darden, L. (2006). *Reasoning in Biological Discoveries*. New York: Cambridge University Press.

Friedman, M. (1974). Explanation and Scientific Understanding. *The Journal of Philosophy* 71(1): 5–19.

Gopnik, A. (1998). Explanation as Orgasm. *Minds and Machines* 8: 101–118.

Hempel, C. (1965). *Aspects of Scientific Explanation and Other Essays in the Philosophy of Science*. New York: Free Press.

Kitcher, P. (1981). Explanatory Unification. *Philosophy of Science* 48(4): 507–531.

Laland, K., Sterelny, K. Odling-Smee, J., Hoppitt, W., & Uller, T. (2011). Cause and Effect in Biology Revisited: Is Mayr's Proximate-Ultimate Dichotomy Still Useful? *Science* 334(6062): 1512–1516.

Lewis, D. (1986). Causal Explanation. in *Philosophical Papers*, Vol. II. Oxford: Oxford UP.

Lombrozo, T. (2011). The Instrumental Value of Explanations. *Philosophy Compass* 6(8): 539–551.

Lombrozo, T. & Carey, S. (2006). Functional Explanation and the Function of Explanation. *Cognition* 99(2): 167–204.

Love, A. (2017). Building Integrated Explanatory Models of Complex Biological Phenomena: From Mill's Methods to a Causal Mosaic. In M. Massimi et al. (eds.), *EPSA15 Selected Papers, European Studies in Philosophy of Science* 5: 221–232.

Mayr, E. (1961). Cause and Effect in Biology. *Science* 134(3489): 1501.

McGill, B. & Nekola, J. (2010). Mechanisms in Macroecology: AWOL or Purloined Letter? Towards a Pragmatic View of Mechanism. *Oikos* 119(4): 591–603.

Miller, J. H., Druckenmiller, P., & Bahn, V. (2013). Antlers of the Arctic Refuge: Capturing Multi-Generational Patterns of Calving Ground Use from Bones on the Landscape. *Proceedings of the Royal Society B: Biological Sciences* 280(1759): 20130275.

Mitchell, S. (2003). *Biological Complexity and Integrative Pluralism*. Cambridge: Cambridge University Press.

Potochnik, A. (2010). Explanatory Independence and Epistemic Interdependence: A Case Study of the Optimality Approach. *The British Journal for Philosophy of Science* 61(1): 213–233.

Potochnik, A. (2013). Defusing Ideological Defenses in Biology. *BioScience* 63(1): 118–123.

Potochnik, A. (2015). Causal Patterns and Adequate Explanations. *Philosophical Studies* 172(5): 1163–1182.

Potochnik, A. (2017). *Idealization and the Aims of Science*. Chicago: University of Chicago Press.

Railton, P. (1981). Probability, Explanation, and Information. *Synthese* 48(2): 233–256.

Reutlinger, A. & Saatsi, J. (eds.) (2018). *Explanation Beyond Causation*. Oxford: Oxford University Press.

Sterelny, K. (1996). Explanatory Pluralism in Evolutionary Biology. *Biology and Philosophy* 11: 193–214.

Strevens, M. (2004). The Causal and Unification Approaches to Explanation Unified – Causally. *Noûs* 38(1): 154–179.

Strevens, M. (2008). *Depth: An Account of Scientific Explanation*. Cambridge: Harvard University Press.

Trams, Eberhard G. (1969). Carotenoid Transport in the Plasma of the Scarlet Ibis. *Comparative Biochemistry and Physiology* 28(3): 1177–1184.

Woodward, J. (2003). *Making Things Happen*. Oxford: Oxford University Press.

3 What Is Biological Knowledge?

KEVIN MCCAIN

3.1 The Genus to which Biological Knowledge Belongs

As biologists can certainly appreciate, it can be helpful to think of *biological knowledge* as a species belonging to the genus *knowledge*. A good starting point for understanding the features of biological knowledge is therefore to ask what features all instances of knowledge have in common.

First of all, it is important to recognize that knowledge is not something that exists independently of knowers. Perhaps the easiest way to get a handle on this idea is to distinguish between knowledge and facts. It is a fact that chimpanzees and gorillas have a common ancestor. However, not everyone knows this fact. Biologists know this because they believe this fact on the basis of very good evidence. Many laypeople do not know this fact, either because they lack evidence or because they simply lack the belief that this fact is indeed a fact. Consider another example: suppose that a particular allele is correlated with the development of cancer in humans. Suppose further that there is no available evidence in support of this correlation. In such a case it is a fact that there is such a correlation, but this fact is not known. Not only this, the fact is not simply knowledge waiting to be found. In order for knowledge of this fact to exist, it must be constructed. A knower, or group of knowers, must gather sufficiently strong evidence in support of this fact and form the corresponding belief on the basis of that evidence. Only after this has occurred is there genuine knowledge of the fact. Similarly, skill knowledge is something that must be constructed rather than discovered. One does not discover that she always knew how to extract DNA, say. Rather, she learns techniques for doing so. That is to say, she develops, or constructs, the appropriate knowledge of how to extract DNA from a sample.

As our discussion has already suggested, there are at least two kinds of knowledge that seem importantly different. Roughly, the first kind of knowledge, what is commonly called "propositional knowledge,"[1] concerns knowledge of facts, and the other, called "knowledge how,"[2] concerns skills or abilities.[3] It seems clear that these are different kinds of knowledge. For example, someone might know many true propositions about how to sing, say from studying musical theory and listening to many great performances, and yet not be able to sing in tune. In such a case, the person has propositional knowledge, but she lacks the know-how of a singer. Alternatively, it also seems that someone could have know-how without having much propositional knowledge. Think about the case of someone who is naturally talented and an amazing singer but entirely uneducated in music. She is able to sing beautifully, but she cannot read music, nor does she have any knowledge of musical theory. It may be that such a person cannot express how she sings in tune to others. Maybe she cannot even express it to herself if she were to think about how she does it. This seems possible, and it would be an instance where someone has know-how without corresponding propositional knowledge. Although there is much more that could be said about the relationship between these two kinds of knowledge, enough has been said to make the point that they are different and to illustrate ways that they may come apart (Ryle 1949; Squire 1987; Lewis 1990; Schacter et al. 2000; Adams 2009; Poston 2009; Devitt 2011; and cf. Stanley & Williamson 2001 and Stanley 2011).

A natural question arises at this point: What is required for a person to have either of these two kinds of knowledge? Knowledge-how is perhaps the simplest to get a clear handle on. In most cases, in order to

---

[1] This is the common phrasing among philosophers. Psychologists sometimes refer to this as "semantic" or "declarative" knowledge. See Schacter et al. 2000 and Squire 1987.

[2] Schacter et al. (2000) and Squire (1987) refer to this as "procedural knowledge."

[3] There is a third sort of knowledge – *acquaintance knowledge* – that seems to be distinct from the other two discussed in the text. Acquaintance knowledge is the sort of knowledge that one has when one knows a particular person. For example, Jack might know a large number of facts about Bill Clinton – when he was born, the years he was president of the United States, and so on without having ever met or interacted with Clinton. Jill might fail to know some facts about Clinton, but she might be personal friends with him. In such a case we would say that Jill has acquaintance knowledge, but Jack does not. Simply put, Jack only *knows about* Clinton whereas Jill *knows him*. For more on this, as well as the other kinds of knowledge discussed in the text, see McCain 2016, ch. 2.

know-how to do something one must have the ability to do it. For example, one knows how to swim, if one can swim. If one cannot swim, one does not know-how to do it. Admittedly, one might have a lot of *propositional* knowledge about how to swim, but one does not really *know-how* without being able to actually swim.[4] Knowledge-how to do something amounts to having certain abilities – namely, the ability to do that thing.

Propositional knowledge is a bit more complex to spell out. Fortunately, it has been a key area of study for many philosophers, so the ground has been covered fairly well. Although there is disagreement about the specifics, it is widely accepted that three primary requirements of knowledge are truth, belief, and justification. This is sometimes referred to as the "traditional account of knowledge" or the "justified true belief account of knowledge" (McCain 2016). The general idea is that knowledge is made up of, or at least entails, justified true belief.[5] That is to say, when you know a proposition is true, you believe the proposition is true, it is true, and your belief is justified (it is based on sufficiently strong evidence).

Why think that knowledge requires justified true belief? To perhaps oversimplify, knowledge is information that we can go on – the things that we know can be relied upon when deciding how we should act (Fantl & McGrath 2009, ch. 3). If someone wants orange juice and she knows that there is orange juice in the fridge, she can use that knowledge to guide her actions. This entails that knowledge requires belief. Only if the person holds the belief that there is orange juice in the fridge can this belief serve as knowledge that can guide this person to open the fridge to get the orange juice when she wants to drink it.

---

[4] There are complications here. One might know-how to swim but recently broke a leg. In such a case, it seems that one still knows how to swim even though one cannot at this time swim. This sort of general concern will be set aside as the primary focus here is on typical cases of know-how. For an excellent discussion of the current state of scholarship on knowledge how, see Carter & Poston 2018.

[5] The reason for this qualification is that while the traditional (and still predominate) view is that knowledge can be analyzed into components which include justified true belief, following Williamson (2000) a significant number of philosophers take knowledge to be an unanalyzable primitive concept. That being said, those who, like Williamson, think that knowledge is primitive in this sense are often willing to grant that knowledge entails justified true belief – i.e., if you know some claim, that entails that the claim is true, you believe the claim, and that you have sufficiently strong evidence for the claim.

Continuing with the thought that knowledge is information that we can rely upon, knowledge requires truth. We cannot really rely upon false information – it is very likely to lead us astray. While at times things are said that suggest that knowledge can be false – "people knew the earth was flat," for instance – this is best understood as simply loose talk. People did not know that the earth was flat because it was not, and is not, flat. Rather, people *thought* they knew, or even had good reasons to think that they knew, that the earth was flat without actually knowing this. Sometimes we think we know, and we do; other times we think we know, but we are mistaken. Holding beliefs of various kinds is possible, but only those that are true are candidates for knowledge.

Finally, knowledge requires justification because this is what separates knowledge from lucky guesses. There is a very big difference between someone who sees the result of a coin toss and someone who simply guesses that the coin landed "heads" up without seeing it. Even if the coin did in fact land "heads" up, only the first person has knowledge. What is the difference? The person who sees the coin is justified in believing this – she has good reasons/evidence in support of the claim that the coin landed "heads" up. The second person does not know how the coin landed. True, she guessed correctly, but she simply guessed. Getting the right answer by luck is not sufficient for knowledge. For these sorts of considerations, it is widely held that in order to know that some claim is true one must have a justified true belief that that claim is true.

Biological knowledge is a species of knowledge. Consequently, biological knowledge has these same features. Know-how in biology (e.g., using a population model to make predictions) requires having certain skills and abilities. Propositional biological knowledge involves believing a true biological proposition on the basis of sufficiently strong evidence. Hence, when it comes to the general nature of knowledge there is nothing particularly special about biological knowledge as such. Nonetheless, there *are* features of how biological knowledge is generated and transmitted as well as additional distinctions between kinds of biological knowledge that make biological knowledge a special kind of knowledge.

3.2 Kinds of Propositional Biological Knowledge

Although both know-how and propositional knowledge are important features of biological knowledge and play key roles in biological practice, from this point on the focus will be on propositional knowledge. When it comes to propositional biological knowledge there is an important distinction that should be drawn. It is useful to distinguish between direct/observational knowledge on the one hand and theoretical knowledge on the other. Direct/observational knowledge is knowledge gained by way of observations. For example, one learns the size of the population of lions in a particular area of Africa, say, by observing the area and keeping track of the number of lions. Such direct/observational knowledge can be readily transmitted, of course. Once someone has observational knowledge of the population size, they can easily share this knowledge with others via testimony – they can tell others or publish the results in a journal, for example.

Theoretical knowledge is different. It is not gained simply by way of making observations. Instead, theoretical knowledge often starts with observational knowledge and builds from there. Theoretical knowledge arises from the attempt to explain the observational knowledge, which is descriptive. As Kampourakis and Niebert (2018, p. 237) have aptly put the point, "While a description [observational knowledge] aims to answer *what* has happened, explanations try to give us an answer to *why* it took place." We gain theoretical knowledge by explaining the observations we make both in and out of the lab. Similar to observational knowledge, we can readily share theoretical knowledge via testimony – common mediums for this are academic publications and professional conference presentations. But, how do we come to have knowledge through the act of attempting to explain what has been observed? This is achieved through inference to the best explanation (IBE).

3.3 Theoretical Knowledge and Inference to the Best Explanation

Theoretical biological knowledge arises via inference to the best explanation (IBE). This prompts three questions. The first is simply what exactly is an explanation? Many answers to this question have been proposed, and it is still a matter of philosophical debate (McCain 2016, ch. 9). For our purposes here, it is enough to think of an explanation as an answer to the question of why (or how) a particular phenomenon occurred (see Chapter 2 for more in-depth discussion). The other two questions, which will be the focus of the rest of this section, are: How can we gain knowledge via IBE? And, is IBE a reliable way to generate knowledge?

3.3.1 Getting Theoretical Knowledge from Explanations[6]

Inference to the best explanation (IBE) is familiar and ubiquitous in our lives. When a veterinarian determines what is wrong with the family pet, she uses IBE. She considers the symptoms the pet has displayed and the potential explanations of those symptoms. She then infers that the explanation that best explains those symptoms is correct. Similarly, when you try to determine why your car will not start, you employ IBE. It was running fine earlier today, the battery is new, it has plenty of fuel, and the alternator has been making a sound for weeks. A reasonable inference to draw here is that the alternator is bad. The bad alternator best explains the data you have about your car. Philosophers have argued that not only do we use IBE to come to knowledge via inferences of this sort, we also employ it any time we gain knowledge via testimony (Fricker 1994; Lipton 1998). Their thinking is that we can reasonably accept what someone tells us (whether this testimony comes orally or via some written medium) only if the best explanation of why they are telling us what they say is that they know what they are talking about.

Given the ubiquity of IBE, it may not be surprising that scientific reasoning employs it in a careful and refined form. It will be helpful to make the general form of IBE more precise here. McCain & Poston (2019) have formalized it this way (though I have changed the phrasing slightly here):

Inference to the Best Explanation (IBE)

(1) There is some data, d, and some background evidence, k.
(2) E explains d better than any available competing potential explanation.
(3) E is a good explanation given k.
(4) Therefore, E is true.[7]

---

[6] For more on how explanatory reasoning leads to knowledge in science in general, see McCain 2016, 2019.

[7] A key difference between McCain & Poston's formulation and others is their inclusion of (3). In essence, the inclusion of (3) means that IBE in terms of a slogan should be "infer the best sufficiently good explanation" rather than, as it is typically understood,

The most famous instance of IBE in biology comes from Charles Darwin. In *The Origin of Species* he justified natural selection on the grounds that it provides the best explanation of a wide variety of biological facts:

> It can hardly be supposed that a false theory would explain, in so satisfactory a manner as does the theory of natural selection, the several large classes of facts above specified. It has recently been objected that this is an unsafe method of arguing; but it is a method used in judging of the common events of life and has often been used by the greatest natural philosophers. (1872/1962, p. 476)

Before moving on, there are three points about IBE that need clarification. The first is that IBE involves inferring that the best explanation of a field of competing explanations is the one that is true. Since at most one of these competing (mutually exclusive) explanations can be true, it is important to realize that the explanations being compared are "potential explanations" – they are each such that *if* true, they would explain the relevant data. Hence, IBE involves inferring that the best *potential* explanation is the *actual* explanation of the data (or phenomenon).

The second point needing clarification is what makes an explanation the "best." A large number of explanatory virtues have been proposed by both scientists and philosophers of science. Common explanatory virtues include: *simplicity* (as Newton 1999, p. 794 put it, "No more causes of natural things should be admitted than are both true and sufficient to explain their phenomena ... For nature is simple and does not indulge in the luxury of superfluous causes"); *explanatory power* (the amount of data explained); *conservatism* (consistency with background knowledge); and *predictive power* (making accurate predictions). Of course, much more could be said about these various virtues. For instance, explanatory power concerns not only the individual points of data explained but also the kinds of data explained. So, an explanation that explains seemingly disparate phenomena is more explanatorily powerful, all else equal, than an explanation that only explains one kind of phenomena. Also, there are many additional explanatory virtues that have been proposed (McMullin 2008; Beebe 2009). Finally, there are a

---

"infer the best explanation." Lipton (2004) and Musgrave (1988) have also argued in support of including similar restrictions on IBE.

number of questions and debates about these virtues (e.g., what kinds of simplicity are relevant? [Sober 2015]) and how they relate to one another (e.g., if one hypothesis has more explanatory power but another is simpler, which is best?). As a result, disagreement about which hypothesis is best can arise because of differing views as to which particular virtues are (most) important and how the virtues should be understood. Such disagreement may also arise because there is debate about the degree to which a particular hypothesis possesses a given virtue. For instance, judgments of the explanatory power of a hypothesis may be affected by the sorts of data that one takes to be most important. Plausibly, some of the controversy in the "units of selection" debates arises because of disagreements concerning which data are most in need of explaining. As a result of this disagreement it is unsurprising that there are differing opinions as to the comparative explanatory power of various hypotheses. Fortunately, we do not need to enter into the intricacies of any of these debates here. Rather, it is enough to have a grasp of some of the most common explanatory virtues.

The third point to clarify is how good an explanation needs to be in order to be sufficiently good given the available evidence. This is an area of IBE that is vague. Plausibly, the reason for this vagueness is that how good an explanation needs to be will vary depending on the context. For example, in everyday life the amount of evidence that is required for making an inference is significantly lower than it is in the biological sciences. Before a scientific journal publishes a result purporting to support a new hypothesis, the researchers involved need to have garnered very good evidence for the hypothesis, passed peer review, and so on. This ensures that the knowledge produced by professional biologists is the sort that can be defended from objections and can withstand significant scrutiny.[8] So, where is the line drawn for IBE in biology? Presumably, a hypothesis that conflicts with well-established biological facts is not going to be good enough. Whereas a hypothesis that fits well with all of our current best biological theories and known biological facts is clearly good enough. Unfortunately, there is a lot of space between these two extremes. Most every instance where IBE is

---

[8] For more on the distinction between high-grade and low-grade knowledge, see Sosa 2011, 2017 and Zagzebski 1996. See Kornblith 2012 for criticism of this distinction as Sosa draws it.

employed in biology, or science in general, is going to fall in this gray area lying between the extremes. It is for this reason that many hypotheses (especially those that are newer on the scene) require a lot of testing and debate before they are generally accepted. This testing and debate is invaluable for making sure that the hypothesis meets the standards for being a "good" explanation. That being said, in many cases there is disagreement about whether a particular hypothesis is good enough to infer. Rather than being a detriment to biology, this is a driving force for increasing our biological knowledge because it spurs on additional research and debate.[9]

3.3.2 Can IBE Be Trusted?

IBE has played a significant role in a number of major scientific advances. As previously noted, Darwin's defense of natural selection relied on IBE. It was also central to the discovery of Neptune, the discovery of the electron, the general adoption of the heliocentric model of the solar system, and many other advances.[10] Despite this success, a number of objections have been leveled at IBE in the philosophical literature. Here we will discuss two of the most important ones.[11]

3.3.2.1 Multiple Plausible Rivals

In clarifying the nature of IBE in Section 3.3 we noted that it is often difficult to determine whether the best explanation is sufficiently good to warrant inference. There are a number of cases where there is an explanation that is the best, but many plausible rival explanations are

---

[9] Kampourakis & Niebert (2018) have persuasively argued that in biology before an explanation is *complete* it needs to not only supply information about causes but also about the causal process by which those causes bring about the phenomenon being explained. This raises the interesting question of how close to complete an explanation needs to be in order to be legitimately inferred. Plausibly, one does not have to have a complete explanation before the explanation is sufficiently good to warrant inference (though, of course, until the explanation is complete, inquiry on the issue is not closed). However, drawing the line demarcating when an explanation is close enough to complete for this purpose is very difficult. Fortunately, we do not need to attempt to do so here.

[10] See Kampourakis 2014 and Trout 2016 for additional examples of IBE in science. Trout argues that modern science's success on the whole is due to IBE.

[11] See McCain & Poston (Forthcoming) for discussion of additional challenges to IBE.

available. For example, Bernhardt (2012) has argued that the RNA world hypothesis is a better explanation of the origin of life on Earth than its rivals (chief among these are the hypotheses that proteins preceded RNA or that proteins and RNA evolved together). Grant for the sake of argument that Bernhardt is correct that the RNA world hypothesis is the best. Should we then infer that it is true? Arguably the answer is "no." The rival hypotheses (protein and protein & RNA) seem to suggest that it would be illegitimate to infer the truth of the RNA world hypothesis. While it might be that the RNA world hypothesis is best, and so most probable, it may be that this hypothesis is less probable than the probability that one or the other of its rivals is true. For example, it could be that the probability of the RNA world hypothesis is .4, and the protein hypothesis and protein & RNA hypothesis are each .3 probable. In this case, the RNA world hypothesis is the best individual hypothesis, but it should not be inferred to be true because the probability of one or the other of the rivals being true is significantly higher (.6). Similarly, consider the event that caused the extinction of the dinosaurs. Assume that the best explanation is that it was the result of a large meteorite striking the earth. Does this license the inference that this purported explanation is in fact the truth of the matter? It seems not because there is a plausible rival – that the event leading to the extinction of the dinosaurs was volcanic activity – in fact recent evidence suggests that both were probably important (see Chapter 10).

Dellsén (2017) has argued that consideration of cases where there are multiple plausible rivals to the best explanation exposes a serious flaw of IBE. The problem, according to Dellsén, is that IBE says that we should infer that the best explanation is true, but in such cases we should not do so. For instance, Dellsén pointed out that when it comes to the RNA world or the meteorite hypothesis for dinosaur extinction, the fact that they are the best explanations is not enough to make it reasonable to infer that they are true. Hence, he claims that IBE yields the wrong result in any cases like these.

The response to this objection to IBE is straightforward, though. It seems that (3) "E is a good explanation given k" rules out the sort of situation that Dellsén discusses. In order for (3) to be satisfied, it cannot be that the disjunction of rival explanations has a higher probability than the best explanation. In other words, as discussed in the case of the RNA world hypothesis, in order for (3) to be satisfied it cannot be that the probability of the best explanation is significantly lower than the probability that one or the other of its rivals is true (McCain & Poston 2019). So, it seems that the problem of multiple plausible rivals does not get off the ground as long as we recognize that IBE requires that we look at the relevant information – the data and our background evidence, which in this sort of case gives us good reason to doubt that the best explanation is true.

Given how straightforward the response to this objection is, one might wonder why Dellsén, or anyone, would think that multiple plausible rivals is a problem for IBE. The reason for this is that often IBE is not presented as carefully as it should be. In fact, it is not uncommon for it to be simply presented in the form of a slogan such as "the best explanation is true" or "infer the best explanation." And, even when it is given more precise formulations it is not uncommon for those formulations to leave out the sort of restriction encapsulated in (3) (see Lycan 2002, p. 413). Against these formulations and the bare slogan, the multiple plausible rivals objection is devastating. Nevertheless, once IBE is properly understood to include (3), the multiple plausible rivals objection loses its effectiveness. Hence, while appreciating the impact that multiple plausible rivals can have on whether the truth of a potential explanation should be inferred is very helpful for understanding how IBE should be construed, the possibility of such cases fails to undermine IBE's legitimacy.

3.3.2.2 Explanatoriness and Truth

The next objection to IBE is a bit more abstract. It questions the existence of a relationship between explanatory virtues and the truth of a hypothesis. Peter Lipton (2004, p. 144) has termed this objection "Voltaire's Objection." The challenge that this objection poses for IBE is one of giving good reasons for thinking that the fact that a potential explanation beats out its rivals is a reason to think that that explanation is true. In other words, the supporter of the legitimacy of IBE is pressed to show that being more explanatorily virtuous (being simpler, having more explanatory power, and so on) makes a potential explanation more likely to be true. For instance, recall the RNA world hypothesis. This time let us assume that this hypothesis is the best explanation of the origin of life on Earth, and let us further assume that it is a really good explanation without plausible rivals. Should we now infer that the RNA world hypothesis is true? In other words, what reason do we have for thinking that being the best explanation of the data we have makes the RNA world hypothesis likely to be true?

One route to responding to this objection to IBE is to attempt to argue on theoretical grounds that individual explanatory virtues, such as simplicity, are truth conducive, or on the same sorts of grounds to argue that explanatory considerations in general are evidence of the truth (See Swinburne 1997 and White 2005 on simplicity; see McCain 2018 and McCain & Poston 2014, 2017 on explanatory considerations in general). Of course, there are those who argue that such attempts are unsuccessful (Roche & Sober 2013, 2014; Sober 2015; Rinard 2017; and Roche 2018). Another route, which will be briefly pursued here, is to simply point out that reasoning via IBE has been and continues to be successful in leading us to true conclusions. While such a route may not satisfy critics' deep philosophical concerns about the legitimacy of IBE, it is sufficient for demonstrating that IBE can be a route to biological knowledge as well as knowledge more generally.

Recall, from earlier, the small sampling of times that IBE has led to scientific conclusions that were later confirmed via observation. While it is true that IBE does not always lead to a true conclusion (something that no supporter of IBE would reasonably deny), it has led to a number of important scientific discoveries. This gives us some reason to think that the sort of explanatory virtues that make one explanation superior to another are apt to lead us to true hypotheses. In addition to examples from the history of science, we have a vast number of examples of successful uses of IBE in our everyday lives. For example, after a bad storm, one might find shingles in the yard and notice a leak in the ceiling. Inferring to the best explanation would lead to the conclusion that the storm damaged the roof. This conclusion would be easily confirmed by examining the roof. This sort of everyday use of IBE is quite common (Douven & Mirabile 2018). In fact, IBE is so ubiquitous that some have argued that it is a basic belief-forming method for humans (Enoch & Schechter 2008). It is one that we use all of the time, and it has a very good track record of success in our everyday lives (Douven & Schupbach 2015 and Douven & Wenmackers 2017).

Of course, one might worry that IBE as used in biology differs from IBE used in everyday life. This is true enough. But, the difference is a matter of degree, not a difference in kind. Biologists are more careful in drawing inferences on the basis of explanatory reasoning, as they should be (Novick & Scholl Forthcoming). Nevertheless, IBE is used successfully in biology. For example, in addition to the case of Darwin discussed earlier, John Snow relied on IBE to determine that cholera was spread by drinking contaminated water (Tulodziecki 2011). Similarly, Ignaz Semmelweis's discovery that cadaveric material on the hands of doctors was the cause of high mortality rates among women who recently gave birth was the result of IBE (Lipton 2004).

Admittedly, the examples of successful IBEs just canvased may not give us the sort of demonstration that specific explanatory virtues are linked to truth that the person pressing Voltaire's objection might want. But, it does give us good reason to think that explanatory virtues can, and do, successfully guide us to true hypotheses. Thus, as with the multiple plausible rivals objection, this challenge to IBE fails to give us reason to think that IBE is not to be trusted.

In sum, IBE is our method for generating theoretical biological knowledge. When we properly employ IBE to come to accept true hypotheses, we thereby have sufficient evidence to be justified in believing those hypotheses. Judicious use of IBE, thus, provides us with the evidential grounds that we need in order to construct knowledge of theoretical biological facts.

3.4 Misunderstandings about Biological Knowledge

Now that we have discussed different kinds of biological knowledge and how we come to have such knowledge, it is worth pausing to discuss two common misunderstandings about the nature of biological knowledge. Both of these are misunderstandings that plague the sciences in general. The first is the myth that scientific knowledge is simply discovered by great thinkers. The second is the misunderstanding that real biological knowledge requires absolute certainty. Let's start with the first misunderstanding.

Often, people tend to think that scientific advances are the result of a scientist working on her own to discover knowledge that is lying preformed in nature. But, this is not true of even purportedly paradigmatic examples. For example, Charles Darwin, Gregor Mendel, and Isaac Newton did not really work in isolation, nor did they simply discover knowledge (see Olesko 2015 on Darwin and Newton; see Kampourakis 2015 on Mendel). They each made extensive use of data collected by other people, and they, as any good scientist would, relied on the scientific work of others in developing their own theories. In other words, they did not discover knowledge that had been waiting for them in the world. Instead, they constructed knowledge by looking at the data and developing explanations that best explained the data. And, even in constructing the knowledge in question, they did not succeed on their own. Biological knowledge in particular, and scientific knowledge in general, is often constructed by a team of scientists working together rather than an individual. In fact, as of 2015 the average number of authors on articles in science journals was 4.4.[12] Furthermore, even single-authored articles will cite and make use of the work of other scientists. Hence, although up to this point in the present chapter, biological knowledge has been discussed as if it were the solitary activity of individuals, it is important to be clear that it is not. While the picture so far developed has been of how an individual biologist might come to possess various kinds of biological knowledge and is the correct picture in as far as it goes, it is limited and should not be understood to be a description of how advances in biological knowledge typically come about. Although it can be easier to get a handle on the nature of biological knowledge by thinking about it in the case of individuals, biological knowledge, like scientific knowledge in general, arises from a thoroughly collaborative activity (De Ridder 2019). Recognizing the collaborative nature of scientific knowledge is key for realizing that knowledge is constructed, not discovered. After all, if knowledge were merely discovered, there would be no need for collaboration. If knowledge is simply out there in the world waiting to be discovered, one only needs to be looking in the right place at the right time. However, if knowledge is constructed, then one not only needs to make the right observations, one needs others to help gather data, debate and refine hypotheses, establish evidence, and so on.

The second misunderstanding is more troubling than thinking that advances in biology are the result of isolated great thinkers simply discovering knowledge. This is the misunderstanding that biological

---

[12] www.economist.com/news/science-and-technology/21710792-scientific-publications-are-getting-more-and-more-names-attached-them-why (Accessed May 1, 2018).

knowledge requires certainty. The thought here is that in order for something to count as biological *knowledge*, we must be absolutely certain that it is true. That is to say, the evidence in support of the claim must be so strong that it is *impossible* (in the broadest sense of the term) to have that evidence and be mistaken. Obviously, this is a high standard, and one that cannot be met in biology (or any domain of science, or any aspect of life!). While biologists can readily recognize that this level of evidence is not required for something to be accepted as true or known, it is worth emphasizing this misunderstanding here for at least three reasons.

The first is that this misunderstanding of what is required for biological knowledge is likely to play a role in misguided resistance to evolution. People may erroneously think that since the details of a theory, such as evolution, are not absolutely certain, it is "just a theory" in the sense that it is not true or known (McCain & Weslake 2013).

Second, in many cases people fail to understand that biological knowledge (like all scientific knowledge) is filled with uncertainties in the details, and as a result they are apt to expect certainty in explanations and predictions when this is impossible to have. A clear example of this is laypeoples' views on genetic testing. We know that various alleles are linked to diseases such as cancer. However, the link in these cases is probabilistic, not deductive. So, for instance, if a woman finds out that she has mutated *BRCA* alleles, she may be apt to think she is definitely going to develop breast cancer. While it is true that she is at an increased risk for breast cancer, it is far from certain that she will develop it even without treatment. Appreciating the uncertainty in cases like this can help one properly weigh the costs and benefits of pursuing various kinds of treatments (Kampourakis 2017 and Kampourakis & McCain 2019, ch. 9).

Third, and finally, when people mistakenly think that biological knowledge requires certainty, they may be misled in ways that are harmful. Two examples of this (aside from the resistance to evolution already mentioned) are doubt about humans' role in climate change and resistance to having children vaccinated. One thing that has led people astray with respect to both of these issues is a misunderstanding of the nature of uncertainty in science, specifically discounting very well-established scientific knowledge because it is not absolutely certain. Plausibly, coming to understand that while biological knowledge (and scientific knowledge more generally) requires strong evidence, it does not require certainty, could go a long way toward rectifying dangerously mistaken views of science and have a profound impact on socioscientific issues (Kampourakis & McCain 2019). In light of this, when discussing with laypeople or reporting findings, biologists would be well served by carefully explaining uncertainties and emphasizing that such uncertainties do not undermine knowledge.

3.5 Conclusion

There are various kinds of biological knowledge, though theoretical knowledge is the sort that we are most apt to think of when discussing things like the "current state of biological knowledge." While the particular models and explanations generated in biology are specialized, the general way that biological knowledge is attained is common to all the sciences, and everyday life. In biology, as in the rest of the sciences, we make use of explanations to generate knowledge via IBE. There are many questions about biological knowledge that remain: How close to complete does an explanation have to be in order to be known to be true? What is the relationship between biological knowledge and understanding? Can one truly appreciate truths in biology with only theoretical knowledge, or does one also need to know how to use various biological models? – just to name a few. Nevertheless, here we have clarified the concept of biological knowledge, especially its theoretical variety, so that it is reasonably clear. Biological knowledge is a matter of accepting a true biological hypothesis on the basis of strong (though short of certain) evidence – evidence that establishes the hypothesis as the best explanation of various biological phenomena. In sum, biological knowledge is high-grade explanatory knowledge of biological phenomena.

References

Adams, M. P. (2009). Empirical Evidence and the Knowledge-That/Knowledge-How Distinction. *Synthese* 170(1): 97–114.

Beebe, J. (2009). The Abductivist Reply to Skepticism. *Philosophy and Phenomenological Research* 79(3): 605–636.

Bernhardt, H. S. (2012). The RNA World Hypothesis: The Worst Theory of the Early Evolution of Life (Except for All the Others). *Biology Direct* 7: 23.

Carter, J. A. & Poston, T. (2018). *A Critical Introduction to Knowledge-How*. New York: Bloomsbury.

Darwin, C. (1962 [1859]). *The Origin of Species*. New York: Collier.

Dellsén, F. (2017). Abductively Robust Inference. *Analysis* 77(1): 20–29.

De Ridder, J. (2019). How Many Scientists Does It Take to Have Knowledge? In K. McCain & K. Kampourakis (eds.), *What Is Scientific Knowledge? An Introduction to the Epistemology of Science*. New York: Routledge.

Devitt, M. (2011). Methodology and the Nature of Know How. *Journal of Philosophy* 108(4): 205–218.

Douven, I. (2017). Abduction. *The Stanford Encyclopedia of Philosophy*. In E. N. Zalta, ed., *The Stanford Encyclopedia of Philosophy*, Summer 2017 Edition. https://plato.stanford.edu/archives/sum2017/entries/abduction/

Douven, I. & Mirabile, P. (2018). Best, Second-Best, and Good-Enough Explanations: How They Matter to Reasoning. *Journal of Experimental Psychology: Learning, Memory, and Cognition* 44(11): 1792–1813.

Douven, I. & Schupbach, J. N. (2015). Probabilistic Alternatives to Bayesianism: The Case of Explanationism. *Frontiers of Psychology* 6: 459. doi: 10.3389/fpsyg.2015.00459

Douven, I. & Wenmackers, S. (2017). Inference to the Best Explanation versus Bayes's Rule in a Social Setting. *British Journal for the Philosophy of Science* 68(2): 535–570.

Enoch, D. & Schechter, J. (2008). How Are Basic Belief-Forming Methods Justified? *Philosophy and Phenomenological Research* 76(3): 547–579.

Fantl, J. & McGrath, M. (2009). *Knowledge in an Uncertain World*. New York: Oxford University Press.

Fricker, E. (1994). Against Gullibility. In B. K. Matilal & A. Chakrabarti (eds.), *Knowing from Words*, pp. 125–161. Netherlands: Kluwer.

Hobbs, J. R. (2004). Abduction in Natural Language Understanding. In L. Horn & G. Ward (eds.), *The Handbook of Pragmatics*, pp. 724–741. Oxford: Blackwell.

Kampourakis, K. (2014). *Understanding Evolution*. Cambridge: Cambridge University Press.

Kampourakis, K. (2015). That Gregor Mendel Was a Lonely Pioneer of Genetics, Being Ahead of His Time. In R. L. Numbers & K. Kampourakis (eds.), *Newton's Apple and Other Myths about Science*, pp. 129–138. Cambridge, MA: Harvard University Press.

Kampourakis, K. (2017). *Making Sense of Genes*. Cambridge: Cambridge University Press.

Kampourakis, K. & McCain, K. (2020). *Uncertainty: How It Makes Science Advance*. New York: Oxford University Press.

Kampourakis, K. & Niebert, K. (2018). Explanation in Biology Education. In K. Kampourakis & M. Reiss (eds.), *Teaching Biology in Schools: Global Research, Issues and Trends*, pp. 236–248. New York: Routledge.

Kornblith, H. (2012). *On Reflection*. Oxford: Oxford University Press.

Lewis, D. (1990). What Experience Teaches. In W. Lycan (ed.), *Mind and Cognition*, pp. 29–57. Malden, MA: Blackwell Publishing.

Lipton, P. (1998). The Epistemology of Testimony. *Studies in History and Philosophy of Science* 29(1): 1–31.

Lipton, P. 2004. *Inference to the Best Explanation*. 2nd ed. New York: Routledge.

Lycan, W. (2002). Explanation and Epistemology. In P. Moser (ed.), *The Oxford Handbook of Epistemology*, pp. 408–433. Oxford: Oxford University Press.

McCain, K. (2014). *Evidentialism and Epistemic Justification*. New York: Routledge.

McCain, K. (2016). *The Nature of Scientific Knowledge: An Explanatory Approach*. Switzerland: Springer.

McCain, K. (2018). Explanatory Virtues Are Indicative of Truth. *Logos & Episteme* 9(1): 63–73.

McCain, K. (2019). How Do Explanations Lead to Scientific Knowledge? In K. McCain & K. Kampourakis (eds.), *What Is Scientific Knowledge? An Introduction to the Epistemology of Science*. New York: Routledge.

McCain, K. & Poston, T. (2014). Why Explanatoriness Is Evidentially Relevant. *Thought* 3(2): 145–153.

McCain, K. & Poston, T. (2017). The Evidential Impact of Explanatory Considerations. In K. McCain & T. Poston (eds.), *Best Explanations: New Essays on Inference to the Best Explanation*, pp. 121–129. Oxford: Oxford University Press.

McCain, K. & Poston, T. (2019). Dispelling the Disjunction Objection to Explanatory Inference. *Philosophers' Imprint* 19(36): 1–8.

McCain, K. & Poston, T. (Forthcoming). Explanation and Evidence. In C. Littlejohn & M. Lasonen-Aarnio (eds.), *Routledge Handbook of Evidence*. New York: Routledge.

McCain, K. & Weslake, B. (2013). Evolutionary Theory and the Epistemology of Science. In K. Kampourakis (ed.), *The Philosophy of Biology: A Companion for Educators*, pp. 101–119. Dordrecht: Springer.

McMullin, E. (2008). The Virtues of a Good Theory. In *The Routledge Companion to Philosophy of Science*, pp. 526–536. New York: Routledge.

Musgrave, A. 1988. The Ultimate Argument for Scientific Realism. In R. Nola (ed.), *Relativism and Realism in Science*, pp. 229–252. Dordrecht: Kluwer.

Newton, I. (1999 [1687]). *The Principia: Mathematical Principles of Natural Philosophy*. I. B. Cohen & A. Whitman, trans. Berkeley, CA: University of California Press.

Novick, A. & Scholl, R. (Forthcoming). Presume It Not: True Causes in the Search for the Basis of Heredity. *British Journal for the Philosophy of Science*.

Olesko, K. M. (2015). That Science Has Been Largely a Solitary Enterprise. In R. L. Numbers & K. Kampourakis (eds.), *Newton's Apple and Other Myths about Science*, pp. 202–209. Cambridge, MA: Harvard University Press.

Poston, T. (2009). Know-How to Be Gettiered? *Philosophy and Phenomenological Research* 79(3): 743–747.

Rinard, S. (2017). External World Skepticism and Inference to the Best Explanation. In K. McCain & T. Poston (eds.), *Best Explanations: New Essays on Inference to the Best Explanation*, pp. 203–216. Oxford: Oxford University Press.

Roche, W. (2018). The Perils of Parsimony. *Journal of Philosophy* 115(9): 485–505.

Roche, W. & Sober, E. (2013). Explanatoriness Is Evidentially Irrelevant, or Inference to the Best Explanation Meets Bayesian Confirmation Theory. *Analysis* 73(4): 659–668.

Roche, W. & Sober, E. (2014). Explanatoriness and Evidence: A Reply to McCain and Poston. *Thought* 3(3): 193–199.

Ryle, G. (1949). *The Concept of Mind*. Chicago: University of Chicago Press.

Schacter, D. L., Wagner, A. D., & Buckner, R. L. (2000). Memory Systems of 1999. In E. Tulving & F. I. M. Craik (eds.), *The Oxford Handbook of Memory*, pp. 627–643. Oxford: Oxford University Press.

Sober, E. (2015). *Ockham's Razors: A User's Manual*. Cambridge: Cambridge University Press.

Sosa, E. (2011). *Knowing Full Well*. Princeton, NJ: Princeton University Press.

Sosa, E. (2017). *Epistemology*. Princeton, NJ: Princeton University Press.

Squire, L. R. (1987). *Memory and Brain*. New York: Oxford University Press.

Stanley, J. (2011). *Know How*. New York: Oxford University Press.

Stanley, J. & Williamson, T. (2001). Knowing How. *Journal of Philosophy* 98(8): 411–444.

Swinburne, R. (1997). *Simplicity as Evidence of Truth*. Milwaukee, WI: Marquette University Press.

Trout, J. D. (2016). *Wondrous Truths: The Improbable Triumph of Modern Science*. Oxford: Oxford University Press.

Tulodziecki, Dana. (2011). A Case Study in Explanatory Power: John Snow's Conclusions about the Pathology and Transmission of Cholera. *Studies in History and Philosophy of Science* 42(3): 306–316.

van Fraassen, B. (1989). *Laws and Symmetry*. Oxford: Oxford University Press.

White, R. (2005). Why Favour Simplicity? *Analysis* 65(287): 205–210.

Williamson, T. (2000). *Knowledge and Its Limits*. New York: Oxford University Press.

Zagzebski, L. (1996). *Virtues of the Mind: An Inquiry into the Nature of Virtue and the Ethical Foundations of Knowledge*. Cambridge: Cambridge University Press.

4 What Is the Nature of Theories and Models in Biology?

EMILY C. PARKE AND ANYA PLUTYNSKI

4.1 Introduction

Biologists try to understand the living world at large by zooming in on tractable portions of it and by constructing and studying models. If you want to study long-term evolution in real time, you can follow finch populations in their natural habitats in the Galápagos (Grant & Grant 1993) or you can evolve microbial models in flasks in a laboratory (Lenski et al. 1991). If you want to understand how multicellularity first emerged in eukaryotes, you cannot go back in time to examine that event directly. Instead, you can construct and study a variety of models: for example, agent-based computer simulations, model organisms such as yeast or volvox, or model-based phylogenetic reconstructions (Bonner et al. 2016).

Biologists study models to generate predictions and to better understand the natural world. While the living world itself is often a natural starting point for generating such understanding through experiments and observations, models allow researchers to abstract away from the complex causal processes in nature and thus are typically simpler, more tractable objects of study. When a scientist cannot directly examine or intervene in the natural system or phenomenon she is ultimately interested in – because it occurred in the deep past, because it is too widespread or invisible or moves too slowly or quickly, or because doing so is unethical or not allowed – she can construct and study models instead.

One persistent view of modeling in biology, and indeed in science generally, is that it is a legitimate but ultimately second-rate way to study the natural world, when the "real" system or phenomenon of

We are grateful to Kostas Kampourakis and Tobias Uller for helpful comments on drafts of this chapter.

interest is too intractable or inaccessible. (This view is also reflected in commonly heard remarks such as, "It's only a model.") However, there are good reasons to think otherwise about modeling in biology. Models allow scientists to study tractable, idealized systems that isolate the key features, or causal variables, that they are ultimately interested in. Mathematical models, especially, have a long and successful track record in service of these ends, in both evolutionary biology and ecology (McIntosh 1986; Kingsland 1995; Provine 2001).

This chapter is about the relationship between theories and models in biology and between models and the world. We start in Section 4.2 with some philosophical issues about modeling, representation, and the relationships between theories, models, and the natural world. Then we discuss some examples of long-term microbial evolution, which drive discussion of two questions in Sections 4.3 and 4.4: What is the status of model organisms, as models particular to the life sciences? And what is the difference between models and experiments? How best to answer these questions is a matter of ongoing debate for philosophers of science. The views in this chapter touch on only some issues in philosophy of modeling, on which there is a huge literature. For further discussion, with a particular eye to biology, interested readers might start with Lloyd (1994); Godfrey-Smith (2006); Weisberg (2013).

4.2 Scientific Theories and Models

A scientific theory explains something in the natural world. To be genuinely explanatory, a theory must be well supported by repeated tests, experiments, or observations. Philosophers have debated how we should think about the relationship between scientific theories and models and other related aspects of scientific reasoning, such as laws. In the following sections we discuss these relationships with an eye to what is particular about the biological sciences.

4.2.1 What Is the Relationship between Theories and Models?

One long-standing view among philosophers is that we should think of scientific theories as collections of laws, or general statements about the world that can be true or false (such as, "Water boils at 100 degrees Celsius," or "Copper conducts electricity"). This view was developed and popularized at a time when philosophers of science focused mostly on physics and did not pay much attention to biology; it made sense to them to spotlight the formal and axiomatic side of theorizing. That focus has since changed. Alternative views of scientific theories are that theories are families of models, or that theories are collections of models, laws, and other less formal assumptions and ideas (see, e.g., van Fraassen 1980; Suppe 1989; Giere 1999).[1] These latter views are more popular today.

It may help to pause briefly here and reflect on what a model is. We will go into more detail in Section 4.2.2, but for now: Models are interpreted structures that scientists take to represent systems or phenomena in the natural world. These structures come in various media. They can be *mathematical*, such as Muller's ratchet or the Lotka–Volterra or Price's equations; they can be *computational*, such as cell lattice models or evolutionary agent-based models; or they can be *material* (also known as *concrete* or *physical*), such as physical scale models (e.g., the DNA double helix) or model organisms.[2]

When philosophers initially spoke of "theories as families of models," they typically had in mind an abstract, formal idea of a "state space" with various "dimensions," such that systems occupy positions in that space given their properties (mass, energy, etc.). This formal way of representing scientific theories, however, is just one of several ways philosophers have thought about the idea that theories might consist of families of models. Perhaps a more intuitive way to grasp this idea is in terms of the typical body of mathematical theory for a given domain; for example, population genetics or population ecology. The "families of models" view would consider these mathematical structures, together with their interpretations (or specifications of the values of variables in the system, or what they are taken to represent), as what make up the theory of evolutionary or ecological population dynamics.

---

[1] Philosophers call (roughly) these three views the "syntactic," "semantic," and "pragmatic" views of scientific theories.

[2] Some people argue about whether mathematical and computational models are really different kinds of models. For example, you might think computational models are just a special kind of mathematical model, because both are fundamentally "made of" mathematics. That is true in a sense, but Weisberg (2013) has argued that these two kinds of models differ in the key features that ground model-based explanation and prediction: In mathematical modeling it is *mathematical structures*, while in computational modeling it is *algorithms*, that are front-and-center in representing (and justifying inferences about) target systems in the natural world.

Many people also think that *laws* figure centrally in scientific theories, where laws have universal scope (they apply to everything) and express necessary (non-accidental) relationships between variables or quantifiable measures of properties such as mass, force, or energy. A famous toy example of this universality and necessity is "All spheres of uranium-235 have a diameter of less than a mile." This universal generalization describes not just how things happen to be, but how they have to be, given what we know about the world: A sphere of uranium-235 even approaching this size could not exist because it would explode. Contrast that example with "All spheres of gold have a diameter of less than a mile." This generalization is true in our universe, as far as we know. But in contrast to the uranium example, it is accidentally true: It is a contingent fact about what happens to be true. There is no special reason it could not have been otherwise. Some people think this way of understanding a law, in terms of universal generalizations that are non-accidentally true, applies nicely to at least some scientific fields (but see Cartwright 1983 for a famous counterpoint).

While understanding laws as central to theories might work nicely for some scientific fields or subfields, it is arguably impossible to find these sorts of laws in biology. There are some potential contenders. For example, Kleiber's law says that an animal's metabolic rate is proportional to the 3/4 power of its body mass (see Savage et al. 2007 for further discussion of mass/metabolic rate scaling). As another example, a population's intrinsic rate of increase is proportional to its constituents' body size (Fenchel 1974). While these and other generalizations about scaling relationships apply to many species, they do not apply as strictly as generalizations about properties of chemical elements in the periodic table, and they come with exceptions (for example, on mass and metabolic rate, see Agutter & Wheatley 2004).

Try to think of another contender for a universal generalization or law in biology. Mendel's "laws" often come to mind; but they, too, have exceptions. For example, the "law of segregation" is routinely violated; in fact, the failure of genes to assort randomly was an important clue to the realization that how frequently genes are transmitted together may have something to do with how close together they are on a chromosome. Even apparently more straightforward biological "laws" come with deviations and exceptions; for example, "living things need water to survive." This is at least in part because the nature of the subject matter of biology is variation. There is no periodic table of biology. Taxa and other biological categories do not have "essences" or fixed properties to universally generalize across. This does not mean that biology offers fewer opportunities for prediction or explanation than other sciences do: Most living things need water to survive, and the discovery of extremophiles that do not is a fascinating anomaly to assimilate into our broader understanding of physiology and evolution, rather than a crisis for those theories. The lack of universal generalizations just means that traditional accounts of scientific theories and laws do not seem to fit so well with biology (see Beatty 1997 for further discussion).

Given that biological generalizations have so many exceptions, it makes sense in many cases to speak of biological generalizations as describing conditions of "idealized" systems. Various kinds of models are used to describe such idealized systems. Models are not intended to represent the natural world with perfect fidelity, but to represent a world much like it, though far less complicated. For instance, Mendel's laws describe a "Mendelian" system of inheritance, which is far simpler and more tractable than the actual patterns and processes of inheritance in living things. This is in part why it is not uncommon in biology to use the terms "theory" and "model" interchangeably. For example, along with many other principles in population genetics, Muller's ratchet is commonly called both a theory and a model of mutational accumulation. Another example might be the Dobzhansky–Muller model (or, sometimes, theory) of the evolution of hybrid incompatibility (Dobzhansky 1936; Muller 1942). In this model, an early stage of speciation — the origins of new species — is represented as a very simple one where hybrid incompatibilities are caused by the interaction between only two genes that have functionally diverged in the respective hybridizing species, such that interbreeding leads to depressed fitness. Even though most speciation events are caused by many causal factors in interaction, this simple model can be used as a starting point to explore various scenarios, either likely causes of speciation in this or that lineage, or processes of speciation more generally.

As we have seen, a key aspect of how biologists go about theorizing is by building and studying models. So despite some philosophers' insistence on treating theories and models as separate sorts of things, this intertwining of concepts and interchangeability of terms is, at least in biology, unsurprising and unproblematic.

4.2.2 What Exactly Is a Model?

A popular way to think about models is that they are interpreted, idealized structures, which scientists construct to represent some target system in the natural world (Giere 1988; Godfrey-Smith 2006; Weisberg 2013). Here, "target system" refers to whatever system(s) or phenomena in the world a modeler ultimately aims to understand, or to explain or predict something about. Target systems are typically not particular biological systems (such as this specific population of fish, plants, or volvox), but instead, general kinds of systems. For instance, we might be interested in general features of speciation, rather than speciation in this or that particular lineage; or, we might be interested in the evolution of altruism in general, rather than the evolution of altruism in this or that lineage.

Models' targets are thus *abstract* with respect to the natural world: Specifying a model's target is a process of abstraction, or leaving out details that are not salient to addressing the scientific question at hand. For example, if a modeler aims to understand something about predator–prey dynamics, her target will include population-level details about predators and prey and their interactions. But she will probably not care, for her purposes, about the particular types of predation typical of lynxes with respect to hares. Or, if she is modeling the evolution of sexual selection, she will leave out details about salinity of the water a particular type of fish lives in, or the compositions of the individual fish microbiomes. Models are also *idealized* with respect to their targets: A modeler distorts or ignores certain facts about her target system, representing it in her model differently from how it is, in all its fine-grained detail. For example, many models in population genetics represent populations as infinite in size. This assumption is important for understanding the mathematical consequences of the model, but nobody thinks the model applies only to populations that are actually infinite in size. Some microbial populations can be impressively huge, but infinitely sized populations do not exist in nature. Abstraction and idealization allow a modeler to isolate and make salient the key factors she is interested in better understanding through her model. It is not so clear that abstraction and idealization are two entirely different processes, but in any case omitting and distorting details of the natural world are important aspects of model building (for further discussion, see Frigg & Hartmann 2012; Weisberg 2013; Potochnik 2017; Levy 2018).

Giere (1999) has argued that scientific models function akin to maps. Maps are partial representations, ignoring some details and including others. They represent the world from a particular perspective, often in service of particular, pragmatic ends — they serve a specific function. In service of these functions, we may give a distorted representation of the world. When we build a map of the New York City subway system, for instance, we might abstract away from the actual physical distances between stops in the subway. Likewise, in science we may want to understand one aspect of a system, and not others, such as causal dynamics operating in some system with respect to only two or three variables, or over one specific time scale. Giere also noted that in mapmaking, one can begin to adopt certain conventions for representations of specific kinds; subway maps around the world typically abstract from physical distances. Similarly, scientists adopt various conventions in building and using mathematical, computational, or even physical models, such as model organisms. As Giere wrote:

> Maps have many of the representational features we need for understanding how scientists represent the world. There is no such thing as a universal map. Neither does it make sense to ask whether a map is true or false. The representational virtues of a map are different. A map may be, for instance, more or less accurate, more or less detailed, of smaller or larger scale. Maps require a large background of human convention for their production and use. Without such they are no more than lines on paper. Nevertheless maps do manage to correspond in various ways with the real world. (Giere, 1999, p. 25)

This way of thinking about modeling — in terms of idealized interpreted structures representing target systems in nature — is popular among contemporary philosophers of science, and it is the default in this chapter. But it is not the only way. Some philosophers of science prefer to think of models as fictions, which modelers construct and explore in roughly the same way as someone would construct and explore other works of fiction (e.g., Godfrey-Smith 2009; Frigg 2010). Others prefer the view that models "mediate" between theories and the world, which suggests that theories and models are closely related but different kinds of things (e.g., Morgan & Morrison 1999).

Many people default to thinking about mathematical models when discussing models in biology. But as mentioned previously, models come in various media. They can be mathematical, computational, or material (also known as concrete or physical). An obvious difference between these three kinds of models is what they are "made of": Mathematical models are composed of mathematical structures and equations; computational models are made of mathematics, algorithms, and sets of instructions and are typically studied on computers; and material models are physical constructs, which in biology include living organisms.

A related difference among the three kinds of models is in the basis for their capacity to successfully stand in for the natural world; for example, their success in helping a scientist explain or predict something about her target system. In each of these three kinds of models, there is a different aspect of the model standing in for the world (Weisberg 2013). In the Price equation, two mathematical terms, w and z, stand in respectively for fitness and the value of a trait in a population. In many biomedical studies, mice stand in for humans as material models of particular health or disease states. In computational cell lattice models such as those discussed in Goel et al. (1970), cells in a two-dimensional grid in the model stand in for a range of complex morphogenetic phenomena in living multicellular organisms. While it is often relatively easy to point at which features of a model are meant to stand in for features of the natural world outside of the model, it is relatively trickier to spell out how this standing-in works. Moreover, one might wonder whether, say, mathematical models and material models are very different, not only in that one is abstract and the other concrete and physical, but also insofar as the ways in which the two are used look rather different in service of making inferences about the natural world.

For instance, model organisms often seem to have a different epistemic role from mathematical models. Model organisms might be used to test hypotheses, for instance, about whether knocking this gene out results in this or that effect. The reasoning at work in such knock-out experiments is that insofar as such organisms and their mechanisms of gene expression are relevantly similar to humans, given their shared ancestry, we might expect similar effects to occur in humans. The reasoning deploying mathematical models makes no such empirical assumptions about similarity of mechanism or shared ancestry. Mathematical models are highly idealized and abstract representations of classes of systems; inferences about the world or target systems are not as direct (though of course, we cannot expect model organisms to have perfect fidelity to their target systems, either). Computational models and inferences from them are rather different, as well; we might use them to generate hypotheses about how system dynamics might play out, given certain assumptions. Or, we might use them to test these same hypotheses, given certain ranges of initial conditions, or inclusion or exclusion of various alternative variables.

Despite these apparent differences, there is a sense in which all three kinds of models are taken to represent target systems, even though the aims and inferences might vary widely, even within each of these categories. We now turn to the question of what allows a model to represent something in the world.

4.2.3 What Is the Relationship between Models and the World?

The model/world relationship is generally understood in terms of representation. But what makes a successful representation? On the surface, there might seem to be an obvious answer: Models successfully represent when they are sufficiently similar to their target systems in nature, and the more the similarities, the better the representation. But there are immediate problems here; similarity is a tricky notion (Downes 1992). Say an ecologist is studying predator–prey dynamics; she might think of her target system as a generalized dynamic across interacting predator/prey populations, or as something more specific, say, the interactions between amoebae and bacterial populations. Consider the differences between two candidate models: the Lotka–Volterra equations, or two piles of toy plastic amoebae and bacterial cells. There is a straightforward sense in which the plastic toy amoebae and bacteria are more similar to (share more features with) the target than a couple of differential equations do. But the key point here is that the features they have in common are not the features that matter to our hypothetical ecologist. This example illustrates that it is not just some intuitively high degree of model–target correspondence that matters for successful scientific representation. Rather, it is capturing the points in common, between a model and a target, that are *relevant* to a scientist's research agenda and aims (explanatory, predictive, or otherwise).

There are different ways to think about what it means to capture the relevant points in common. Some people think it is about models and targets being partially or nearly identical in structure. Some people think it is about measuring the relevant similarities between a model and target and have tried to fill in more details of what this looks like. Michael Weisberg has suggested a formal measure for judging relevant similarities in terms of "weighted feature matching." This is based on a relationship between features shared by the model and target and features that differ between the two, which are assigned weights according to how salient they are to the causal relationships that matter to a modeler (Weisberg 2013, ch. 8). For instance, model organisms differ in their "closeness" to target systems in various respects; some model organisms may be better models for human metabolism, and others may be better models for human cognition. However, it is less clear how this works in cases where the target is less accessible to observation than humans' metabolic or cognitive functions. Say the target system occurred in the deep past, like the first appearance of multicellularity in eukaryotes or the origin of life itself. In these cases, model–target similarity cannot be assessed as directly as it can in the cases of the amoeboid predators and bacterial prey, or the organismal models of human metabolism, because nobody knows for sure which features the target has or lacks — that is why scientists are building and studying models of it.

Moreover, sometimes modelers use multiple models, because no single model alone can capture all the relevant features of the target. For very complex systems — such as the global climate — representing all the causal interactions and feedback processes in a single model may be too computationally challenging. This is, for instance, why climate modelers often build multiple models and use these together to generate predictions. This strategy of using several different models (some of which might capture some features of the system better than others) to determine if they arrive at similar predictions is sometimes described as robustness analysis. In cases where many different models, similar in some respects but different in others, arrive at the same predictions, we might take the prediction or outcome of such models to be more well confirmed (Lloyd 2015).

How do models successfully represent the world, given that they are idealized, and often highly so, in the sense outlined in Section 4.2.2 – they omit and distort the facts about the target? Here is an example: A mathematical model from population genetics, the pseudohitchhiking model, sheds light on important stochastic evolutionary forces beyond genetic drift (Gillespie 2000). This model can represent how substitutions of one allele or gene variant at one locus can lead to patterns that resemble genetic drift at a closely linked neutral locus. It does so by zooming in on a single genetic locus and assuming that population size is infinite. These idealizations allow for isolating the causal factors most important for understanding the particular theoretical issue at stake for the modeler: the effects of linkage on patterns of distribution of alleles in a population subject to selection.

Angela Potochnik has argued that models successfully represent *in virtue of* being highly idealized, counter to a popular way of thinking that they successfully represent *despite* being highly idealized (Potochnik 2017). The hitchhiking example illustrates this, because, according to Potochnik, "idealizations' positive representational role is to indicate the nature of a factor's relevance to the focal causal pattern." It is by "representing as if" that such idealized models help scientists better understand the focal pattern of interest. For instance, the notion of effective population size represents a population as if it were a size it is not, which is "justified by the functional similarity between how the actual population and an ideal population of that size experiences genetic drift" (p. 54).

Perhaps another example that illustrates the useful role of idealization even more clearly is the history of studying the evolution of cooperation. A long-standing puzzle for evolutionary biology has been: How on earth could altruism or cooperation ever evolve via natural selection in a world of organisms whose primary interest is their own survival and reproduction? Early and crucial traction on answering this question was gained by studying abstract and idealized models: specifically, computational models of iterated prisoner's dilemma tournaments, in which two model "agents" (representing organisms) have to decide whether to cooperate with one another or defect, and where defecting looks like the "safer" move but mutual cooperation is in fact overall better for both (Axelrod 1984). These models helped clarify the formal structure of the problem and a possible way to make sense of the key causal factors involved in cooperation and its evolution, and thus allowed modelers to isolate the causal variables they saw as salient to the problem.

A key feature (and often, challenge) of model building relates to these points about idealization. This is the issue of *trade-offs*. There is a constant balancing act between generality, realism, and precision (Levins 1966; Matthewson & Weisberg 2009). Though this classification has been contested (see, e.g., Orzack & Sober 1993), it seems that some models (what Levins, 1966, called "Type I" models), sacrifice generality for precision and realism. Such models might allow us to make numerically precise predictions, but only for a narrow range of systems, under a narrow range of conditions. Other models (what Levins called "Type II" models) might sacrifice realism for generality and precision. For instance, we might leave out a great deal of detail but generate a general model that can allow for predictions across an array of contexts. Still other models (what Levins called "Type III" models) might sacrifice precision for generality and realism. In these cases, we might be interested in general qualitative outcomes, and developing a model that can be used across a range of cases, without precise predictive ability. Levins claimed that we cannot maximize all three virtues simultaneously. That is, there is a three-way trade-off between generality, realism, and precision in model building in population biology. Whether or not one agrees that these trade-offs are inevitable, it does seem clear that one may strategically choose to develop a model that falls short in one respect in service of other goals. Levins also argued that if we treat the same problem with several different models, with a common biological assumption and convergent results, we have a robust theorem.

In broad strokes, plenty of modeling in biology is not so different from modeling in other sciences. Mathematical models play central roles in theorizing in physics, chemistry, and economics, as well as in biology. Mathematical, computational, or material models can stand in for target systems that cannot be observed directly because they are inaccessible in space or time (think of distant planets, starts, and galaxies; black holes; or the first life forms that emerged on Earth around four billion years ago). However, there is one important aspect of modeling that is particular to the life sciences: studying model organisms. In the remainder of this chapter, we discuss an example of experimental microbial modeling, and use it as a launching point to address some further questions about modeling. This example offers a lens through which to expose assumptions and challenge standard views about the roles of model organisms and modeling in biology more generally.

4.3 Lenski's Long-Term Evolution Experiment

Richard Lenski's laboratory at Michigan State University used an initial population of genetically identical E. coli to found 12 populations in

12 identical environments (flasks of bacterial growth medium). They began propagating the populations via serial transfer: Every 24 hours, as the populations start to use up their environmental resources, a representative sample is transferred from each well-mixed flask to a fresh flask, and thus allowed to continue evolving. This experiment began in 1988 and is still going today, an impressive 30+ years and 70,000+ generations later.

A key motivation was to study the long-term dynamics of adaptation and diversification by watching evolution "as it happens" (Lenski 2011). Because the populations started out identical and are evolving in identical environments, any differences that arise over time in fitness, physiology, or morphology are due to new mutations and the effects of adaptation, chance, or the history of individual lineages (Travisano et al. 1995). Bacteria are especially tractable organisms for this sort of work: They have rapid generational turnover; they are easy to manipulate; we know a lot about their genomes; and they have large populations that can be contained in tiny spaces. Also, populations of them can be frozen, defrosted, revived, and "rerun" from any previous evolutionary timepoint, allowing for easy comparisons between evolved populations and their ancestors. So individual lineages can be quite literally replayed, by regrowing them from the "frozen fossil record" (Blount et al. 2008). This allows researchers to do something methodologically they could previously do only with computer simulations: Back up a lineage's evolutionary history and "replay" it an arbitrary number of times (see Beatty 2006; Blount et al. 2008). This ability to replay evolutionary trajectories is just one sense in which microbes can be thought of as especially epistemically tractable, in addition to practically tractable, objects of study (O'Malley & Parke 2018).

4.3.1 Lessons from Lenski's Work

Lenski's long-term experiment has led to an impressive range of insights. *E. coli* can represent populations and phenomena outside the laboratory. Here are a few cases to briefly illustrate this range, which we use as steering examples in the rest of the chapter:

- **High mutation rates.** Most new mutations are not beneficial, so one might think that in well-adapted populations in unchanging environments, the genomic mutation rate would stay the same or even decrease. In just three of Lenski's populations, they saw the opposite: Over the first few years of the experiment, a few populations evolved genomic mutation rates orders of magnitude higher than the wild type mutation rate. The researchers explained this theoretically in terms of mutator hitchhiking, a phenomenon whereby mutator alleles (which raise the entire genomic mutation rate) can "hitch-hike" to high frequencies if, in non-recombining genomes such as Lenski's *E. coli*, they arise tightly linked to a beneficial mutation. In other words, in this organism, genes do not recombine in every generation. So, we might find that genes "come along for the ride" if they are closely linked to other genes that increase fitness. Against the background of that theoretical conceptual model, they propose that this explanation for evolved high mutation rates in the laboratory populations could explain observations of mutator alleles in *E. coli* and other pathogenic bacteria outside the laboratory (Sniegowski et al. 1997).

- **Punctuated evolution.** Over the first few thousand generations of Lenski's experiment, the *E. coli* cells got bigger, and they did so in jumps: Cell size would remain static for hundreds of generations, then dramatically increase, remain static for a while, and dramatically increase again. The researchers explained this in terms of the arrival of beneficial mutations that rapidly sweep to fixation (Elena et al. 1996). The authors also suggested that this short-term evolutionary dynamic in the laboratory might tell us something about punctuated equilibrium — where evolution occurs in short periods of stasis punctuated by rapid bursts of change — which was a hot topic in contemporary discussions of macroevolutionary trends in the fossil record (Gould & Eldredge 1977).

- **Emergence of novel traits.** A number of productive surprises have also arisen over the course of Lenski's research. Most notable was the emergence of a completely novel trait, the ability to use environmental citrate as an energy source. Lenski's populations live in a glucose-limited growth medium that also contains citrate. Citrate is an additional energy source in principle, but *E. coli* cannot use it as an energy source in oxic conditions ... or at least could not, until just over 30,000 generations in to the experiment when one population was found to have evolved the ability to do so (Blount et al. 2008).

- **Model-model comparisons: *E. coli* and Avida.** Lenski's group has also branched out into studying digital evolving systems, especially Avida. Avida is an agent-based model in which "digital organisms," whose genomes are strings of computer code, compete in a digital environment for resources in the form of computer memory. Avida is like a sort of digital petri dish, and Lenski's group has used it alongside the long-term evolution experiment to explore evolutionary dynamics and to think about the outcomes of their *in vitro* and in silico evolutionary models. For example, a paper on the evolutionary origin of complex features explores the dynamics of a broader class of events exemplified by the emergence of citrate utilization discussed previously (Lenski et al. 2003).
- **Evolutionary models of cancer.** Another interesting case, building on Lenski's and related work, is thinking of the laboratory populations as an evolutionary model of cancer. Kathleen Sprouffske and colleagues (2012) have discussed how we can gain traction on studying the evolutionary dynamics of cancer cells in a human body by studying the evolutionary dynamics of microbial models in the laboratory. Both are populations of cells that originate from a single ancestor, evolving with mutation but without recombination; they noted key differences as well, such as spatial structure (cancer cells exist in the relatively stable three-dimensional structure of tumors; experimental populations are in well-mixed flasks so are effectively zero-dimensional, where everyone can come into contact with everyone else). In any case, as the authors discussed, we can use bacterial populations to model the evolution of clonal cell lines generally, and hence as the basis for at least cautious inferences about cancer as an evolutionary phenomenon.

This was just to briefly illustrate the breadth of the findings from decades of long-term microbial evolution in the laboratory, and how it has been linked to a wide range of elements of theoretical background, and to other models along the way. We use this as a basis for discussing some further questions about modeling in Section 4.4. But first, an interlude about the status of model organisms as models.

4.3.2 Are Model Organisms Really Models?

Throughout the chapter so far, we have included model organisms, along with constructs such as physical scale models, as examples of material models. There is debate about whether or not model organisms are really models, and if so, in what sense. This debate is about more than just semantics, or what to call the organisms studied in biology laboratories: It is about how we should understand those organisms' roles as bases for inferences to the world beyond the laboratory. By "model organisms" people typically have in mind the organisms that biologists and biomedical researchers propagate and study in their labs, usually with the goal of having something to say about some particular target entities or phenomena (such as human disease, similar organisms in the wild, or evolutionary or ecological dynamics more broadly; see Ankeny & Leonelli 2011). Classic model organisms in this sense include *Drosophila*, mice, and nematodes.

Some philosophers categorize model organisms as a kind of material model, alongside things such as MONIAC (The Monetary National Income Analogue Computer; Bissell 2007), physical scale models and model airplanes in wind tunnels (Frigg & Hartmann 2012; Weisberg 2013). Model organisms are not "built from scratch" in quite the way other material models typically are, but they can be thought of as at least partially constructed by modelers, in virtue of being bred and modified to make these organisms tractable objects of laboratory study (Weisberg 2013). On this view, model organisms are interpreted structures that biologists study as analogs for their chosen target(s) of inquiry in the natural world. In this way, some have argued, model organisms function in ways akin to other theoretical models.

However, some contend that it is a mistake to classify model organisms as models akin to theoretical models. Arnon Levy and Adrian Currie have defended this view (2014). They argue that model organisms are not representations of targets in the natural world, but rather specimens of those targets. On their view, inferences from model organisms are not justified by directly comparing features known to be shared by the organism and some target(s) outside the laboratory. Instead, inferences are justified by what they call "empirical extrapolation," based on evidence from phylogeny, or other reasons to think that model and target organisms will share particular conserved traits of interest. The basic idea here is that model organisms are living representatives of broader classes of organisms, which gives inferences made from model organism research a special kind of justification, unique to the life sciences.

The point of saying that model organisms are not theoretical models is to separate their role in (biological) theorizing from the role played by mathematical and computational models. But it does not seem quite right to say that model organisms are not theoretical models (see O'Malley & Parke 2018). Consider the cases from the Lenski experiment in Section 4.3.1. Some of them look like studying a specimen; for example, the high mutation rates case, where the twelve *E. coli* populations in the lab can be thought of as specimens of pathogenic bacteria (including *E. coli* and others) more generally, outside the lab. Some other cases look like studying an indirect representation, often quite abstractly. The cancer case illustrates this most starkly: Nobody thinks Lenski's *E. coli* are specimens of cancer cells, in any sense resembling the way that they are specimens of *E. coli* outside the lab. In that case, they are taken as tractable representations of cancer as an evolutionary phenomenon, in virtue of sharing some crucial relevant similarities (such as being large populations of cells evolving clonally). This is not to suggest that *every* case could be designated either studying a specimen or studying a representation — these are not two categories with a clear line dividing them; at best there is a spectrum here. And to be fair to Levy and Currie, they do suggest that some cases of experimental microbial evolution look like what they call theoretical modeling, more than other studies of model organisms.

In any case, while it seems right to say that not *all* biological models are theoretical models, it seems wrong to carve up which models are theoretical and which are not in terms of what the model is made of (mathematics, computational algorithms, organisms, or other material constructs). Model organisms are sometimes the basis for what Levy and Currie call empirical extrapolation and sometimes function like theoretical models for the purposes of explanation or prediction (and this certainly does not exhaust their roles; for example, models are also used for exploratory purposes, or to demonstrate what is possible). A model's relationship to theory and to the world depends centrally on the modeler and her intentions. Furthermore, it is hard to say which sorts of models count as "theoretical" and which ones do not, because all sorts of models (material, mathematical, computational, and experimental) play important roles in developing, articulating, and testing theoretical claims. As we suggested at the beginning of the chapter: Models are not of one kind, nor do they all play the same predictive or explanatory role.

4.4 What Is the Difference between Experiments and Models?

To highlight an aspect of modeling unique to biology – studying model organisms – we have focused on Lenski's long-term evolution *experiment*. Perhaps some eyebrows were raised at that choice. (Why) is Lenski's experiment an example of modeling? Is it just because its subjects, *E. coli*, are called "model organisms," or because of some other features of the research?

"Experimental system," "model," "experimental model," and a handful of related terms are commonly used interchangeably in reference to model organisms. On the surface, this suggests that experiments and material models are interchangeable in biology. On the other hand, a common view is that experiments are somehow "closer to the world" than models are. Many people think something like the following: When you learn something in an experimental context, you are learning something about the "real world," whereas when you learn something in a modeling context, you are (at least initially) learning something only about your model. In this and other senses, models are thought of as lower down a hierarchy than experiments, in terms of their quality as bases for generating knowledge (or explanations, or predictions, etc.) about the natural world.

These sorts of views about experiments and models come up all the time, especially this general sense that experiments are on the top of a methodological hierarchy. Think of how common it is to hear people say things like "It's only a model," or "Nice simulation, but where are the data from the *real* world?" Students are taught that there is this thing called "The scientific method" and that it involves coming up with a hypothesis and designing an experiment to test it (see Chapter 9). Modeling is often left out of this picture, in its simplest versions. These simplest versions portray scientific inference as a matter of straightforward deduction from sets of hypotheses. This fails to capture an important part of biological practice. As we have seen, models are used in a diverse array of ways, to represent abstractly systems of interest, to generate hypotheses, and to test them. When modeling is included in discussions of scientific method, it is often treated as (purely) subservient to experiments: Modeling is what you do with your experimental data to better understand or analyze them; then you design and repeat your experiment.

Some philosophers of science think there are important differences between experiments and models regarding the value of what scientists can learn from them. Most of these arguments have focused on differences between how experimental systems and models relate to scientists' target systems in the natural world. In particular, the claim is that experimental systems are specimens of scientists' targets in the natural world and correspond closely and materially to those targets; whereas models only represent targets in the natural world and correspond to them only distantly and formally. There is an intuitive pull to this sort of view. If a scientist is interested in ultimately having something to say about humans, there seems like a straightforward sense in which poking and prodding actual humans in a laboratory gets her closer than watching dots move around on a screen in a computer model. In defense of this sort of view, Mary Morgan has said, "The same materials are in the experiment and the world ... [they] share the same stuff" (2005). So on the basis of these differences, experiments are argued to generate more reliable scientific knowledge (for two versions of this sort of view, see Guala 2002; Morgan 2005; for responses and more recent discussion of these issues, see Parker 2009; Parke 2014; Currie & Levy 2019).

Here, briefly, are some problems with these claims about experiments being overall superior to models (see Parke 2014 for further discussion). Recall the first two cases of inferences from Lenski's work discussed in Section 4.3.1: high mutation rates and punctuated evolution. On the view outlined above, there might be good reasons to think that the high mutations rate case is experimental, while the punctuated evolution case is modeling. In the former case, the lab *E. coli* with high mutation rates bear a clear and close relationship to *E. coli* and other pathogenic bacteria outside the lab, which are the targets of the inference. In the latter case, the *E. coli* in the lab do not bear a clear relationship to the targets of the inference: Long-dead lineages that have left evidence of punctuated evolutionary trends in the fossil record (apart from the trivial point that both are biological organisms). On the one hand, it may be reasonable to take the former case to be an experiment and the latter a model. On the other hand, it might seem counterintuitive to label the exact same lab work as either experimenting or modeling, depending only on the inference in question, if one subscribes to the view that these are methodologically distinct.

In any case, it is not the terminological difference that is at stake, but instead: making clear what inferences are warranted. The punctuated evolution example in particular illustrates the difficulties in assessing whether objects of study are representations of, versus specimens of (or materially similar to), their targets. The claim there is about a long-term evolutionary dynamic; the identity of any specific natural populations that instantiate that dynamic is not at stake. Neither is their phenotypic or phylogenetic closeness to *E. coli*. The question of whether the model organisms and the target in nature are "made of the same stuff" stops making sense in cases like this. Stepping back from these particular cases: The same model organisms in flasks in a lab can look relatively much "closer to" or more "distant from" the natural world, depending on what inference about the world is at stake. The difference between specimens and representations does not track any in-principle facts about the medium of an object of study. Being comprised of organisms or physical materials, or mathematics or computational algorithms, does not tell us, in and of itself, about an object of study's quality as an inferential basis.

Carving science at its joints in this way, and using the resulting categories as bases for judgments about inferential or predictive power, focuses on the wrong issues. There is no hierarchy of explanatory or inferential power, such that experiments (or models, for that matter) are automatically at the top. There are, of course, clear methodological differences between studying mathematical equations on a computer and studying organisms in the laboratory or field, such as the resources and timescales involved. But there is no real difference between experiments and studies of material models. And when you learn something in an experimental context or in *any* modeling context, you are learning something (at least initially) only about your experimental system or your model. Any further inferences about target systems in the natural world need further justification. For instance, we can study the evolution of cooperation in a highly idealized computer simulation, but before we draw any inferences about how to apply such models, for example, to human behavior, we need to check whether and how the conditions described in the simulation fit with the natural world. Of course, this justification does not have to start from scratch on a case-by-case basis. There are plenty of disciplinary traditions of taking one system to reliably and robustly tell us about another, for good reasons including track records (for example, well-designed clinical trials telling us about human patients at large, or climate models telling us about the weather tomorrow, at least in some locations). But there is no principled sense in which experiments should automatically license more confidence in inferences about the world, compared to models, purely in virtue of being experiments (for further discussion, see Parke 2014; for alternative views, see Roush 2018; Currie & Levy 2019).

4.5 Conclusion

Models play a variety of roles and serve a variety of aims in biology. We have only scratched the surface of these in this chapter (for example, exploratory and heuristic roles of modeling are important, in addition to their roles in generating explanations and testing hypotheses, which we have focused on here). As we have seen, theories and models in biology present interesting opportunities to challenge some standard views about scientific theories and models. For instance, we have argued that theories in biology are by and large not sets of universal laws, but instead consist largely of families of models, suitably interpreted. Interpretations might provide us with general information about, e.g., the conditions under which altruism might evolve, or the evolution of biodiversity. However, the aims of modeling are not (or not only) in service of general theory. They are also used as tools for discovery, and may also function effectively as experimental systems. Model systems can help us arrive at new insights about the likely dynamics of evolutionary processes in the distant past, just as computer simulations can teach us new things with surprising outcomes. Hypotheses, theories, and models relate to one another in dynamic, multifaceted ways. False or deliberately idealized models can provide useful information about the causal dynamics of systems of interest. Model organisms perform a variety of functions as models for both representing general dynamic processes, as well as in service of addressing particular experimental questions.

References

Agutter, P. S. & Wheatley, D. N. (2004). Metabolic Scaling: Consensus or Controversy? *Theoretical Biology and Medical Modelling* 1(13).

Ankeny, R. A. & Leonelli, S. (2011). What's So Special about Model Organisms? *Studies in History and Philosophy of Science Part A* 42(2): 313–323.

Axelrod, R. (1984). *The Evolution of Cooperation*. New York: Basic Books.

Beatty, J. (1997). Why Do Biologists Argue like They Do? *Philosophy of Science* S64: 231–242.

Beatty, J. (2006). Replaying Life's Tape. *The Journal of Philosophy* 103(7): 336–362.

Bissell, J. (2007). The Moniac: A Hydromechanical Analog Computer of the 1950s. *IEEE Control Systems Magazine* 27(1): 69–74.

Blount, Z. D., Borland, C. Z., & Lenski, R. E. (2008). Historical Contingency and the Evolution of a Key Innovation in an Experimental Population of *Escherichia Coli*. *Proceedings of the National Academy of Sciences of the United States of America* 105(23): 7899–7906.

Bonner, J. T. (2016). *Multicellularity: Origins and Evolution*. Cambridge, MA: MIT Press.

Brodland, G. W. (2015). How Computational Models Can Help Unlock Biological Systems. *Seminars in Cell & Developmental Biology* 47–48: 62–73.

Cartwright, N. (1983). *How the Laws of Physics Lie*. Oxford: Clarendon Press.

Currie, A. & Levy, A. (2019). Why Experiments Matter. *Inquiry* 62(9–10): 1066–1090.

Dobzhansky, T. (1936). Studies on Hybrid Sterility. II. Localization of Sterility Factors in Drosophila Pseudoobscura Hybrids. *Genetics* 21(2): 113–135.

Downes, S. M. (1992). The Importance of Models in Theorizing: A Deflationary Semantic View. *PSA: Proceedings of the Biennial Meeting of the Philosophy of Science Association* 1992(1): 142–153.

Elena, S. F., Cooper, V. S. & Lenski, R. E. (1996). Punctuated Evolution Caused by Selection of Rare Beneficial Mutations. *Science* 272(5269): 1802.

Fenchel, T. (1974). Intrinsic Rate of Natural Increase: The Relationship with Body Size. *Oecologia* 14(4): 317–326.

Frigg, R. (2010). Fiction and Scientific Representation. In R. Frigg & M. Hunter (eds.), *Beyond Mimesis and Nominalism: Representation in Art and Science*, pp. 97–138. Dordrecht: Springer.

Frigg, R. & Hartmann, S. (2012). Models in Science. In E. Zalta (ed.), *The Stanford Encyclopedia of Philosophy* (Summer 2018 ed.). https://plato.stanford.edu/arch ives/sum2018/entries/models-science/

Giere, R. N. (1988). *Explaining Science: A Cognitive Approach*. Chicago: University of Chicago Press.

Giere, R. N. (1999). *Science without Laws*. Chicago: University of Chicago Press.

Gillespie, J. H. (2000). Genetic Drift in an Infinite Population: The Pseudohitchhiking Model. *Genetics* 155(2): 909–919.

Godfrey-Smith, P. (2006). The Strategy of Model-Based Science. *Biology and Philosophy* 21(5): 725–740.

Godfrey-Smith, P. (2009). Modes and Fictions in Science. *Philosophical Studies* 143(1): 101–116.

Goel, N., Campbell, R. D., Gordon, R., Rosen, R., Martinez, H., & Yčas, M. (1970). Self-Sorting of Isotropic Cells. *Journal of Theoretical Biology* 28(3): 423–468.

Gould, S. J. & Eldredge, N. (1977). Punctuated Equilibria: The Tempo and Mode of Evolution Revisited. *Paleobiology* 3(2): 115–151.

Grant, B. R. & Grant, P. R. (1993). Evolution of Darwin's Finches Caused by a Rare Climatic Event. *Proceedings of the Royal Society of London B* 251(1331): 111–117.

Guala, F. (2002). Models, Simulations and Experiments. In L. Magnani & N. J. Nersessian (eds.), *Model-Based Reasoning: Science, Technology, Values*, pp. 59–74. Dordrecht: Kluwer.

Kingsland, S. E. (1995). *Modeling Nature*. Chicago: University of Chicago Press.

Kingsland, S. E. (2005). *The Evolution of American Ecology, 1890–2000*. Baltimore: John Hopkins University Press.

Lenski, R. E. (2011). Evolution in Action: A 50,000-Generation Salute to Charles Darwin. *Microbe* 6(1): 30–33.

Lenski, R. E., Ofria, C., Pennock, R. T., & Adami, C. (2003). The Evolutionary Origin of Complex Features. *Nature* 423: 139–144.

Lenski, R. E., Rose, M. R., Simpson, S. C., & Tadler, S. C. (1991). Long-Term Experimental Evolution in *Escherichia Coli*. I. Adaptation and Divergence during 2,000 Generations. *The American Naturalist* 138(6): 1315–1341.

Levins, R. (1966). The Strategy of Model Building in Population Genetics. *American Scientist* 54(4): 421–431.

Levy, A. (2018). Idealization and Abstraction: Refining the Distinction. *Synthese* 1–18.

Levy, A. & Currie, C. (2014). Model Organisms Are Not (Theoretical) Models. *The British Journal for the Philosophy of Science* 66(2): 327–348.

Lloyd, E. A. (1994). *The Structure and Confirmation of Evolutionary Theory*. Princeton, NJ: Princeton University Press.

Lloyd, E. A. (2015). Model Robustness as a Confirmatory Virtue: The Case of Climate Science. *Studies in History and Philosophy of Science Part A* 49: 58–68.

Matthewson, J. & Weisberg, M. (2009). The Structure of Tradeoffs in Model Building. *Synthese* 170(1): 169–190.

McIntosh, R. P. (1986). *The Background of Ecology: Concept and Theory*. Cambridge: Cambridge University Press.

Morgan, M. S. (2005). Experiments versus Models: New Phenomena, Inference and Surprise. *Journal of Economic Methodology* 12(2): 317–329.

Morgan, M. S. & Morrison, M. (eds.) (1999). *Models as Mediators: Perspectives on Natural and Social Science*. Cambridge: Cambridge University Press.

Muller, H. J. (1942). Isolating Mechanisms, Evolution, and Temperature. *Biology Symposium* 6:71–125.

O'Malley, M. A. & Parke, E. C. (2018). Microbes, Mathematics, and Models. *Studies in History and Philosophy of Science Part A* 72: 1–10.

Orzack, S. H. & Sober, E. (1993). A Critical Assessment of Levins's The Strategy of Model Building in Population Biology (1966). *The Quarterly Review of Biology* 68(4): 533–546.

Parke, E. C. (2014). Experiments, Simulations, and Epistemic Privilege. *Philosophy of Science* 81(4): 516–536.

Parker, W. S. (2009). Does Matter Really Matter? Computer Simulations, Experiments, and Materiality. *Synthese* 169(3): 483–496.

Potochnik, A. (2017). *Idealization and the Aims of Science*. Chicago: University of Chicago Press.

Provine, W. B. (2001). *The Origins of Theoretical Population Genetics: With a New Afterword*. Chicago: University of Chicago Press.

Roush, S. (2018). The Epistemic Superiority of Experiment to Simulation. *Synthese* 195(11): 4883–4906.

Savage, V. M., Allen, A. P., Brown, J. H., Gillooly, J. F., Herman, A. B., Woodruff, W. B., & West, G. B. (2007). Scaling of Number, Size, and Metabolic Rate of Cells with Body Size in Mammals. *Proceedings of the National Academy of Sciences* 104(11): 4718–4723.

Servedio, M. R., Brandvain, Y., Dhole, S., Fitzpatrick, C. L., Goldberg, E. E., Stern, C. A., Cleve, J. V., & Yeh, D. J. (2014). Not Just a Theory – The Utility of Mathematical Models in Evolutionary Biology. *PLoS Biology* 12(12): e1002017.

Sniegowski, P., Gerrish, P. J., & Lenski, R. E. (1997). Evolution of High Mutation Rates in Experimental Populations of *E. Coli*. *Nature* 387(6634): 703–705.

Sprouffske, K., Merlo, L. M., Gerrish, P. J., Maley, C. C., & Sniegowski, P. D. (2012). Cancer in Light of Experimental Evolution. *Current Biology* 22(17): R762–R771.

Suppe, F. (1989). *The Semantic Conception of Theories and Scientific Realism*. Chicago: University of Illinois Press.

Travisano, M., Mongold, J. A., Bennett, A. F., & Lenski, R. E. (1995). Experimental Tests of the Roles of Adaptation, Chance, and History in Evolution. *Science* 267(5194): 87–90.

van Fraassen, B. (1980). *The Scientific Image*. Oxford: Oxford University Press.

Weisberg, M. (2013). *Simulation and Similarity: Using Models to Understand the World*. Oxford: Oxford University Press.

5 How Are Biology Concepts Used and Transformed?

INGO BRIGANDT

5.1 Introduction

Scientific knowledge (and its transformation) is often presented in terms of models or overarching theories (Chapter 4). This chapter, in contrast, focuses on *concepts* as units and organizers of scientific knowledge. Concepts, on the one hand, are more fine-grained units in that a scientific theory contains many individual concepts. On the other hand – and this makes a look at concepts in biology particularly interesting – a concept can be used across several theories, and it can persist even when a theory has been discarded. The concept of a species continues to be used well after pre-Darwinian theories about species were abandoned, and this concept is used across all of biology, in such different theoretical contexts as vertebrate development and microbial ecology. The gene concept is likewise used in very different fields; it has survived despite the flaws of the original Mendelian theory of inheritance and a move toward molecular accounts.

A scientific concept is expressed in public discourse by a scientific term. But one should not identify concepts with terms, as the same term (the same word) can be used with different meanings and thus correspond to different concepts. For instance, the term "function" can refer to an evolutionary adaptation, where a trait is understood to have a certain function if in the past there was selection for this function. But there are many contexts, especially in molecular biology, where the same term "function" instead means (current) activity, for example, a protein binding or a gene being expressed in some tissue – a notion of function that is

I am indebted to Kostas Kampourakis and Tobias Uller for detailed comments on a previous version of this chapter. The work on this chapter was supported by the Social Sciences and Humanities Research Council of Canada (Insight Grant 435-2016-0500).

independent of evolutionary considerations, given that such an activity can actually be present even if it is new and in the past there was no selection for it. Conversely, different terms may all happen to express the same concept; and one may ponder which of several possible terms is best to talk about a previously established concept. For example, the way in which development generates phenotypic variation used to be referred to by the term "developmental constraint," but since this term has misleading connotations of variation always being prevented, nowadays other terms, such as "developmental bias" and "evolvability," are more commonly used to express the same basic idea.

Philosophers construe a concept as the mental content associated with a term, and because of its content, the concept plays a distinctive role in reasoning, from theorizing to practical action. A concept may be explained or standardized by means of a definition, but often the content of a concept is more encompassing than a short definition. A definition permits one to *identify* an object, but some concepts (e.g., "allopatric speciation") embody larger causal models, in which case they can be used to *explain* how and why a biological phenomenon occurred. It is because of its rich content that a biology concept can perform various functions in theory and practice. The following sections provide several examples of the different roles that concepts can play.

What matters especially in the present discussion is that concepts are not static objects (embodying knowledge that was obtained long ago and captured by an authoritative definition), but *dynamic* entities. Even once introduced, a biological concept can be revised and undergo transformation upon new empirical findings. Another dynamic aspect of concepts is that they guide scientific activity and investigative practice (Chapter 7; Feest & Steinle 2012). A biology concept does not just figure in abstract theorizing, but figures more general in biological *practice*, for instance, by motivating steps of experimental discovery. Given that philosophers commonly employ the term "epistemic" to refer to anything related to human knowledge, including the formation of knowledge, my discussion can be said to be about the epistemic role of concepts, as long as it is made clear that this also includes biological practice and the methodological role of concepts. Overall, this chapter is not just about what the content of biological concepts is, but how concepts are *used* by biologists.

Illustrated by means of several biology concepts, the following discussion first covers one possible function of a concept – to set an explanatory agenda – a function that is of particular importance because of its forward-looking nature of motivating ongoing and future explanatory efforts. Then I discuss the gene concept as a prime case where a concept has undergone substantial transformation throughout its rich history, without ceasing to evolve. Finally, I turn to instances of conceptual diversity where a term can fruitfully be associated with different meanings, using species concepts as a case where a pluralism about concepts is warranted.

5.2 Setting a Research Agenda: A Forward-Looking Function of a Concept

Like all scientific concepts, biological concepts refer to scientific phenomena, be it specific entities (e.g., synapses) or complex processes (e.g., natural selection). A concept can be articulated by a definition. Such a definition not only specifies what natural objects the concept picks out but also conveys some important characteristics of these objects (and standardizes how a scientific term is to be understood within a scientific community). But besides describing phenomena of the *natural world*, it is also important to consider how concepts are used by *scientists* and how they function in scientific practice. We particularly need to ask for what *scientific purpose(s)* a given concept is used. One biology concept may primarily be used for the purpose of classification, such as classifying organisms into different species (see also Chapter 11). Another concept may instead serve the scientific aim of explanation, for example, by explaining the mechanism of cell-cell signaling (see also Chapter 2). Scientific aims are also called *epistemic aims* by philosophers, although one needs to bear in mind that scientific aims are not exclusively about intellectual understanding ("episteme" meaning knowledge) but often include practical aims (Kitcher 2001; Potochnik 2017). Relevant examples of practical aims are uncovering causes and mechanisms for the purpose of intervening in nature by means of applications (e.g., biomedical ones) or investigating species and ecosystems in order to provide guidance to conservation efforts.

Paying attention to the aims and purposes for which biological concepts are used is important, because it reveals the *forward-looking* nature of concepts. All too often concepts are merely seen as the outcome of science – a term being coined once a new biological entity has been discovered, or a mature definition being established once the relevant scientific knowledge is in. But concepts also continuously undergo transformation, and they function by *guiding ongoing scientific practice*. A biology concept can motivate *future* scientific efforts, and it can also provide a scaffold to direct the generation of new knowledge and the organization of complex knowledge.

One instance of the forward-looking functions of concepts is when a biological concept sets a research agenda. A striking case in point is the concept of *evolutionary novelty* (as it is especially used in evolutionary developmental biology). A novelty is a qualitatively new trait that arose in some biological taxon, such as the evolutionary origin of the fins of fish, or their later transformation into the limbs of tetrapods – which are quite different from fins, among other things due to the presence of digits. There are major scientific efforts devoted to understanding the evolution of particular novelties, and accounting for evolutionary novelty may well be among the major aims of evolutionary biology. Despite the scientific significance of the concept of novelty, there is in fact disagreement about how to *define* novelty. One definition may focus on a morphological trait as such, while another definition will also consider its underlying developmental or genetic basis. Many definitions focus on structure (e.g., defining a trait as novel whenever it is not homologous to any ancestral trait; Müller & Wagner 1991), while some definitions bring in functional considerations such as adaptation and selection (Hallgrímsson et al. 2012). A universally agreed-upon definition would pick out exactly those biological traits that are novelties, while excluding those traits that do not qualify as novel.

However, the root of the definitional disagreement is that there may not be any principled difference between evolutionary changes that are qualitatively distinct (and thereby novel) and those that are mere quantitative variants of existing traits. Indeed, there are many morphological structures that look clearly novel, but where closer biological investigation reveals that various components of such a structure already had precursors in ancestral taxa or that the "novel" structure arose from modifications of previously existing regulatory pathways (Hall & Kerney 2012). In the case the tetrapod limb, the classical example mentioned previously, at the very least the digits of the limb were deemed to be completely novel. Yet *Hox* gene expression patterns also seen in the distal part of fish fins and other data have been used to suggest that even the digits of tetrapods can be homologized with structures in fish (Johanson et al. 2007).

But this disagreement of how to define "evolutionary novelty" does not impugn this concept. Some scientific concepts do serve the purpose of classification (which in the present case would be classifying traits into those that are novel and those that are non-novel), in which case a precise definition is needed. But this is hardly the purpose that the concept of evolutionary novelty serves; instead, this concept fulfills a vital scientific function by setting up an *explanatory agenda* (Brigandt & Love 2012). The point is that rather than debating which particular structures are really novel, it is more fruitful to work toward explanations of the evolutionary origin of various structures.

Philosopher of biology Alan Love (2013b) has called the explanation of novelty a *problem agenda*. The label "agenda" reflects that this scientific problem does not just consist in a single question, but rather in a whole set of interrelated questions and explanatory tasks. The label also highlights that there is an ongoing and forward-looking process. Some scientific concepts embody explanatory theories, and the concept of evolutionary novelty may at some point in the future do this. However, rather than delivering complete explanations, the concept of evolutionary novelty points to explanatory frameworks that yet have to be established. Indeed, evolutionary developmental biologists argue that a traditional neo-Darwinian approach, which focuses on the dynamics of genetic variants within populations, is unable (or at least woefully incomplete) to adequately explain the very origin of novelties. Instead, knowledge about developmental processes is needed in order to understand how the modification of those processes could have given rise to the novel structure. The problem agenda of evolutionary novelty entails that a host of explanatory ingredients are needed, some of which come from different biological disciplines (Love 2013a). This includes phylogenetic trees from systematics, fossil evidence about ancestral structures from paleontology, gene regulatory pathways from developmental genetics, phenomena such as phenotypic plasticity studied in developmental biology (especially ecological developmental biology), and knowledge about selection pressures implicating ecology. As a result, the concept of evolutionary novelty sets an agenda that not only motivates ongoing and future research efforts, but also prompts *interdisciplinary* research (Brigandt 2013; Love 2013b).

A related function of the concept of evolutionary novelty is that it provides intellectual identity to a biological approach. In some cases, a scientific concept can even provide identity to a scientific discipline. Accounting for novelty is certainly one of the core items on the agenda of evolutionary developmental biology. At the same time, there are different possible ways of construing evolutionary developmental biology and its relation to other fields (Brigandt & Love 2012). Some biologists prefer to construe evolutionary developmental biology as an autonomous discipline. But rather than asserting independence from other disciplines, often evolutionary developmental biology is framed as a synthesis of evolutionary biology and developmental biology (among other disciplines), so as to augment the neo-Darwinian Modern Synthesis established in the middle of the twentieth century. However, the label "synthesis" may suggest in a misleading fashion that different disciplines have been merged into one, single discipline (of evolutionary developmental biology), which is certainly not the case. Therefore, a more cautious characterization may be that evolutionary developmental biology is a field that also operates at the intersection of different fields. We fortunately do not have to settle on a unique construal of the disciplinary nature of evolutionary developmental biology. In any case, agenda-setting concepts such as evolutionary novelty provide substantial *intellectual identity* to the approach of evolutionary developmental biology (and coordinate interaction with other fields).[1]

Although the scientific problem agenda set up by the concept of evolutionary novelty (or, more precisely, by how this concept is *used* in biological practice) is quite complex, there is some structure to it. The different component questions of the overall problem agenda stand in systematic relations, which provide clues for how the different explanatory contributions – some coming from different biological fields – are to be coordinated so as to arrive at an explanatory framework (Love 2013a).

---

[1] The concept and explanatory agenda of evolvability have likewise been seen as establishing intellectual or disciplinary identity for evolutionary developmental biology (Hendrikse et al. 2007). Looking at precursors in the 1980s, well before any distinct field of evolutionary developmental biology existed, the concept of developmental constraint (also setting an explanatory agenda) already provided some intellectual identity across different approaches (Brigandt 2015).

For instance, one first needs evidence about the phylogenetic sequence of structural transformations (including which traits are homologous to precursors and which ones are new) before information about regulatory pathways and developmental processes (in different extant taxa) can be used to formulate an explanation of how changes in developmental mechanisms resulted in the novel structure under study. Thereby, in addition to motivating explanatory (and interdisciplinary) efforts – as one *forward-looking* function of a concept – the concept of evolutionary novelty also provides a scaffold for integrating new scientific findings into an emerging explanatory framework – another forward-looking aspect.

Another example is the concept of a *living fossil* (Lidgard & Love 2018). Here it is also controversial which taxa should actually count as living fossils. Various different criteria have been proposed and used: a gross similarity to an ancestral fossil or slow evolutionary change (compared to similar taxa), an unusually long geological presence (compared to similar taxa), a very low current taxonomic richness relative to the past, or a very limited geographic range compared to the taxon's past. One complaint about some criteria is that they are ill-defined. For instance, according to one criterion a living fossil is a taxon that is known from the fossil record before it is discovered alive. But this is not about the characteristics of this taxon. This criterion is about what scientists happen to have discovered first (the fossil or the extant version), which would make the living fossil status relative to the idiosyncrasies and hazards of scientific discovery. Another issue is what to do with cases of morphological stasis where molecular genetic change has occurred (which is in a sense the opposite of what we encountered in the case of evolutionary novelty, where a morphologically novel trait may be due to minor modifications of gene regulatory mechanisms). Generally, employing one rather than another of the available criteria yields a different judgment about whether a given taxon is actually a living fossil. But in this case again, the point is not (or should not be) to be able to classify taxa into living fossils and others. Rather, the concept of a living fossil can still play a fruitful role in biological practice by setting an explanatory agenda. Directing future investigative efforts, this problem agenda includes uncovering the mechanisms responsible for the retention of morphological traits over a longer period of time, investigating the relative rates of change across different traits in one lineage (e.g., molecular as opposed to morphological traits) plus explaining why some traits change faster than others, and accounting for slow evolutionary change in one lineage compared to other lineages (Lidgard & Love 2018).

The main message of this section is that apart from referring to phenomena of the natural world, biology concepts also have important functions for researchers and are used for various aims and purposes in scientific practice. If the primary scientific aim is to classify objects or to precisely characterize phenomena, an accurate definition is in fact needed. But we have seen that definitions of "evolutionary novelty" and "living fossil" are contested, and that these concepts instead have an additional, more fruitful function – to set an explanatory agenda. In this fashion, such concepts are not just the outcome of scientific knowledge obtained in the past, but also have important implications for ongoing and future scientific practice.

5.3 Concept Change and Transformation

So far we have seen that concepts are not only receptacles that contain knowledge about biological phenomena, but are also used for purposes set by scientists, including explanatory and investigative agendas (see also Kindi 2012). And it is because of such scientific purposes and aims underlying the use of concepts that the latter can have a forward-looking dimension with impacts on future research, beyond storing knowledge that was previously acquired. One aspect of the forward-looking dimension of concepts we now turn to is the fact that a concept can *change* and undergo transformation. A concept may be articulated by a definition, which standardizes its meaning for biologists and facilitates communication across different biological fields. A definition may also provide focus to ongoing research in that it delineates the objects or the phenomenon to be investigated in this context, for example, what counts as "speciation." At the same time, the definition of a biological term is often revised (hopefully amounting to an improvement on an earlier definition). And for such a modification of a concept to be possible, it cannot be the case that scientists always adhere to the original definition – which would mean to set aside findings about biological processes that do not count as instances of speciation on some definition of "speciation," or to take as irrelevant for the concept of the gene all cellular structures that happen to not count as genes on some original definition of "gene." Therefore, biology concepts are *open-ended*, in that they permit modification and can come to encompass new and quite different phenomena.

One important example of how a concept has continuously undergone change, and in fact major transformation, is the *gene concept* (Weber 2005; Kampourakis 2017; Rheinberger & Müller-Wille 2017). Based on studies of heredity in the second half of the nineteenth century, the gene concept was established in the early twentieth century during the period sometimes called Mendelian genetics, which from 1920 onwards gave rise to the more mature classical genetics. In classical genetics, genes were also often called *alleles*. What could actually be observed in breeding studies and mutation experiments were phenotypic traits (more specifically, phenotypic differences between individuals and patterns of phenotypic inheritance across generations), where alleles had to be inferred as the entities that were physically passed on to the next generation and had an effect on these phenotypic traits. Consequently, the classical gene concept construed genes in terms of their function, more precisely, their *phenotypic function*. This was obviously not a definition of genes that would specify the internal structure of genes.

Our knowledge about the structure (as well as function) of genes has changed considerably (Carlson 1966). Around 1900, many biologists still maintained that the hereditary material resides in the cytoplasm (as opposed to the nucleus). In contrast, the chromosome theory of inheritance maintained that genes are part of or in any case carried by the chromosomes. It was not until 1920, based on the *Drosophila* studies by the Morgan school, that the chromosome theory was widely accepted. For instance, the inheritance of some phenotypic traits was shown to be linked to a specific sex chromosome. And the crossing-over of chromosomes that could be observed with microscopes explained why the inheritance of some alleles (those on the same chromosome) was linked, but only with a certain probability (that declines the farther apart the alleles are on the chromosome). While alleles were structurally seen to be parts of chromosomes, a definition in terms of phenotypic function still prevailed (and phenotypic impact was the only way to experimentally distinguish different genes and locate them on a chromosome).

Classical geneticists were well aware of the fact the functional relation between genes and phenotypic traits is a *many–many one* relation (Morgan et al. 1915). Not only has a gene an influence on several phenotypic traits, but a phenotype is produced by the interaction of many genes. An allele is gene "for" a phenotype (e.g., a specific eye color as one of the many mutant traits studied in *Drosophila*) only in the sense that relative to an otherwise identical genetic background, this allele results in this phenotype (e.g., the mutant eye color), whereas other alleles at this chromosomal locus (e.g., the wildtype allele) yield a different phenotypic trait.[2] Therefore, a classical gene as such does not produce a phenotypic trait in an individual; rather, a *difference* in classical genes between individuals accounts for different phenotypes.

Later on, biochemical studies in the fungus *Neurospora* suggested that the proximate effect of an allele is the production of a specific enzyme, leading to the one gene–one enzyme hypothesis (Beadle & Tatum 1941), which was later refined to the one gene–one polypeptide hypothesis. The discovery of the molecular structure of DNA and studies with bacteria and bacteriophage viruses (which permitted much more fine-grained genetic studies) led to the advent of molecular genetics and the molecular gene concept. Roughly speaking, the molecular gene concept construes a gene as a linear, continuous segment of DNA, which (because it is preceded by a promoter) is transcribed to RNA and subsequently translated in accordance with the genetic code to a linear sequence of amino acids (a polypeptide). Note that this is more of a *structural* definition: A gene is a specific sequence of nucleotides. Moreover, the function of interest is the coding for a molecular product (a polypeptide). By now focusing on the more proximate function of molecular genes, the traditional many–many relation between classical genes and gross phenotypes was replaced by a *one–one* relation between genes and their molecular products (Griffiths & Stotz 2013).

This was clearly a significant scientific advance. But in the context of *concept* transformation we also need to ask why the very definition

---

[2] "Although there is little that we can say as to the nature of Mendelian genes, we do know that they are not 'determinants' in the Weismannian sense .... All that we mean when we speak of a gene for pink eyes is a gene which differentiates a pink eyed fly from a normal one – not a gene which produces pink eyes *per se*, for the character pink eyes is dependent upon the action of many other genes" (Sturtevant 1915, p. 265).

of "gene" changed.[3] Why not take onboard the new knowledge from molecular genetics (e.g., that genes are strands of DNA and code for a polypeptide before having an impact on development and an organism's phenotype) while still consistently defining genes in the classical way, as something leading to a difference in a specific phenotypic trait between individuals? To appreciate the motivations for the change of gene concept, we yet again have to look at the *scientific aims* for which concepts are used in scientific practice (Brigandt 2010). Whereas the very origin of the gene concept was the study of patterns of inheritance between generations, molecular genetics is hardly concerned with this and instead focuses on processes going on within individual organisms, typically within single cells. The purpose for which the gene concept was primarily used gradually changed during the history of genetics, which explains why the manner in which this concept was defined changed (once empirical information relevant to the new purpose became available). The molecular gene concept is used for the purpose of explaining the molecular mechanisms by which genes code for their molecular products, i.e., RNAs and polypeptides. The classical gene concept (linking genes with phenotypes) does not provide this mechanistic explanation, whereas the molecular gene concept serves this purpose by means of construing a gene as a DNA segment with a specific nucleotide sequence (resulting in corresponding RNA and polypeptide sequences). Thus, the focus on a new scientific aim had the forward-looking impact on the gene concept of motivating the subsequent widespread adoption of a molecular definition.

In its early stages, the molecular gene concept basically construed a gene as an open reading frame, as a continuous segment of DNA delimited by a start codon and a stop codon and preceded by a promoter sequence. To the extent that all molecular genes fit this structural definition and any such DNA structure codes for one molecular product, this definition avoided the many–many relation between classical genes and phenotypes and seemed to yield a stable resting place for the

---

[3] Despite the advent of the molecular gene concept, the classical gene concept continues to be used in some biological contexts, as Section 5.4 discusses. So rather than one definition of "gene" being fully replaced by another one, this conceptual change consists in the addition of a novel definition of the term "gene." But we still have to understand why this new, molecular definition originated in the first place and became the most prominent one for most areas of biology.

hitherto changing gene concept (Griffiths & Stotz 2013). However, the concept transformation did not stop upon the very advent of the molecular gene concept, and the molecular gene concept has kept changing ever since the complexities of gene structure and function in eukaryotic cells were revealed (Portin 1993; Keller 2000), which underscores the open-ended nature of many concepts. In fact, already in the 1980s there were signs that a fairly unified conception of molecular genes had been replaced with diverse conceptions, where the meaning of "gene" employed in a specific situation could only be gathered from the context (Falk 1986).

Here I can mention only some of the complexities that kept driving the change of the gene concept. Most prominently, the relation between continuous DNA segments and polypeptide products is often *many–many*, after all. Because of the mechanism of RNA splicing, a DNA segment is transcribed to a pre-mRNA, but then some chunks of the pre-mRNA (the introns) are spliced out, i.e., removed, before this mature mRNA (now consisting of exons only) is translated to a polypeptide. In the case of alternative splicing, a DNA segment is repeatedly transcribed to identical pre-mRNAs, but these pre-mRNAs may be spliced differently, resulting in different exon combinations and different polypeptide products. Consequently, one cell can produce chemically diverse proteins from one and the same DNA segment. There is also trans-splicing, where two or more DNA segments (possibly located on different chromosomes) are independently transcribed to pre-mRNAs, and chunks of these pre-mRNAs are spliced together to one mature mRNA, resulting in one polypeptide product. Are two such DNA segments two independent genes (which have to collaborate to generate one product at all)? Or do the two rather make up one gene (which simply consists of disjoint nucleotide sequences)? But what to say for those cases where a DNA segment engages in trans-splicing involving another DNA segment, as well as codes for its own product? Generally, the many–many relation between DNA segments and gene products raises difficult questions for how to annotate genes, for how to decide where a gene begins and ends, how many different genes there are in a certain genomic region, and for formulating an account of what genes really are.

Some further complications related to the gene concept are that not only segments of the sense strand of the double-stranded DNA (which used to be called the coding strand) are transcribed, but also that some segments of the antisense strand ("non-coding strand") can be transcribed so as to qualify as a coding gene. There are also many cases of overlapping genes, for example, a gene that codes for a product being situated within the intron of another, larger gene. Finally, due to the processes of RNA editing (modification of individual nucleotides of the mRNA before translation) and translational recoding, the final amino acid sequence may not be fully determined by the DNA nucleotide sequence. As a result, the central dogma of molecular biology that DNA sequence determines RNA sequence which in turn determines the protein's amino acid sequence is not accurate (as other sources of information can be involved in specifying the amino acid sequence), and the genetic code as the mapping from nucleotide to amino acid sequence does not always tell the complete story (Griffiths & Stotz 2013).

Overall, while at the beginning of molecular biology a structural construal of genes seemed possible, the subsequent history has shown that considerations about gene *function* are vital. Ultimately, deciding what genes are and whether some structural entity counts as a gene depends on its functional behavior, in particular whether it is transcribed and whether it codes for some product. This also depends on the larger genomic and even the cellular context, which then provides a more meaningful context for assessing gene function than simply looking at a DNA nucleotide sequence as one structure. The advent of the "postgenomic era," including functional genomics and transcriptomics, has certainly increased the focus on gene function, and keeps providing findings that continue to modify our conceptions of genes. Textbooks may still provide stereotypical characterizations of genes (a molecular gene as an open reading frame coding for a protein); and researchers may use such stereotypical visions as a general starting point, but then immediately employ a more sophisticated account that is geared to a particular genomic case and investigative context, overall resulting in a context-dependent use of the gene concept across the molecular biology community (Keller 2000; Griffiths & Stotz 2013).[4]

---

[4] This can also mean that apart from consistently using a stereotypical or simplistic construal, a textbook may contain different definitions of genes, without necessarily clarifying the reasons for this (Kampourakis & Stern 2018).

5.4 Pluralism and Conceptual Diversity

The previous section has emphasized that not only does the gene concept continue to change, but also that the molecular gene concept exhibits significant variation. There is no unified construal of what genes are, and biologists use the concept depending on their specific research context. One reason for this is the structure of the natural world: the complexity of gene structure and function. Another reason for this conceptual diversity is epistemic: the various scientific considerations and interests that scientists bring to the task. The philosophers Karola Stotz and Paul Griffiths (2004) adopted the label *conceptual ecology* for the project of understanding the reasons for the diversification and transformation of the gene concept. Rather than endorsing proposals that a unified gene concept behind the diversity can still be recovered (e.g., Waters 2000) or that the gene concept had better be eliminated in favor of other terms (e.g., Keller 2000), the agenda of Stotz and Griffiths has been to map out the empirical pressures that move the gene concept in a novel direction and the scientific agendas that diversify the use of the gene concept across different biological communities. Scientific aims matter here again; for instance, some researchers may focus on RNA as the gene product of interest, whereas others on the polypeptides produced (given their biochemical roles as enzymes). Since the relation between DNA segments and RNAs is one–one, whereas from DNA segments to polypeptides there are often many–many relations, focusing on polypeptide rather than RNA as the gene product of interest may result in a different account of whether separate DNA segments each count as an independent gene.[5]

The discussion on conceptual change in Section 5.3 was framed in terms of a move from the classical to the molecular gene concept.

---

[5] Another example of the impact of different scientific aims is how the homology concept diversified once it came to be used in newly formed biological fields, such as evolutionary developmental biology. I have described this as the homology concept undergoing an "adaptive radiation" – borrowing the metaphor from evolutionary biology (Brigandt 2003). Biologists nowadays talk about a phylogenetic as opposed to a developmental concept of homology. It is less important to adjudicate whether these are really distinct concepts or whether they are different variants of one concept. Instead, the philosophically interesting task ("conceptual ecology") is to understand the reasons for this diversity upon the homology concept coming to occupy new conceptual niches, among other things by the concept being used for different specific biological purposes in different fields.

However, the classical gene concept is still used nowadays, routinely in population genetics and in some contexts of medical and behavioral genetics (Griffiths & Stotz 2013). The coexistence of the classical gene concept and the molecular gene concept means that the conceptual plurality is in fact much more pronounced than just the diverse ways in which the molecular gene concept is used. Scientific aims yet again have an impact on why it is better to use a certain gene concept in a given context. The molecular gene concept construes a gene as a specific sequence of nucleotides, which is important for understanding how it codes for molecular products – even in cases where many different polypeptides are being produced from this molecular gene (and the connection to a particular phenotype is unclear). In contrast, population genetics accounts for the dynamics of genes within populations in terms of (more abstract) genes having phenotypic effects – regardless of what the molecular basis of this phenotype may be – and models how the fitness of such a phenotype results in gene frequency changes in the population. Likewise, in medical genetics (in addition to the molecular gene concept) one can also find talk about a gene "for" breast cancer, understood as anything that has this phenotypic effect of a clear breast cancer risk (Moss 2003). This can be of interest even if several different nucleotide sequences lead to this phenotype under study.

Another prominent example of conceptual diversity is the existence of several different *species concepts*. Despite ongoing developments, the basic situation has existed for several decades: taxonomist Richard Mayden (1997, 2002) has provided a list of 22 different species concepts, and philosopher of biology John Wilkins (2018) has given a detailed update that also includes a thorough discussion of views on species in the history of biology, especially how the species problem arose in the nineteenth century. Some species concepts may be related or are different versions of a basic approach, for instance, there are several concepts focusing on reproductive isolation between species, just like there are several evolutionary species concepts as well as several phylogenetic species concepts. Still, the plurality of concepts is hard to do away with. In what follows, I can only present a few examples of species concepts, while my main aim is to give a sense of why it is important for biologists to have several such concepts at their disposal.

One reason for this conceptual diversity is *ontological*, in that nature is too complex that one single account would capture all of it. More precisely, there are different processes that shape biodiversity and provide cohesion within populations and differentiation into different lineages, where different concepts are needed to cover different aspects of this overall biodiversity (Ereshefsky 1992). But another reason for the conceptual diversity is *epistemic* – in line with this chapter's focus on how concepts are used and for what purposes they are used. In a nutshell, the right conceptual tool is needed for each job. One species concept may work well for one biological task, but be unsuitable for other legitimate purposes (Kitcher 1984).

One prominent species concept is the so-called *biological species concept* championed by Ernst Mayr (who used a clever label for his favored concept). It states that a species is a group of potentially interbreeding natural populations that are reproductively isolated from other such groups (i.e., from other species).[6] This species concept is theoretically interesting because it ties into some models explaining speciation. If two populations of what is still one (interbreeding) species become geographically isolated, due to the strongly reduced gene flow between them they may diverge genetically and phenotypically. Thus, they may eventually lose the ability to interbreed and will then be two distinct species according to the biological species concept. Another advantageous aspect of this concept is that when it was applied to the *Anopheles* mosquito genus, it succeeded in distinguishing two populations as different interbreeding species so as to shed light on the distribution of malaria in Europe – given the importance for human health, this was a scientific use of a species concept for clearly practical purposes (Ludwig 2016).[7]

At the same time, the biological species concept has clear drawbacks. One is practical, in that the concept (relying on the ability to interbreed) cannot be used to classify extinct species that are only known as fossils, which matters for the discipline of paleontology. In this domain, instead the *morphological species concept* may be used, which construes

---

[6] To illustrate that even within the approach focusing on reproductive isolation there are different concepts, another species concept is Templeton's mate recognition concept.

[7] More recently, however, it has been argued that adequately fighting malaria requires the use of the phylogenetic species concept, which offers a yet more fine-grained and detailed classification than the biological species concept by splitting certain interbreeding populations into different species (Attenborough 2015).

species as the smallest groups that are consistently and persistently distinct and distinguishable by morphological traits. But there also exist significant theoretical problems for the biological species concept. Hybridization among even distantly related lineages is widespread among plants and animals. Even proponents of a biological species concept agree that what they consider distinct species can undergo hybridization, so that they are able to interbreed after all. Conversely, geographically distant populations from the same species may not even be able to potentially interbreed. (In the case of a so-called ring species, geographically adjacent populations A and B do interbreed, as do B and C, etc., but A and F cannot potentially interbreed.) Moreover, a fundamental trouble with the biological species concept is that only sexual species interbreed, so that this species concept cannot even be theoretically employed for most microbial species. Instead, microbiologists may use a sort of morphological species concept (applied to cellular structures), the *genetic species concept* (which views organisms to belong to different species if their genetic similarity crosses a threshold), the ecological species concept, or a phylogenetic species concept – all of which are also used for sexually reproducing species (Bzovy 2017; Wilkins 2018).

To briefly convey more of this conceptual diversity, according to the *evolutionary species concept*, a species is a lineage evolving separately from others and with its own unitary evolutionary role and tendencies of evolutionary change, in particular long-term phenotypic change. This account is theoretically appealing, but not very useful for (putative) species that have originated recently and that do not yet have a clear evolutionary tendency or fate. And even though this concept was introduced by the paleontologist G. G. Simpson, it is not practically applicable in paleontological cases where a species to be classified (or other related species) is only known from a limited number of fossils. The *ecological species concept* (introduced by Leigh van Valen) assumes that a species is a lineage or a closely related set of lineages that occupies an adaptive zone minimally different from any other lineage in its range (and that evolves separately from all lineages outside its range). This concept focuses on ecological competition and natural selection as factors creating differences between different species. Two populations that interbreed (and are thus one species on the biological species concept) may well occupy different ecological niches, and thus exhibit relevant phenotypic (including behavioral) differences and therefore qualify as two separate species on the ecological concept. Finally, *phylogenetic species concepts* (of which there also are different variants) arose with the advent of phylogenetic systematics (cladistics). They generally conceive of species as the smallest group of organisms forming a phylogenetic lineage that are diagnosable by a unique combination of characters, which can for instance be articulated by a definition that requires the presence of a synapomorphy (i.e., a derived and thus phylogenetically new trait that is shared among these organisms). A phylogenetic species concept will consider two distinguishable phylogenetic lineages as two distinct species, even if the organisms from these two populations can interbreed (Mayden 1997; Wilkins 2018).

It is easy to note that different species concepts happen to be used, but the question is: How should one evaluate this diversity? A few decades ago some biologists insisted that their favored species concept is the only legitimate one (or the most fundamental one), and that the diversity of different species concept will go away (Simpson 1961; Ghiselin 1987; Mayr 1987). But the dominant view these days is that several species concepts are legitimate and needed for biological practice. Philosophers use the label *pluralism* for a stance that endorses a plurality of species concepts. Apart from the ontological complexity of the processes generating biodiversity (Ereshefsky 1992), we have already encountered an epistemic argument for pluralism: Different species concepts fulfill *different legitimate biological purposes*, such as the explanation of speciation, or the classification of fossil taxa (Kitcher 1984). It is certainly true that there are cases where using one rather than another species concept yields a different account of the boundaries and the number of species, so as to result in classifications that cross-cut each other (Conix 2018). This clearly matters in the context of conservation biology. If a population that is likely to become extinct is classified as a distinct species, it is indeed considered as an endangered species; but this is not the case if it is deemed to merely be a population of a larger species (which as a whole is not endangered). Polar bears and Alaskan brown bears can interbreed with each other, but many will argue that the conservation status of polar bears should not be contingent on Alaskan brown bears, which are then to count as a separate species. Likewise, the eventual recognition of the Alabama sturgeon as an independent species resulted in its being listed as a critically endangered species, which was preceded by scientific, political, and legal debates (Scharpf 2000). The fact that a particular species concept can be favored due to conservation biology considerations highlights that a scientific concept can fulfill practical purposes that have sociopolitical ramifications.[8]

The diversity of species concepts and the possibility of cross-cutting classifications sounds certainly challenging. But taxonomists use various local decisions about the aims of classification to arrive at the most legitimate taxonomic account (Conix 2018). Moreover, the fact that different species concepts can serve different specific purposes should not be misunderstood to mean that in one scientific context only one species concept is of relevance (while other species concepts will be used by other researchers). Instead, several species concepts can be jointly used during a research project. A case in point is how yeast taxonomists use the morphological, the ecological, the biological, the genetic, and the phylogenetic species concept to supplement one another so as to arrive at a solid classification (Bzovy 2017). The morphological species concept may be used early on in a study of a new strain of yeast, but then additional concepts provide more fine-grained information, such as the ecological species concept or the biological species concept for sexual yeast taxa. This interplay among different species concepts illustrates again that concepts are relatively fine-grained units of knowledge whose use is subject to change and that dynamically interact, guided by local scientific purposes.

5.5 Concluding Remarks

Science is often described in terms of scientific theories and scientific disciplines, as relevant units of knowledge and knowledge organization. But this discussion has shown that concepts are also epistemic units of interest. On the one hand, concepts can have a longer reach insofar as a concept (e.g., "species") can be used across different disciplines and can persist even if an older theory has been abandoned (the classical gene concept continuing to be used despite the original Mendelian theory of

---

[8] Conservation concerns may often favor the splitting of populations into distinct species (so-called "taxonomic splitting"), but considerations about habitat fragmentation (likewise a conservation concern) can work in the opposite direction, i.e., they favor the viewing of several populations as forming one species (Frankham et al. 2012).

genetics having become obsolete). One the other hand, concepts can be more fine-grained and dynamic than disciplines and theories, in that many different concepts are used by a discipline or figure in a theory, where these individual concepts interact and the use of one such concept can be sensitive to and change across scientific contexts. Indeed, beyond the traditional focus on scientific theory, philosophers of science have come to study scientific *practice*. Rachel Ankeny and Sabina Leonelli (2016) have adopted the notion of "repertoires" to analyze the organization of scientific practice – repertoires do not just contain scientific ideas, but also such things as experimental tools, databases, and model organisms. Yet in addition to such material entities, this chapter has made plain that intellectual entities such as concepts are also an important player in scientific practice, at least if one is mindful of how biology concepts are being *used*.

Such a focus on the practice of science also aligns with current trends in science education, which beyond the content of science – encapsulated in learning outcomes – also endeavors to teach students about the process of scientific investigation (e.g., NGSS Lead States 2013). In the case of biological concepts, we have seen that concepts not only embody scientific knowledge that was obtained in the past, but also actively *guide* ongoing and future scientific practice. One reason for this is that biology concepts are used for specific *purposes* (which are often a combination of intellectual and practical purposes). Such concept use can generate diversity in science. A plurality of species concepts is needed as one species concept may be more suitable for a concrete biological task then another one; and a researcher may first employ one and then another species concept depending on the particular biological aim at hand. Biological concepts are also open-ended, in that definitions once adopted are revised and concepts keep undergoing transformation. In the case of the gene concept, we have seen that the advent of the molecular gene concept (and its ongoing modification) was guided by the concept coming to be used for the purpose of explaining the molecular processes of gene function. A concept such as evolutionary novelty or living fossil may not embody explanatory models (at least not yet), while still significantly influencing biological practice by setting an explanatory agenda. The problem agenda of accounting for the evolutionary origin of evolutionary novelty even motivates interdisciplinary research, and the concept of evolutionary novelty provides an epistemic scaffold that guides how new findings and explanatory contributions are to be inte-grated in an emerging, future scientific account.[9] Teaching about scien-tific investigation, including how open-ended concepts motivate and diversify research, matters for science education not only because of the need to convey how science works (beyond what scientific theories say). It is also impossible to cover the ever increasing and changing body of current knowledge in secondary education, while teaching in the classroom about how concepts can scaffold students' and scientists' problem solving provides a lasting lesson.

References

Ankeny, R. A. & Leonelli, S. (2016). Repertoires: A Post-Kuhnian Perspective on Scientific Change and Collaborative Research. *Studies in History and Philosophy of Science* 60: 18–28.

Attenborough, R. (2015). What Are Species and Why Does It Matter? Anopheline Taxonomy and the Transmission of Malaria. In A. M. Behie & M. F. Oxenham (eds.), *Taxonomic Tapestries: The Threads of Evolutionary, Band Conservation Research*, pp. 129–155. Canberra: ANU Press.

Beadle, G. W. & Tatum, E. (1941). Genetic Control of Biological Reactions in Neuro-spora. *Proceedings of the National Academy of Sciences of the USA* 27(1): 499–506.

Brigandt, I. (2003). Homology in Comparative, Molecular, and Evolutionary Develop-mental Biology: The Radiation of a Concept. *Journal of Experimental Zoology Part B: Molecular and Developmental Evolution* 299(1): 9–17.

Brigandt, I. (2010). The Epistemic Goal of a Concept: Accounting for the Rationality of Semantic Change and Variation. *Synthese* 177(1): 19–40.

Brigandt, I. (2013). Explanation in Biology: Reduction, Pluralism, and Explanatory Aims. *Science & Education* 22(1): 69–91.

Brigandt, I. (2015). From Developmental Constraint to Evolvability: How Concepts Figure in Explanation and Disciplinary Identity. In A. C. Love (ed.), *Conceptual Change in Biology: Scientific and Philosophical Perspectives on Evolution and Development*, pp. 305–325. Dordrecht: Springer.

Brigandt, I. & Love, A. C. (2012). Conceptualizing Evolutionary Novelty: Moving beyond Definitional Debates. *Journal of Experimental Zoology Part B: Molecular and Developmental Evolution* 318(6): 417–427.

Bzovy, J. (2017). *Species Pluralism: Conceptual, Ontological, and Practical Dimen-sions*. Dissertation, Western University. http://ir.lib.uwo.ca/etd/4309

---

[9] Beyond biology and in the case of scientific concepts in general, philosopher of science Nancy Nersessian (2008) discusses how problem situations (and concomitant model-based reasoning) influence the creation and modification of concepts.

Carlson, E. A. (1966). *The Gene: A Critical History*. Philadelphia: W. B. Saunders.

Conix, S. (2018). Radical Pluralism, Classificatory Norms and the Legitimacy of Species Classifications. *Studies in History and Philosophy of Biological and Biomedical Sciences* 73: 27–34.

Depew, D. (this volume). How Is Biology Affected by Conceptual Frameworks? In K. Kampourakis & T. Uller (eds.), *Philosophy of Science for Biologists*. Cambridge: Cambridge University Press.

Ereshefsky, M. (1992). Eliminative Pluralism. *Philosophy of Science* 59(4): 671–690.

Falk, R. (1986). What Is a Gene? *Studies in History and Philosophy of Science* 17(2): 133–173.

Feest, U. & Steinle, F. (eds.) (2012). *Scientific Concepts and Investigative Practice*. Berlin: de Gruyter.

Frankham, R., Ballou, J. D., Dudash, M. R., Eldridge, M. D. B., Fenster, C. B., Lacy, R. C., et al. (2012). Implications of Different Species Concepts for Conserving Biodiversity. *Biological Conservation* 153: 25–31.

Ghiselin, M. T. (1987). Species Concepts, Individuality, and Objectivity. *Biology and Philosophy* 2(2): 127–143.

Griffiths, P. & Stotz, K. (2013). *Genetics and Philosophy: An Introduction*. Cambridge: Cambridge University Press.

Hall, B. K. & Kerney, R. (2012). Levels of Biological Organization and the Origin of Novelty. *Journal of Experimental Zoology Part B: Molecular and Developmental Evolution* 318(6): 428–437.

Hallgrímsson, B., Jamniczky, H. A., Young, N. M., Rolian, C., Schmidt-Ott, U., & Marcucio, R. S. (2012). The Generation of Variation and the Developmental Basis for Evolutionary Novelty. *Journal of Experimental Zoology Part B: Molecular and Developmental Evolution* 318: 501–517.

Hendrikse, J. L., Parsons, T. E., & Hallgrímsson, B. (2007). Evolvability as the Proper Focus of Evolutionary Developmental Biology. *Evolution & Development* 9(4): 393–401.

Johanson, Z., Joss, J., Boisvert, C. A., Ericsson, R., Sutija, M., & Ahlberg, P. E. (2007). Fish Fingers: Digit Homologues in Sarcopterygian Fish Fins. *Journal of Experimental Zoology Part B: Molecular and Developmental Evolution* 308B(6): 757–768.

Kampourakis, K. (2017). *Making Sense of Genes*. Cambridge: Cambridge University Press.

Kampourakis, K. & Stern, F. (2018). Reconsidering the Meaning of Concepts in Biology: Why Distinctions Are so Important. *BioEssays* 40: 1800148.

Keller, E. F. (2000). *The Century of the Gene*. Cambridge, MA: MIT Press.

Kindi, V. (2012). Concept as Vessel and Concept as Use. In U. Feest & F. Steinle (eds.), *Scientific Concepts and Investigative Practice*, pp. 23–46. Berlin: de Gruyter.

Kitcher, P. (1984). Species. *Philosophy of Science* 51(2): 308–333.

Kitcher, P. (2001). *Science, Truth, and Democracy*. Oxford: Oxford University Press.

Lidgard, S. & Love, A. C. (2018). Rethinking Living Fossils. *BioScience* 68(10): 760–770.

Love, A. C. (2013a). Interdisciplinary Lessons for the Teaching of Biology from the Practice of Evo-Devo. *Science & Education* 22(2): 255–278.

Love, A. C. (2013b). Teaching Evolutionary Developmental Biology: Concepts, Controversies, and Consequences. In K. Kampourakis (ed.), *The Philosophy of Biology: A Companion for Educators*, pp. 323–341. Dordrecht: Springer.

Ludwig, D. (2016). Ontological Choices and the Value-Free Ideal. *Erkenntnis* 81(6): 1253–1272.

Mayden, R. L. (1997). A Hierarchy of Species Concepts: The Denoument in the Saga of the Species Problem. In M. F. Claridge, H. A. Dawah, & M. R. Wilson (eds.), *Species: The Units of Biodiversity*, pp. 381–424. London: Chapman and Hall.

Mayden, R. L. (2002). On Biological Species, Species Concepts and Individuation in the Natural World. *Fish and Fisheries* 3(3): 171–196.

Mayr, E. (1987). The Ontological Status of Species: Scientific Progress and Philosophical Terminology. *Biology and Philosophy* 2: 145–166.

Morgan, T. H., Sturtevant, A. H., Muller, H. J., & Bridges, C. B. (1915). *The Mechanism of Mendelian Heredity*. New York: Henry Holt.

Moss, L. (2003). *What Genes Can't Do*. Cambridge, MA: MIT Press.

Müller, G. B. & Wagner, G. P. (1991). Novelty in Evolution: Restructuring the Concept. *Annual Review of Ecology and Systematics* 22(1): 229–256.

Nersessian, N. J. (2008). *Creating Scientific Concepts*. Cambridge, MA: MIT Press.

NGSS Lead States (2013). *Next Generation Science Standards: For States, by States*. Washington, DC: The National Academies Press.

Portin, P. (1993). The Concept of the Gene: Short History and Present Status. *Quarterly Review of Biology* 68(2): 173–223.

Potochnik, A. (2017). *Idealization and the Aims of Science*. Chicago: University of Chicago Press.

Rheinberger, H.-J. & Müller-Wille, S. (2017). *The Gene: From Genetics to Postgenomics*. Chicago: University of Chicago Press.

Scharpf, C. (2000). Politics, Science, and the Fate of the Alabama Sturgeon. *American Currents* 26(3): 6–14.

Simpson, G. G. (1961). *Principles of Animal Taxonomy*. New York: Columbia University Press.

Stotz, K. & Griffiths, P. E. (2004). Genes: Philosophical Analyses Put to the Test. *History and Philosophy of the Life Sciences* 26: 5–28.

Sturtevant, A. H. (1915). The Behavior of the Chromosomes as Studied through Linkage. *Zeitschrift für induktive Abstammungs- und Vererbungslehre* 13(1): 234–287.

Waters, C. K. (2000). Molecules Made Biological. *Revue Internationale de Philosophie* 4(214): 539–564.

Weber, M. (2005). *Philosophy of Experimental Biology*. Cambridge: Cambridge University Press.

Wilkins, J. S. (2018). *Species: The Evolution of the Idea*. Boca Raton, FL: CRC Press.

6   Why Does It Matter That Many Biology Concepts Are Metaphors?

KOSTAS KAMPOURAKIS

6.1 The Metaphorical Nature of Biology Concepts

What does scientific knowledge consist of? At the most basic level it is structured by concepts. Scientific concepts have important representational and heuristic roles in the acquisition and justification of scientific knowledge because they both represent natural entities, properties, and processes and also make their investigation possible (Arabatzis 2019; see also Nersessian 2008; MacLeod 2012). This is why concepts and their meanings have long been the focus of philosophical studies (see for instance the concepts analyzed in The Stanford Encyclopedia of Philosophy, https://plato.stanford.edu; or those biology concepts in edited collections such as Keller & Lloyd 1992; Hull & Ruse 1998; Hall & Olson 2003; Hull & Ruse 2007). This chapter deals with a central feature of scientific concepts: their metaphorical nature.

It may come as a surprise, but most biology (and science in general) concepts are actually metaphors. For example, biologists routinely refer to DNA as "book"; to DNA synthesis as "replication"; to RNA synthesis as "transcription"; to protein synthesis as "translation"; and to RNA modification as "editing" (see Kay 2000, pp. 14–30). Such metaphors not only have facilitated understanding of the respective phenomena but have also influenced scientific discourse and the direction that research has taken. For instance, the metaphor of "information" encoded in DNA set the context for deciphering the genetic code, whereas the metaphor of the genome as the "book of life" was central to the launch of the human genome project. The announcement of the first sequence of the

Many thanks to David Depew, Dan Nicholson, Andrew Reynolds, and Tobias Uller for their very helpful comments on earlier versions of this chapter.

human genome in 2000 as the outcome of the Human Genome Project was presented in BBC under the title "Reading the Book of Life," stating that: "The blueprint of humanity, the book of life, the software for existence – whatever you call it, decoding the entire three billion letters of human DNA is a monumental achievement."[1] The genome was thus perceived both as a book containing information and as a program through which this information is used. In a similar manner, the ENCODE (Encyclopedia of DNA Elements) project in 2012 was presented in CNN under the title "DNA Project Interprets 'Book of Life,'" in which it was stated that: "When the Human Genome Project sequenced the human genome in 2003, it established the order of the 3 billion letters in the genome, which can be thought of as 'the book of life.'"[2] It must be noted at this point that this metaphor is inherent in the notion of the "encyclopedia" employed by the researchers of the ENCODE project themselves. Therefore, it is important to understand the features of these metaphors in detail.

Metaphors are very common in science because in order to represent or explain something that we do not fully understand, it is often easier to refer to something else that is familiar and considered to have similar characteristics. Simply put, a metaphor is understanding and experiencing something in terms of something else (Lakoff & Johnson 2003, p. 5). It can be thought of as a mapping from a source domain of everyday experience to a target domain, with the aim of better understanding the latter in terms of the former. This is done by emphasizing particular features of the source domain and hiding others (Brown 2003, p. 29). Metaphors thus make entities, phenomena, processes, or mechanisms comprehensible because people are guided to think about them in terms of other entities, phenomena, processes, or mechanisms with which they are already familiar. As John Avise has nicely put it: "Evocative metaphors can distill an ocean of information, whet the imagination, and suggest promising channels for navigating uncharted ... waters" (Avise 2001, p. 86).

Let us consider an example in some detail. Cell biologists refer to "signaling" proteins and to "receptor" proteins to describe their chemical interactions that bring about changes in the status of a cell. Whereas

---

[1] http://news.bbc.co.uk/2/hi/indepth/scitech/2000/humangenome/760893.stm
[2] http://edition.cnn.com/2012/09/05/health/encode-human-genome/

the two proteins simply interact chemically, they are described as "signal" and "receptor" because the outcome of this interaction is a change, or a series of changes, within the cell. Therefore, scientists describe the protein that makes the difference in bringing about these changes as a "signaling" protein, *as if* this protein is transferring a signal for initiating these changes. They also describe the other protein as a receptor, *as if* it receives the signal transferred by the other protein. Models of intracellular signaling have thus been dominated by metaphorical language from electrical engineering, with the major background metaphor being that of the "cell is a computer." More recently, researchers have been using metaphors in which the molecular components of signaling pathways are described as dynamic and flexible agents that cooperate in signaling pathways, thus resembling an active team of workers rather than a static computer circuit board (Reynolds 2018, pp. 114–115).

However, two kinds of problems can emerge from the use of metaphors. The first is that it is often very easy to confuse the target domain with the source domain. That is, it is often very easy to forget that the properties and the features of the source domain attributed to the target domain are not really its own properties and features. For instance, in the example of "signaling" and "receptor" proteins previously, one should keep in mind that signaling proteins are not really active agents in intercellular communication and do not have any intentions, and their interactions and subsequent changes depend on several other molecules around them. Most importantly, there is really no signal transfer as it would happen in an electrical circuit. The second problem is that the focus may eventually come to be only on those features of the target domain in which the metaphor is better illustrated, and thus overlook other, perhaps important, aspects that do not fit well in the metaphor. In the case of the signaling-receptor interaction, the problem with focusing on the interaction between the two proteins too much is that it may mask the importance of the broader cellular context in which the chemical interactions between these two proteins take place. This is why it is necessary to be aware both of the specific features of metaphors and of their different uses. If this is not done, then metaphors can cause significant problems both within science and in science communication.

Genes can be considered as an exemplar case of the negative impact that the bad use of metaphors and the use of bad metaphors can have.

Two metaphors have been mainly used by biologists and non-biologists in order to account for what genes "do": The genome has been described as a book and as a genetic program. However, these metaphors may have blinded biologists to genomic imperfections (see Morange 2002, pp. 22–24). Perhaps more influential than all has been the metaphor of "gene action," introduced in the beginning of the twentieth century. This metaphor has been unquestionably productive, but also problematic because it facilitated the attribution of agency, autonomy, and causal primacy to genes. By not knowing what a gene is and by talking at the same time about gene action, it became possible to consider genes as the basis of life. Genes are thus often described as autonomous agents producing phenotypes, which is another anthropomorphic metaphor that may imply that genes do things on their own (Keller 1995, pp. 8–11).[3] A set of different metaphors instead draw attention to how genes interact, and not act, in the context of developmental processes (see Kampourakis 2017).

In this chapter I discuss the metaphorical nature of biology concepts, with the following main aims: (1) analyze the metaphorical nature of scientific concepts; (2) explain the value and the problems of the use of these metaphors in scientific research, with two concrete examples; and (3) argue that because these metaphors are indispensable for doing and communicating science, scientists had better be aware of, and explicit about, the metaphorical nature of the concepts that they use in their discourse.

6.2 The Interaction View of Metaphor

In a now classic analysis of metaphor, philosopher Max Black argued against the view that metaphor is something that is used for purely rhetorical purposes, and that all metaphorical elements of a sentence can be replaced by literal elements. According to what he described as the *substitution view of metaphor*, a metaphorical expression M is considered as simply a substitute for a literal expression L; in this view,

---

[3] Several other analyses of metaphors exist (e.g., Ruse 2000, 2005; Thompson 2000; Cuddington 2001; Griffiths 2001; Ouzounis & Mazière 2006; Kaplan 2008; Pigliucci 2010; Boudry & Pigliucci 2013; Mindell 2013; Nicholson 2013; Fusco, Carrer, & Serrelli 2014; Kampourakis 2019).

M has only substituted L, which could have been used instead for the same purpose. In this case, in other words, a metaphor is used in order to express something that could have also been expressed literally. This is done either for stylistic reasons with the aim of entertaining the reader, or because a literal equivalent is not available in that particular language (this latter case, in which a word is used metaphorically in order to fill a gap in the vocabulary, is described as catachresis). According to Black, the substitution view sees metaphor as having no philosophical interest, as it only relates to language issues. A special case of the substitution view of metaphor is what Black called the *comparison view*, according to which a metaphorical expression M is considered as similar or analogous to a literal expression L (Black 1955, pp. 278–283). This reduces metaphor to simile. The comparison view does not attribute much philosophical interest to metaphors ("it is just semantics!"), and simply pointing out that a metaphor can replace a literal description does not offer any understanding of why this should be the case.

However, it is possible that a metaphor does not simply express some similarity that already existed, but, rather, that the metaphor itself creates the similarity. Black thus proposed what he called the *interaction view of metaphor*, according to which, "when we use a metaphor we have two thoughts of different things active together and supported by a single word, or phrase, whose meaning is a resultant of their interaction" (Black 1955, p. 285). Black called the word or phrase that expressed the metaphorical idea the *focus* of the metaphor, and the rest of the sentence in which that word or phrase occurs the *frame*. For Black, the focus of a metaphor is not simply analogous to some literal meaning; rather, the frame imposes extension of meaning upon the focus words, which thus obtains a new meaning that is not the same with the meaning of the literal sense.

To illustrate the contrast between the two views, let us use Black's example, the metaphor "man is a wolf." According to the comparison view of metaphor, this particular metaphor presents a comparison between men and wolves, without explicitly specifying what this is about. In contrast, according to the interaction view of metaphor, our thoughts about men and wolves are active together and interact to produce a meaning. In this case, how man (or humans to be more precise) is seen through the metaphorical expression depends on what people may consider to be true about wolves, which in turn makes understanding a metaphor a culture-dependent issue. Therefore, and in contrast to the comparison view, the metaphorical expression cannot remain unaffected by the metaphor. Rather, the interaction view of metaphor entails that humans are not simply compared to wolves, but that, through the metaphor, we come to see both humans and wolves differently than before, as similarities and analogies are created by the metaphor itself. This has important implications, as some aspects will be emphasized and others will be neglected. This is a feature that is very important to keep in mind for understanding the role of metaphors in science (Black 1955).

We can now summarize the components of the interaction view of metaphor, that is, what is going on when we say that "P is S" (Bradie 1999, p. 160; Hesse 1966, pp. 157–166):

1) a principal subject P (in Black's example "man");
2) a secondary subject S (in Black's example "wolf");
3) a set of implications I associated with S (what entails for humans to be considered as wolves);
4) a set of attributions A that P acquires in virtue of our looking at P through the lens of S (what properties humans are now seen to have as a result of considering them as wolves).

Before I illustrate why this summary can be a very valuable tool for the analysis of metaphors that helps us understand them better, I would like to add one more point:

5) a set of properties Z of P that are masked due to our looking at P through the lens of S (what properties of humans are now overlooked as a result of considering them as wolves).

Let us call this framework *structure-of-metaphor* scheme, or SoM scheme. This scheme is summarized in Box 6.1 and I apply it to several metaphors that are presented in the subsequent sections.

Before analyzing metaphors with this scheme, it is important to also consider the functions of metaphors. That is, for the analysis of any metaphor, it is important to consider what this metaphor is used for and what one intends to achieve by using it. It seems that we can

---

BOX 6.1 **The Structure of Metaphor Scheme**

Metaphor: P is S
  consists of:

1) a principal subject P;
2) a secondary subject S;
3) a set of implications I associated with S;
4) a set of attributions A that P acquires in virtue of our looking at
   P through the lens of S;
5) a set of properties Z of P that are masked due to our looking at
   P through the lens of S

---

distinguish between three different, but of course interrelated, functions for metaphors (Bradie 1999):

- The *heuristic* function: This is about the use of metaphors within science. In this case, metaphors are useful to scientists themselves in order to explore new phenomena with reference to other phenomena they already understand well. For example, one might think of a membrane protein that binds molecules as a receptor in order to explore its function.
- The *theoretical* function: This is also about the use of metaphors within science. In this case, metaphors serve for explanations and for theoretical understanding, and are thus indispensable for scientific understanding. For example, once the function of a membrane protein was found to be like that of a receptor, that protein could thereafter be described as a receptor-protein to indicate its function and properties.
- The *rhetorical* function: This is about the use of metaphors in science education and communication. In this case, metaphors serve to explain science concepts, processes, and mechanisms to non-experts by referring to other concepts, processes, and mechanisms with which they are more familiar.

Let us call this scheme the function-of-metaphor scheme, or FoM scheme. Figure 6.1 presents the SoM and the FoM scheme for the receptor-protein metaphor.

In what follows, I present an analysis of the machine metaphor in biology, based on an explicit application of the SoM and the issues around

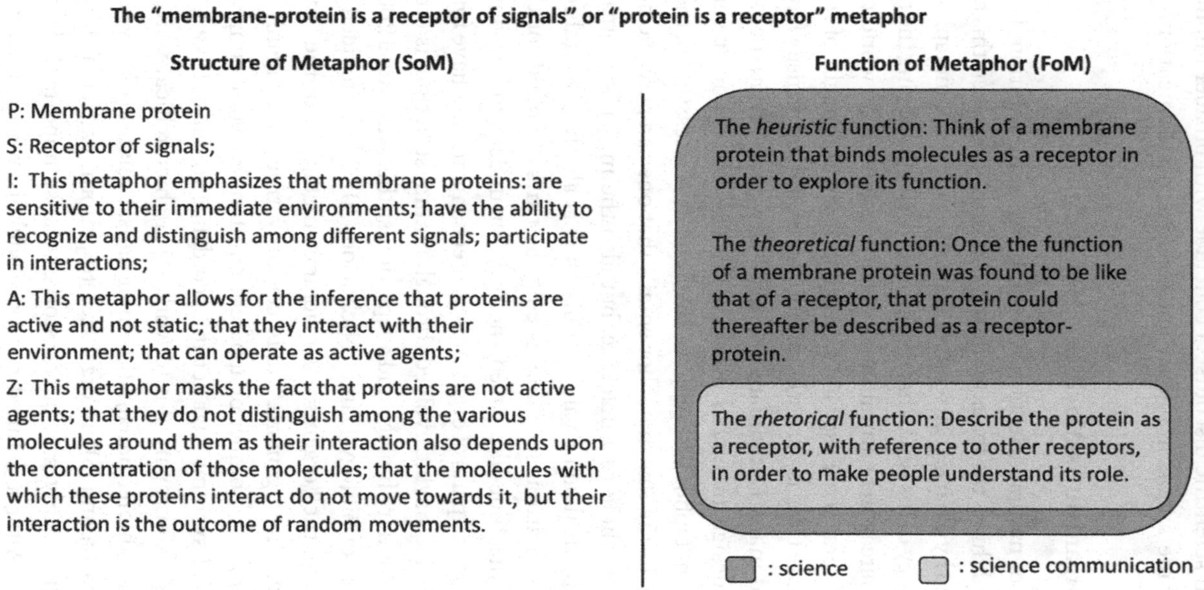

The "membrane-protein is a receptor of signals" or "protein is a receptor" metaphor

Structure of Metaphor (SoM)

P: Membrane protein

S: Receptor of signals;

I: This metaphor emphasizes that membrane proteins: are sensitive to their immediate environments; have the ability to recognize and distinguish among different signals; participate in interactions;

A: This metaphor allows for the inference that proteins are active and not static; that they interact with their environment; that can operate as active agents;

Z: This metaphor masks the fact that proteins are not active agents; that they do not distinguish among the various molecules around them as their interaction also depends upon the concentration of those molecules; that the molecules with which these proteins interact do not move towards it, but their interaction is the outcome of random movements.

Function of Metaphor (FoM)

The *heuristic* function: Think of a membrane protein that binds molecules as a receptor in order to explore its function.

The *theoretical* function: Once the function of a membrane protein was found to be like that of a receptor, that protein could thereafter be described as a receptor-protein.

The *rhetorical* function: Describe the protein as a receptor, with reference to other receptors, in order to make people understand its role.

⬛ : science   ⬜ : science communication

FIGURE 6.1 The structure of metaphor and function of metaphor schemes applied for the "protein-is-receptor" metaphor.

the rhetoric, the heuristic, and the theoretical functions of metaphor. Then I apply the SoM scheme and the FoM scheme to my own analysis of the metaphor of natural selection as used by Darwin in the *The Origin of Species*. The aim of this exercise is to reveal the potential and the limitations of these metaphors, as the latter are more often than not overlooked in scientific discourse. I conclude with broader implications for the use of metaphors in science and science communication.

6.3 The "Organism-Is-Machine" Metaphor

The concept of natural selection, a main process in evolution, is itself a metaphor (and this is the focus of the next section). However, this concept was developed by Charles Darwin as a response to another metaphor of the natural theology of his time: that of the organic world as being the product of design. Whereas the natural theologians considered the world to be an artifact in a literal sense, as God's creation, Darwin thought of it *as if it were* designed, that is, strictly in the metaphorical sense. However, Darwin – and many Darwinians and neo-Darwinians ever since – never rejected the design metaphor (Ruse 2000). This metaphor fit very well with another metaphor first proposed by René Descartes: that of the world as a machine. Using this metaphor entails that one can perceive not only the world as a whole as a machine, but also the individual, organismal parts as mechanical contrivances. This view is absolutely vital for Darwin's theory of natural selection: Seeing the parts of orchids or those of barnacles as machine-like paved the way for understanding how they worked. Seeing the parts of organisms as mechanisms that have functions toward some end, having evolved through natural selection, is crucial to modern evolutionary biology and adaptationist thinking (Ruse 2005).

Philosopher of biology Daniel Nicholson (2013) has provided a thorough analysis of the "organism is machine" metaphor, or the "machine conception of the organism." It is certainly true that organisms and machines exhibit a number of apparent similarities. For example, both: (1) are physical systems that act in accordance to natural laws; (2) use energy and transform part of it to work; (3) are hierarchically structured and internally differentiated; (4) exhibit causal relations between their interacting parts; and (5) are organized to perform particular functions. However, this is only part of the story, because important differences exist as well. After a detailed analysis, Nicholson concluded that the

TABLE 6.1 *Some major differences between organisms and machines (based on Nicholson 2013, p. 674)*

Features	Organisms	Machines
Purposiveness	Intrinsic	Extrinsic
Organization and production	System itself	Maker
Maintenance and repair	System itself	Maker and/or user
Functional determination	System itself	Maker and/or user
Properties of parts	Dependent on whole	Independent from whole
Ontogenic priority	First whole, then parts	First parts, then whole

most fundamental difference between organisms and machines, which accounts for many other differences, is that whereas they both are purposive systems, organisms are internally purposive, whereas machines are externally purposive. This means that organisms have functions that are self-serving, whereas machines have functions that serve an external agent. This entails a number of other differences, as shown in Table 6.1 (Nicholson 2013, p. 671).

Why has the organism-is-machine metaphor been so prevalent in biology even though it clearly misrepresents fundamental features of organisms? To answer this question, Nicholson applied the Function-of-Metaphor scheme, described previously, to shed light on the heuristic, the rhetorical and the theoretical function of the machine metaphor. He argued that the heuristic value of this metaphor is that if the parts of an organism are considered independently from the whole organism, they constitute extrinsically purposive systems that function for the benefit of the whole organism and depend upon it, and they thus resemble machines. Therefore, thinking about the parts of organisms as machines makes it possible to formulate hypotheses about their functions and roles (heuristic function). If this works out well, considering the parts of organisms as machines can have a high theoretical value for the communication among scientists and for investigating their properties (theoretical function). Finally, there is also an important rhetorical function, because non-experts can come to understand the functions and roles of the parts of organisms if they think of them as machines.

An interesting feature of the machine metaphor is that it has many different instantiations in particular domains of inquiry. For instance, in developmental biology it has assumed the form of the genetic program of development, whereas in evolutionary biology it is central in the view of evolution as a forward-looking, problem-solving, optimizing design-process. Both of these instantiations are problematic, however. An organism does not develop in the way that a computer runs a program, that is, as the programmed execution of an algorithmic sequence of predetermined steps. Similarly, organisms do not actually evolve according to any design specifications, in the same sense that an object is designed by an engineer, and so the interpretation of natural selection as an engineer promotes a distorted view of how evolution proceeds (Nicholson 2014). Another example is cell and molecular biology where the machine metaphor is embodied in the conception of protein complexes as rigidly structured and highly coordinated and efficient molecular machines. This view implies that the organization of a cell can be explained in a reductionist manner, as well as the idea that its molecular pathways can be construed as deterministic circuits. However, again this view is misconceived because it fails to convey the dynamic, self-organizing nature of cells, the fluidity and plasticity of their components, and the stochasticity and non-linearity of the intracellular processes (Nicholson 2019).

Therefore, whereas the organism-as-machine metaphor can be quite useful, it also has some potential unwanted consequences. For example, in thinking about organisms as machines, both biologists and laypeople may focus too much on the similarities between machines and organismal parts. As a result, they may not consider organismal parts *as if they were* machines only, but actually think that they are machines. This is actually something that Intelligent Design proponents have exploited (Nicholson 2013, pp. 674–676). Other philosophers have also noted this problem and have convincingly argued that for this reason organisms should not be portrayed as machines (Pigliucci & Boudry 2011; Brigandt 2013).

Let me now apply the Structure of Metaphor scheme (Box 6.1) to the organism-as-machine metaphor:

P: organisms;

S: machines;

I: this metaphor emphasizes that organisms: are physical systems operating according to natural laws; use or modify energy and transform part of it to work; are hierarchically structured and internally differentiated; consist of causally related parts; exhibit purposiveness and functions;

A: this metaphor allows for the inferences that organisms: exhibit design; exhibit complexity; that they might fail to operate and die out if some parts malfunction; that they are the product of an intentional and intelligent designer; and

Z: this metaphor masks all the differences between organisms and machines presented in Table 6.1.

All in all, the "organism-is-machine" metaphor provides the grounds for a natural/physical conception of organisms, independent of any vitalistic or other unnatural metaphysical ideas. However, at the same time, it appears not only compatible with, but even pointing toward, the existence of intelligent design in nature. As Richard Lewontin nicely summarized the issue:

> While we cannot dispense with metaphors in thinking about nature, there is a great risk of confusing the metaphor with the thing of real interest. We cease to see the world *as if* it were *like* a machine and take it to *be* a machine. The result is that the properties we ascribe to our object of interest and the questions we ask about it reinforce the original metaphorical image and we miss the aspects of the system that do not fit the metaphorical approximation. (Lewontin 2000, p. 4, emphases in the original)

However, there is a sense in which the failure of metaphors in science is not exclusively bad. Whereas metaphors such as the "organism-is-machine" metaphor can certainly be misleading, there is a lot that we can learn from its failure in this particular theoretical role. The reason for this is that this failure can turn our attention to particular aspects in which the target, in this case the organism, fails to live up to the expectations we project onto it on the basis of the source domain, in this case the machine. Therefore, by understanding why the organism is not a machine, we may eventually learn a lot about what the organism actually is (Nicholson 2018, p. 162).

6.4 The "Natural Selection" Metaphor

Let us now consider another well-known concept, natural selection. This concept was developed by Charles Darwin as a response to the natural–theological metaphor of the organic world being the product of design and its implications about the existence of a divine designer. Here is how Darwin defined natural selection in the 1st edition of *The Origin of Species:*

> Owing to this struggle for life, any variation, however slight and from whatever cause proceeding, if it be in any degree profitable to an individual of any species, in its infinitely complex relations to other organic beings and to external nature, will tend to the preservation of that individual, and will generally be inherited by its offspring. The offspring, also, will thus have a better chance of surviving, for, of the many individuals of any species which are periodically born, but a small number can survive. I have called this principle, by which each slight variation, if useful, is preserved, by the term of Natural Selection, in order to mark its relation to man's power of selection. We have seen that man by selection can certainly produce great results, and can adapt organic beings to his own uses, through the accumulation of slight but useful variations, given to him by the hand of Nature. But Natural Selection, as we shall hereafter see, is a power incessantly ready for action, and is as immeasurably superior to man's feeble efforts, as the works of Nature are to those of Art. (Darwin 1859, p. 61)

In this passage, Darwin explicitly referred to the analogy between natural selection and artificial selection. However, this is not the only analogy that Darwin used. He developed his theory based on three analogies: (1) between the struggle for existence in human societies and the struggle for existence in nature, (2) between artificial selection and natural selection, and (3) between the (physiological) division of labor and ecological specialization. Darwin was primarily inspired by his cultural milieu, as his influences were distinctive features of the Victorian era: Artificial selection was done in the context of animal and plant breeding that was a form of Victorian technology, whereas the ideas of struggle for existence and division of labor came from the political economists Thomas Malthus and Adam Smith, respectively, who had developed their theories in the English context (Kampourakis 2014, ch. 4; see also Waters 2009). Therefore, analogies are predominant in the concept of natural selection.

Darwin was of course aware of that. In the 3rd edition of *The Origin of Species,* he wrote:

In the literal sense of the word, no doubt, natural selection is a misnomer; but who ever objected to chemists speaking of the elective affinities of the various elements? – and yet an acid cannot strictly be said to elect the base with which it will in preference combine. It has been said that I speak of natural selection as an active power or Deity; but who objects to an author speaking of the attraction of gravity as ruling the movements of the planets? Every one knows what is meant and is implied by such metaphorical expressions; and they are almost necessary for brevity. So again it is difficult to avoid personifying the word Nature; but I mean by Nature, only the aggregate action and product of many natural laws, and by laws the sequence of events as ascertained by us. With a little familiarity such superficial objections will be forgotten. (Darwin 1861, p. 85)

Darwin thus acknowledged not only that natural selection is a metaphor, but also that it has limitations. Darwin intended to refer to a process of *unconscious selection* taking place in nature. Therefore, he struggled for some time to figure out what was the factor that drove selection, in order to distinguish natural selection from artificial selection, which is a *conscious process of selection* that animal breeders perform. Artificial selection requires an intelligent external selector who picks variants according to particular aims or goals. Natural selection, in contrast, is an unmediated, unintentional natural process of struggle that results in some individuals surviving and reproducing whereas others die out, where no external selector exists. Darwin explained that the competition among individuals of the same species that takes place simultaneously with competition among individuals of different species plays the role of the "selector." Thus, it is the external environment, and the different types of competition that it entails, which is the cause of natural selection. Individuals of the same species interact with one another but also with others from different species. In the long run, those individuals of a species that can compete more effectively in their environment with individuals of their own or of other species will be those that will survive and reproduce (Kampourakis 2014, ch. 4; see also Kohn 2009, pp. 93–94).

We can thus understand natural selection in terms of artificial selection, with the important difference that in the former case there is no agent selecting. However, it is exactly this lack of someone doing the selection that makes natural selection an ambiguous metaphor. It is actually more intuitive to think of natural selection as a process in which a conscious agent (e.g., God, Nature) selects something, rather than an unconscious process (see Blancke et al. 2014). Darwin later decided to adopt a different metaphor to resolve the problems with natural selection: Herbert Spencer's concept of the survival of the fittest, which we find for the first time in the 5th edition of *The Origin of Species*:

> I have called this principle, by which each slight variation, if useful, is preserved, by the term Natural Selection, in order to mark its relation to mans power of selection. But the expression often used by Mr. Herbert Spencer of the Survival of the Fittest is more accurate, and is sometimes equally convenient. We have seen that man by selection can certainly produce great results, and can adapt organic beings to his own uses, through the accumulation of slight but useful variations, given to him by the hand of Nature. But Natural Selection, as we shall hereafter see, is a power incessantly ready for action, and is as immeasurably superior to mans feeble efforts, as the works of Nature are to those of Art. (Darwin 1869, pp. 72–73)

Darwin noted that the concept of the survival of the fittest was more accurate, and sometimes more convenient, than natural selection.

However, the metaphors of "natural selection" and of the "survival of the fittest" are different in one very important aspect, as philosopher David Depew (2013) has shown. Darwin initially thought that adaptation through natural selection occurs through "the accumulation of slight but useful variations" (Darwin 1859, p. 61). This means that for Darwin adaptation was possible only if natural selection operates for many generations on some variants that may confer an advantage to their bearer's survival and reproduction. This is different from the idea of the "survival of the fittest," where there is no accumulation of slight, advantageous variations but simply elimination of the individuals that bear the non-advantageous characters and survival of those individuals that possess the advantageous ones. There is thus an important difference between the two views: "natural selection" is the case of *selection for* particular variants; adaptation is the cumulative effect of selection on the available variation for many generations. In contrast, "survival of the fittest" is a case of *selection against* particular variants; in other words, it is a process through which variation is quickly eliminated because only one or very few variants remain. In short, Darwin replaced a conceptualization of selection as a "creative" process with one of selection as an "eliminative" process. We see in this case that which metaphor we use also makes a big difference for how a concept is understood.

Let me now apply again the structure of metaphor scheme (Box 6.1) to analyze the metaphor of natural selection. According to this, we can make the following inferences for the "natural selection" metaphor:

P: differential survival in nature;

S: artificial selection by breeders;

I: this metaphor emphasizes that: some features are selected whereas others are not; the selected features serve some purpose; the apparent design in organisms is the outcome of such a selection process; features that do not serve a purpose may not be selected;

A: this metaphor allows for the inferences that selection: can be a purposeful process; can be a conscious process; can involve an external selector who guides it; can result in intelligent design; and

Z: this metaphor masks the ideas that: natural selection is an unconscious process taking place in nature; the apparent design that we see in organisms is incidental; the purposiveness of features is internal (self-serving) and not external (other-serving).

The metaphor of natural selection has an obvious heuristic value. Although Darwin was not able to show that natural selection had indeed brought about adaptation (Kampourakis 2014, pp. 121–125), it has been perhaps the most important evolutionary concept of the twentieth century, which guided much related research (Mayr 2002). Some biologists have gone as far as to consider natural selection as THE most important process in evolution, with Richard Dawkins famously describing it as the Blind Watchmaker, invoking images of an external agent (Dawkins 2006). Perhaps the importance of natural selection and adaptation has been exaggerated (Gould & Lewontin 1979). Nevertheless, its heuristic value is unquestionable and it was already evident for Darwin who noted that:

> When we no longer look at an organic being as a savage looks at a ship, as at something wholly beyond his comprehension; when we regard every production of nature as one which has had a history; when we

contemplate every complex structure and instinct as the summing up of many contrivances, each useful to the possessor, nearly in the same way as when we look at any great mechanical invention as the summing up of the labour, the experience, the reason, and even the blunders of numerous workmen; when we thus view each organic being, how far more interesting, I speak from experience, will the study of natural history become! (Darwin 1859, pp. 485–486)

In contrast, the theoretical value of the metaphor is limited. Selection is literally an active and conscious process, and so it is more intuitive to think about selection in these terms. Therefore, the metaphor of natural selection emphasizes the purposeful character of this process and masks the fact that it is a natural, unconscious process. The situation became worse with the metaphor of the survival of the fittest, which represents selection as an eliminative process where all variants, except for one, die out. However, this does not seem to be how the process takes place.

Finally, one should probably have mixed feelings about the rhetorical function of natural selection. Whereas the analogy with artificial selection was supposed to make the process comprehensible, Darwin did not succeed in convincing most people about the importance of natural selection. It was not until the early twentieth century, and after the emergence of genetics, that natural selection started to become considered as an important evolutionary process, both within and outside science. Still, even today, many people do not accept, and/or do not understand, how a slow and gradual process can result in organs such as the eye that resembles intelligently designed artifacts such as the telescope. As Richard Lewontin (again) has nicely put it:

Darwin, quite explicitly, derived this understanding of the motivating force underlying evolution from the actions of plant and animal breeders who consciously choose variant individuals with desirable properties to breed for future generations. "Natural" selection is human selection writ large. But of course, whatever "nature" may be, it is not a sentient creature with a will, and any attempt to understand the actual operation of evolutionary processes must be freed of its metaphorical baggage. (Lewontin 2010)

Again, this is a case where the failure of the metaphor of natural selection can help us better understand what natural selection is not: It is not a conscious process directed toward some kind of perfection or optimality. This then allows us to better understand what natural selection is, which is better described as a process of differential survival of individuals in a particular environment. This opens up the way for the consideration of other metaphors that may be useful, such as the metaphor of environmental filtration (Rosenberg & McShea 2008, p. 18). But we should keep in mind that this metaphor, too, like any metaphor, has both value and limitations.

6.5 Conclusion

In this chapter, I have argued that the fact that particular concepts are metaphors has important implications for conceptual understanding in science. Metaphors are based on an analogy between a source and a target domain. However, this analogy may end up emphasizing some of their similarities only, while at the same time masking important differences between the two domains. The similarities may be important for the heuristic function of metaphors, but at the same time there are important issues related to their theoretical and rhetorical function. Therefore, it is important for scientists to reflect upon the meanings and the uses of metaphors they use, as well to be aware of and explicit about their limitations. As it is not possible to entirely refrain from using metaphors, we had better use them in well-defined frameworks, rather than inappropriately building the frameworks on those metaphors. As all language and thought is to some extent metaphorical, then the question is not whether we should rely on metaphors or not. Rather the question is which different metaphors we can use in order to manage to capture several different aspects of the target domain. All metaphors have value and limitations, and what is important is to keep that in mind when using them in science research, teaching, and public communication.

References

Arabatzis, T. (2019). What Are Scientific Concepts? In K. McCain & K. Kampourakis (eds.), *What Is Scientific Knowledge? An Introduction to Contemporary Epistemology of Science*. New York: Routledge.

Avise, J. C. (2001). Evolving Genomic Metaphors: A New Look at the Language of DNA. *Science* 294(5540): 86–87.

Black, M. (1955). Metaphor. *Proceedings of the Aristotelian Society* 55: 273–294.

Blancke, S., Schellens, T., Soetaert, R., Van Keer, H., & Braeckman, J. (2014). From Ends to Causes (and Back Again) by Metaphor: The Paradox of Natural Selection. *Science & Education* 23(4): 793–808.

Boudry, M. & Pigliucci, M. (2013). The Mismeasure of Machine: Synthetic Biology and the Trouble with Engineering Metaphors. *Studies in History and Philosophy of Biological and Biomedical Sciences* 44: 660–668.

Bradie, M. (1998). Explanation as Metaphorical Redescription. *Metaphor and Symbol* 13(2): 125–139.

Bradie, M. (1999). Science and Metaphor. *Biology and Philosophy* 14: 159–166.

Brigandt I. (2013). Intelligent Design and the Nature of Science: Philosophical and Pedagogical Points. In K. Kampourakis (ed.), *The Philosophy of Biology: A Companion for Educators*. Dordrecht: Springer.

Brown, T. L. (2003). *Making Truth: Metaphor in Science*. Urbana and Chicago: University of Illinois Press.

Cuddington, K. (2001). The "Balance of Nature" Metaphor and Equilibrium in Population Ecology. *Biology and Philosophy* 16: 463–479.

Darwin, C. R. (1859). *On the Origin of Species by Means of Natural Selection, or the Preservation of Favoured Races in the Struggle for Life* (1st ed.). London: John Murray (available at http://darwin-online.org.uk)

Darwin, C. R. (1861). *On the Origin of Species by Means of Natural Selection, or the Preservation of Favoured Races in the Struggle for Life* (3rd ed.). London: John Murray (available at http://darwin-online.org.uk)

Darwin, C. R. (1869). *On the Origin of Species by Means of Natural Selection, or the Preservation of Favoured Races in the Struggle for Life* (5th ed.). London: John Murray (available at http://darwin-online.org.uk)

Dawkins, R. (2006 [1986]). *The Blind Watchmaker*. London: Penguin Books.

Depew, D. (2013). Conceptual Change and the Rhetoric of Evolutionary Theory: "Force Talk" as a Case Study and Challenge for Science Pedagogy. In Kampourakis, K. (ed.), *The Philosophy of Biology: A Companion for Educators*. Dordrecht: Springer.

Fusco, G., Carrer, R., & Serrelli, E. (2014). The Landscape Metaphor in Development. In A. Minelli & T. Pradeu (eds.), *Towards a Theory of Development*, pp. 114–128. Oxford: Oxford University Press.

Gould, S. J. & Lewontin, R. C. (1979). The Spandrels of San Marco and the Panglossian Paradigm: A Critique of the Adaptationist Programme. *Proceedings of the Royal Society of London Series B. Biological Sciences* 205(1161): 581–598.

Griffiths, P. E. (2001). Genetic Information: A Metaphor in Search of a Theory. *Philosophy of Science* 68(3): 394–412.

Hall, B. K. & Olson, W. M. (2003). *Keywords & Concepts in Evolutionary Developmental Biology*. Cambridge, MA: Harvard University Press.

Hesse, M. B. (1966). *Models and Analogies in Science*. Notre Dame, IN: University of Notre Dame Press.

Hull, D. L. & Ruse, M. (eds.) (1998). *The Philosophy of Biology*. Oxford: Oxford University Press.

Hull, D. L. & Ruse, M. (eds.) (2007). *The Cambridge Companion to the Philosophy of Biology*. Cambridge: Cambridge University Press.

Kampourakis, K. (2014). *Understanding Evolution*. Cambridge: Cambridge University Press.

Kampourakis, K. (2017). *Making Sense of Genes*. Cambridge: Cambridge University Press.

Kampourakis, K. (2019). Genetics Makes More Sense in the Light of Development. In Giuseppe Fusco (ed.), *Perspectives on Evolutionary and Developmental Biology: Essays for Alessandro Minelli*, pp. 115–122. Padova: University of Padova Press.

Kaplan, J. (2008). The End of the Adaptive Landscape Metaphor? *Biology and Philosophy* 23: 625–638.

Kay, L. (2000). *Who Wrote the Book of Life? A History of the Genetic Code*. Stanford: Stanford University Press.

Keller, E. F. (1995). *Refiguring Life: Metaphors of Twentieth Century Biology*. New York: Columbia University Press.

Keller, E. F. & Lloyd, E. A. (1992). Keywords in Evolutionary Biology. Cambridge, MA: Harvard University Press.

Kohn, D. (2009). Darwin's Keystone: The Principle of Divergence. In M. Ruse and R. J. Richards (eds.), The Cambridge Companion to the "Origin of Species," pp. 87–108. Cambridge: Cambridge University Press.

Lakoff, G. & Johnson M. (2003 [1980]). *Metaphors We Live By*. Chicago: The University of Chicago Press.

Lewontin, R. C. (2000). *The Triple Helix: Gene, Organism, and Environment*. Cambridge, MA: Harvard University Press.

Lewontin, R. (2010). Not So Natural Selection. *The New York Review of Books*, www.nybooks.com/articles/2010/05/27/not-so-natural-selection/, accessed April 3, 2019.

MacLeod, M. (2012). Rethinking Scientific Concepts for Research Contexts: The Case of the Classical Gene. In Uljana Feest & Friedrich Steinle (eds.), *Scientific Concepts and Investigative Practice*, pp. 47–74. Berlin: De Gruyter.

Mayr, E. (2002). *What Evolution Is*. London: Weidenfeld and Nicolson.

Mindell, D. P. (2013). The Tree of Life: Metaphor, Model, and Heuristic Device. Systematic Biology 62(3): 479–489.

Morange, M. (2002). *The Misunderstood Gene*. Cambridge MA: Harvard University Press.

Nersessian, N. (2008). *Creating Scientific Concepts*. Cambridge, MA: The MIT Press.

Nicholson, D. J. (2013). Organisms ≠ Machines. *Studies in History and Philosophy of Biological and Biomedical Sciences* 44: 669–678.

Nicholson, D. J. (2014). The Machine Conception of the Organism in Development and Evolution: A Critical Analysis. *Studies in History and Philosophy of Biological and Biomedical Sciences* 48: 162–174.

Nicholson, D. J. (2018). Reconceptualizing the Organism: From Complex Machine to Flowing Stream. In D. J. Feest & J. Dupré (eds.), *Everything Flows*, pp. 139–166. Oxford: Oxford University Press.

Nicholson, D. J. (2019). Is the Cell Really a Machine? *Journal of Theoretical Biology* 477: 108–126.

Ouzounis, C. & Mazière, P. (2006). Maps, Books and Other Metaphors for Systems Biology. BioSystems 85: 6–10.

Pigliucci, M. (2010). Genotype–Phenotype Mapping and the End of the "Genes as Blueprint" Metaphor. *Philosophical Transactions of the Royal Society B* 365: 557–566.

Pigliucci, M. & Boudry, M. (2011). Why Machine-Information Metaphors are Bad for Science and Science Education. *Science & Education* 20(5–6): 453–471.

Rosenberg, A. & McShea, D. W. (2008). *Philosophy of Biology: A Contemporary Introduction*. New York: Routledge.

Reynolds, A. S. (2018). *The Third Lens: Metaphor and the Creation of Modern Cell Biology*. Chicago: The University of Chicago Press.

Ruse, M. (2000). Metaphor in Evolutionary Biology. *Revue Internationale de Philosophie* 54, 214(4): 593–619.

Ruse, M. (2005). Darwinism and Mechanism: Metaphor in Science. *Studies in History and Philosophy of Biological & Biomedical Sciences* 36: 285–302.

Thompson, N. S. (2000). Shifting the Natural Selection Metaphor to the Group Level. *Behavior and Philosophy* 28(1–2): 83–101.

Waters, C. K. (2009). The Arguments in The Origin of Species. In J. Hodge & G. Radick (eds.), *The Cambridge Companion to Darwin* (2nd ed.), pp. 120–143. Cambridge: Cambridge University Press,

7   How Do Concepts Contribute to Scientific Advancement?

Evolutionary Biology As a Case Study

DAVID J. DEPEW

7.1 Introduction

In 1962 Thomas Kuhn published one of the twentieth century's most fruitful and most scandalous books, *The Structure of Scientific Revolutions* (Kuhn 1962). He argued that the form of inquiry we call natural science periodically undergoes more or less sudden shifts in what he called paradigms. His paradigmatic examples were the displacement of Ptolemaic astronomy by Copernican heliocentrism in the sixteenth century; and in the late eighteenth century the discrediting of alchemy, which sought to turn one element into another, by Lavoisier's chemistry, which took the elements to be atomic and chemical change to consist of compounding these elements in particular ways. These are not especially disconcerting cases, since they helped give birth to modern science in the first place. What *was* disconcerting was Kuhn's contention that, even after it was up and running, modern science – institutionalized practices of inquiry in which hypotheses and theories are rigorously tested by careful observation and experimentation – shows the same pattern. Disciplinary communities, he claimed, typically rally around a new paradigm until enough anomalies pile up to invite an even newer one.

But these shifts are far from inevitable. They don't happen unless and until someone is creative enough to think up something new. Kuhn argued that this process is so jagged and paradigms are so "incommensurable" with each other that belief in science's cumulative progress is unwarranted. This was not just disconcerting, but scandalous, especially because Kuhn maintained that a scientist who shifts paradigms is doing something comparable to undergoing a religious conversion. Without enough evidence, he or she just begins to see things differently.

Reading *Structure* with undergraduates is likely to expose teachers to a common reaction to it. Since elementary school, students have had it drummed into their heads that what makes modern science reliably predictive and explanatory is its commitment to the idea that scientific concepts, hypotheses, and theories arise from carefully collected and sifted sensory experience. If they are not confirmed by observations, ideas, hypotheses, and theories, they are peremptorily rejected. On this view, experimental testing cannot succeed unless the process of hypothesizing is guided by observation. It is commonly believed that this is why science's results keep accumulating and why its technological applications are so powerful.

So deeply ingrained is this so-called "Baconian empiricist" way of understanding science that students tend to retain its basic ideas.[1] As a result, when they read Kuhn they often fail to revise their preconceptions about how science works. Instead, they conclude that Kuhn is rejecting science and embracing relativism, skepticism, or, once they have become acquainted with this shibboleth, "social constructionism." Truth be told, many scientists themselves react the same way.

Kuhn's fellow philosophers and historians of science responded differently to *Structure*. They, too, were doing what Kuhn was doing: studying historical cases in order to test whether this or that philosophical analysis of scientific method succeeds or fails to explain past cases of scientific discovery and so has a good chance of explaining new ones. But they did it with more nuance. They paid less attention than Kuhn to the psychology of discovery and commitment and more to lines of communication in and between field, laboratory, scholarly journals, and public debates, that is, to what Bruno Latour called "science in action" (Latour 1987). They suggested that, although there is a place for paradigms, research programs and in the longer run research traditions are the units of scientific creativity, recruitment, hypothesis testing, problem solving, disputation, and theory replacement (Lakatos 1970; Laudan 1977; Latour 1987; Zammito 2004, pp. 133–136, on diverging uptakes of Kuhn).

In this process, concept formation and its deployment in theorizing play roles that simplistic empiricism is bound to miss or misconstrue.

---

[1] This is a bit of a slur on Francis Bacon, since it is more indebted to Jonathan Swift's satire of the Laputans in *Gulliver's Travels* than to anything that might be found in *The Advancement of Learning* or *Novum Organum*.

Defining key theoretical terms, making mathematical models of how the variables named by these terms interact, measuring how well a model fits particular cases, and using metaphors to extend models to novel cases and new fields – all these activities are integral to the ongoing work of science. Far from importing arbitrary elements into science, metaphors, models, and conceptual frames are necessary if confirmation and falsification are to be possible.

In this chapter, I press these claims by reference to the history of evolutionary biology. In doing so, I suggest their relevance to the process of science generally, as well as its public understanding and pedagogy.

7.2 The Role of Metaphors in Scientific Theories: Darwin's Gradualist Concept of Natural Selection As a Case Study

Darwin's *Origin of Species* was written during an intense period of philosophical scrutiny into scientific methodology. There is little or no drama in the immense logbook of his researches, but there is plenty of it lurking in the *Origin* itself. The drama in *Origin* springs from Darwin's fear that if he remained silent his lifework would be "forestalled" by a priority dispute with Wallace, while if he came forward as having been working with the idea long before Wallace came up with it he would expose himself and his family to condemnation from learned clerics who opposed species transmutation, including some of his own teachers. Accordingly, he wrote *Origin* in a way calculated to get some cover from the three most prominent philosophers of science of his time, William Whewell, John Herschel, and John Stuart Mill.

Of the three, Mill was the most empiricist. He judged natural selection to be a potentially fruitful hypothesis, but at present no more than that. It was certainly not (yet) a theory strong enough to displace the view that each species is specially created by God, an early version of what is now called intelligent design. Mill was not alone in thinking that the claim that natural selection takes the place of an intelligent designer is itself compromised by Darwin's appeal to what animal and plant breeders do. Is not natural selection's power to design adapted organisms, he asked, little more than creationism without a creator (Forest 2018)?

Herschel was potentially more open to Darwin because he held that concepts such as Newton's gravity might not be directly observable but can acquire support from observable consequences that follow from adopting them. Still, noting that the great Newton himself drew attention to gravity as a real phenomenon when he described the way sloshing water systematically climbs up the side of a spinning barrel, Herschel required that true causes (*verae causae*) must be more than deductions from a hypothesis (Herschel 1831). They must in some sense be independently observable. Darwin tried to meet this demand by describing the heritable differences induced by the selective breeding of pigeons (Darwin 1859, ch. I). When direct evidence for Malthusian population dynamics in all species, not just man (ch. III), is combined with the observable modifying power of artificial selection, discriminative selection in nature appeared to him to count as a *vera causa*.

Accordingly, Darwin was distraught when he learned that Herschel, to whom he had sent a copy of *Origin*, was unpersuaded. "I have heard by a roundabout channel," he wrote to his mentor Charles Lyell, whose methodology in the science of geology Herschel had praised and Darwin took himself to be extending to biology (Hodge 1982), that "Herschel says my book 'is the law of higgledy-piggledy.' What exactly this means I do not know, but it is evidently very contemptuous. If true, this is a great blow and discouragement" (Darwin to Lyell, December 10, 1859, DCP, #2575).[2] What Herschel meant was that, unlike the homogeneous, quantifiable, and hence computable factors that lawfully and predictably determine, say, the rate of fall of a body as it enters a gravitational field, what makes for survival and reproductive success in Darwinism is a mixed bag of non-computable traits. It was a good point.

For his part, Whewell construed modern science as inductive, but he did not construe induction as empiricism (Whewell 1840; Snyder 2006). Instead, he believed that the hypothetical deduction of a number of propositions from a theoretical concept, such as gravity and its associated laws, for example, is itself an inductive process that can offer support, sometimes even decisive support, for a theory. It does so, however, only if what Whewell called a "consilience [jumping together] of inductions" reveals the power of a theory's defining concept to simplify and unify an entire field. It is hard not to see Darwin trying hard in

---

[2] All selections from Darwin's correspondence are taken from the Darwin Correspondence Project (DCP) (www.darwinproject.ac.uk). The project assigns a number to each letter. I include these in the references.

the last third of *Origin* to conform to this conception (Ruse 1975). Nonetheless, it soon became clear that Whewell thought that the meaning of the term "species" was already fixed in modern biology and that this meaning ruled out species transmutation as self-contradictory. No help there.

What help Darwin did get came from Wallace, Spencer, and Huxley. All of them wanted to use his book to help liberate science from clerical influence by restricting it to purely natural causes. Hence, they were quite aware that Darwin's analogy between natural and artificial selection imparted a suspiciously designing cast to natural selection. Accordingly, in 1866 they ganged up to urge Darwin to replace "natural selection" with "survival of the fittest" in future editions of *Origin*. The purely *ex post facto* connotations of this process would undercut the creationist objection that natural selection is so haunted by intentional design that it merely substitutes one theology for another (Wallace to Darwin, July 2, 1866, DCP #5140). In the 5th and subsequent editions of *Origin*, Darwin halfheartedly complied. He was in no position to lose friends.

With friends like this, however, he didn't need enemies. Wallace never fancied the analogy with the breeder's art in his way of formulating natural selection, so he had no objection to scotching it. Huxley was more interested in descent from a common ancestor than in natural selection – in fact, a good case can be made out that he never even understood it. He urged Darwin not to rule out the possibility that new species come about by sudden, single-generation jumps (saltation), thereby revealing his indifference to Darwin's confession that he took his theory to stand or fall on its multigenerational gradualism about how natural selection evolves adaptations (Huxley to Darwin, November 23, 1859, DCP #2544; Darwin 1859, pp. 189, 194, 471).

Even more ominously for the future of Darwinism is that the phrase "survival of the fittest" allowed Spencer, who coined it, to assimilate natural selection to his own theory of evolution. In his 1864 *Principles of Biology*, he used "natural selection" to assign to Malthusian population pressure the role of forcing embryos either to die or adjust to their environments during their development. This usage had the effect of narrowing the difference between the struggle for existence and Darwinian natural selection. Not only did it claim that individuals either acquire or fail to acquire fitness or adaptedness in a single generation, which Darwin's gradualism ruled out in principle; it also confined natural selection to selection against the unfit. Spencerism thereby ran roughshod over the creative power of natural selection working gradually over many generations to turn variations initially uncoupled from their utility (and in that respect arising by chance) into adaptations. That was the point of Darwin's comparison of natural with artificial selection. In consequence Spencer lacked a concept of natural selection as a creative process. Instead, he ascribed the innovative power to adapt to the direct, molding effect of environmental pressures on the plasticity of organisms *in utero* and even postnatally.

It is not a pretty spectacle to see Darwin backtracking on this point. He, too, was impressed by the plasticity of early development. He told Lyell that he had underestimated "the direct action of external conditions in producing varieties" because evidence for it was published only after 1859 (Darwin to Lyell, October 4, 1867, DCD, #5640). Many of the objections that creationists, including proponents of intelligent design, still raise against Darwinism are actually aimed at Spencerism in the mistaken belief that it is Darwinism. Examples are the objection that natural selection is tautologous, since by definition only the fit survive, and the moral objection that natural selection is (no more than) a bloody, wasteful, process of elimination that is inconsistent with the goodness of God (so-called Social Darwinism).

I have retold this unhappy tale in order to suggest how essential Darwin's analogy between natural and artificial selection was to his theory. The comparison is indeed metaphorical. Darwin himself says so. In the 3rd edition of *The Origin* he wrote:

> It has been said that I speak of natural selection as an active power or deity; but who objects to an author speaking of the attraction of gravity as ruling the movement of planets? ... Whoever objected to chemists speaking of the elective affinities of various elements? ... An acid cannot strictly be said to elect the base with which it will in preference combine ... With a little familiarity such superficial objections will be forgotten ... Every one knows what is meant and is implied by such metaphorical expressions. They are almost necessary for brevity. (Darwin 1861, ch. IV, rearranged)

Aristotle said long ago that a metaphor is the transfer of a property to something to which it does not properly belong, as when Homer refers to old age as stubble (Aristotle, *Poetics* 21.1457b1–30; *Rhetoric* III.10.1410b13–14). On this view we simply get used to seeing things in terms of metaphors. The more accustomed we get, the more literally we take them (see also Chapter 6). For some writers, in fact, the process of moving from live to dead metaphors, to speak metaphorically, is how language itself evolved (Lakoff & Johnson 1980). Maybe so. But in the passage I have just quoted, this is not what Darwin is getting at. A remarkable feature of his comparison of artificial and natural selection is that, rather than overlooking differences, as dead metaphors do, it calls attention to the disanalogy between artificial and natural selection and turns the difference in natural selection's favor. Animal and plant breeders, Darwin writes, can focus only on superficial, externally visible traits, whereas nature, in addition, "acts on every internal organ, every shade of constitutional difference, and the whole machinery of life" (Darwin 1859, ch. IV). It can do so, moreover, over tracts of time vast enough to give rise to new races, species, genera, and higher *taxa*. In fact, the creative power of gradual natural selection explains the very possibility of artificial selection, and why breeders can do so much even with superficial traits.

It was by reflexively turning this metaphor back on itself that Darwin's natural selection became a scientific concept whose power to shape nature unifies an array of discoveries in fields from embryology to taxonomy.[3] Having shown its explanatory power in this way, the concept of natural selection, born from a metaphor, became the basis of a full-blown theory of natural selection.[4]

Darwin implied that it was in just this way, too, that Newton developed his concept of gravity and used it to unify terrestrial with celestial physics. The concept of gravity had persuasive uses for Darwin. Just as Franklin and others extended the Newtonian model of gravitational dynamics to electricity, so too Darwin appropriated it when he imagined Malthus's inherent rate of reproduction as an inertial tendency analogous to

---

[3] Linguists and philosophers who assign a key role to metaphor in scientific theorizing stress the interactive nature of conceptual metaphors: The source of the metaphor and the target to be comprehended throw light on each other. See Black 1962; Hesse 1966; Lakoff & Johnson 1980. Darwin's metaphors exhibit this quality. They are not mere decorations or aids to comprehension by non-scientists.

[4] See Gayon 1997 on Darwin's recognition of the difference between the phenomenon, power, and theory of natural selection.

Newton's straight-line motion that is bent into dynamic equilibrium with the environment by the gravity-like force of resource scarcity (Depew & Weber 1995). That is perhaps why the carefully constructed last sentence of *Origin* links Darwin's theory with Newton's achievement. It exclaims, "Whilst this planet has gone on cycling on according to the law of gravity ... endless forms [of life] most beautiful have been and are being evolved" by natural selection (Darwin 1859, ch. XIII).

7.3 Conceptual Clarification As Scientific Problem-Solving: Neo-Darwinism As Case Study

7.3.1 The Statistical Turn Resolves Conceptual Issues in Darwinism

It was hard for Darwin's early supporters to miss the core of his theory of evolution by natural selection, but, undervaluing as they did his comparison to the art of breeding, miss it they indeed did. Why didn't Darwin fight harder for his brainchild? In a way he did. He wrote carefully observed monographs designed to exhibit the power of gradual natural selection to evolve adaptations in orchids, climbing plants, insectivorous plants, and worms (Darwin 1862, 1875a, 1875b, 1881). But as his extended correspondence with his botanical American friend Asa Gray demonstrates, Darwin never tackled the interpretive problem head on because as long as he lived he toiled as much as his friends, including Gray, within the bounds of scientific assumptions in which quasi-Newtonian dynamical models and intelligent design seemed the only available alternatives. All he could come up with, he told Gray, was that God created the laws by which natural selection evolved well-designed "contrivances," but in doing so necessarily left much to chance. "Not that this notion *at all* satisfies me," he continued. "The more I think the more bewildered I become" (Darwin to Gray, May 22, 1860, DCP, #2814). It is not too much to say that if new models of dynamics had not emerged in the twentieth century, the inner-driven, goal-oriented, vitalist, sometimes panpsychic, and decidedly non-Darwinian theories of evolution that became ascendant in the last third of the nineteenth century might have continued to dominate the field long enough to instigate and assimilate the insights of the molecular revolution in genetics that began in the 1950s (Bowler 1983, 2013).

That this did not happen, and instead a form of natural selection relying on genetics became dominant by the middle of the twentieth century, is due to a series of often public but highly productive controversies that unfolded over many decades. First, in 1888 August Weismann showed to the satisfaction of most experts that acquired traits are not inheritable. This fact discredited Spencerism, which as we have seen passed as Darwinism, because it depended on "soft inheritance," as it has come to be called. In the United Kingdom there were genuine Darwinians around who might have taken advantage of this moment. The so-called "biometricians" were using increasingly powerful statistical methods to validate and measure gradual natural selection in the wild. But they were restricted to observable (phenotypic) traits. Thus, they were unable to withstand an appeal to hard inheritance that, reinforced by the rediscovery (actually, reformulation) of Mendel's laws of inheritance, took sudden genetic mutations, not gradual selection, to be the cause of new traits and species. It took until 1918 for R. A. Fisher, who was simultaneously the most important figure in the development of advanced statistical science and an ardent Darwinian, to demonstrate that the probability that genetic mutations of the sort postulated by the new Mendelians will take hold in populations is exceedingly small (Fisher 1918, 1930). On the other hand, genetic mutations of small effect, if any happen to be available that lead a population to reproduce more effectively in a specific environment, can readily be amplified by natural selection across many generations to gradually evolve adaptations. Suddenly the fortunes of Darwinism, or rather of genetic neo-Darwinism, began to shine more brightly.

This happened because Fisher used models of system dynamics taken from statistical rather than classical physics (Fisher 1930; Depew & Weber 1995). The distribution of gases in statistical mechanics and thermodynamics is measured by looking at ensembles of molecules whose trajectories are probabilistic rather than deterministic. There is nothing that says that all the molecules of oxygen in a room cannot suddenly move to one corner, but it is very improbable. Generally, the distribution of mixtures of gas molecules in our atmosphere will remain in stable equilibrium unless an impinging force dictates otherwise. Similarly, population-genetic models show that distributions of genotypes in a freely interbreeding, and infinitely large, population whose reproduction conforms to Mendel's laws will remain in the same equilibrium distribution generation after generation unless and until natural selection or other external factors affect reproduction rates and evolutionary trajectories. This means that natural selection and other evolutionary "forces" act on, or at the very least are only visible in, organisms seen as members of closely related Mendelian populations. The philosopher of biology Elliott Sober has argued that a Laplacean or Maxwellian demon whose eyes were trained only on the paths of individual atoms or molecules, or by analogy genes and the individual organisms that contain them, will fail to notice not only the causes that figure in evolutionary explanations, but even the phenomena to be explained (Sober, 1984).[5]

The new statistical population dynamics led to solutions to a range of conceptual problems that had long dogged both Darwinism and its rivals. For example, it discredited non-Darwinian theories such as "aristogenesis" or "orthogenesis" because all such accounts were based on the assumption that phylogenetic evolution is development or ontogeny writ large, and so is inner driven and end-oriented in a grand sense (Gould 1977). Darwin himself was almost captured by this picture. His self-proclaimed disciple Ernst Haeckel fully embraced it. But the new models of evolutionary dynamics taken from statistical physics made it a matter of definition that organisms develop but don't evolve, while populations evolve but don't develop. Ontogeny–phylogeny recapitulation is a conceptual confusion.

The change in models also undermined the assumption of early statistical Darwinians such as Darwin's cousin Francis Galton that the traits of successive generations have an inherent tendency to regress to a less than well-adapted mean. Galton and his followers had been put on the defensive by accusations that natural selection is generally impotent to evolve optimal adaptations to new environments in the face of this regressive tendency. Within a few generations, any novel variation would be "swamped" by less adapted, or even maladapted, traits. By contrast, the new statistical models made it a matter of principle that the default condition of Mendelian populations is that genotypic

---

[5] There are good reasons to say that phenomena, not data, are the proper *explananda* of scientific hypotheses and theories. Data are observable. But only by torturing the notion of observation beyond all measure (under the influence of simplistic empiricism) can we say that phenomena are observed or observable. Only after much processing (including discussion) does "what the data say" identify a phenomenon (Bogen & Woodward 1988). Quibbling creationists may say that natural evolution, selection, and genetic drift are not observable, but this doesn't count for much (Hofmann & Weber 2003).

distributions will show no such tendency. Unless affected by selection or other factors, they will remain in genetic equilibrium indefinitely. Since these forces are constantly at work, fitter variations can spread easily through populations and in this way evolve adaptations.

Additional conceptual consequences of the changed models soon followed. For example, Fisher pointed out that, contrary to Darwin's own brand of Darwinism, evolution by natural selection need not occur only under Malthusian conditions of scarcity, such as the manufactured threat of imminent starvation in Ireland in the 1840s. It will ensue wherever there is a difference between available genetic variation and comparative reproductive advantage. This means that it is probably occurring all the time.

Another realization was that "genetic drift" constitutes an evolutionary factor distinct from adaptive natural selection. In small enough populations, genotypes can become prevalent by sheer chance, just as every experienced gambler knows that in playing roulette on a fair wheel a ball might land on red 10 or more times in a row without in the least contravening the long-run tendency of black and red to even out. There is a law of small numbers as well of large ones. The population geneticist Sewall Wright used this statistical principle to explain how and in what sorts of Mendelian populations speciation occurs (Wright 1932). Genetic drift is a phenomenon that could not even be noticed in older ways of modeling evolutionary dynamics, let alone explained by them.

Although it took a longer time (and assistance from the first generation of analytic philosophers of biology) before it became clear, the new models were also able to liberate Darwinism from three old accusations.

The first of these conceptual clarifications takes the wind out of the sails of the old claim that natural selection is causally empty because its explanations are tautologies. Models taken from statistical physics distinguish between the macrostate of a system and the microstates of its constituents. Applied to evolutionary theory, fitness is a not a single characteristic like physical robustness, but a macroscopic propensity to reproduce more effectively that supervenes on myriad changing and interacting properties of organisms (Mills & Beatty 1979). From this perspective, Darwin need not have been discouraged by Herschel's accusation that natural selection is a "law of higgledy-piggledy." Still, he could not have jumped out of his time any more than Herschel could. Both were confined to working with models adapted from Newtonian physics.

Second, population thinking allowed species to be reconceived as geographically distributed populations within which reproduction occurs freely, but between which genetic barriers evolve to prevent it (Dobzhansky 1937). This meant that species are recast as historical particulars, not abstract classes defined by essential, that is, necessary and sufficient, properties (Hull 1978). This redefinition removed a perennial objection to the very possibility of descent with modification from a common ancestor, namely that species can't change except perhaps saltationally.[6] It also made it clear that species are very real things.

Finally, the new models alleviated the conflict between mechanistic determinism and teleology that vexed Gray, Darwin, and their contemporaries. Darwinism takes seriously the idea that adaptations evolve in order to perform environment-relative life-enhancing functions. In Darwin's milieu, intelligent design creationism recommended itself because it seemed that blind mechanisms could never explain biological functions. But probabilistic models show that genetic mutations with small effects will predictably, if gradually, spread through interbreeding populations if they amplify the representation of traits that enhance differential reproductive success. These favored traits are adaptations. They can be said quite literally to come to be in order to perform specific functions or attain certain goals. They exist for the sake of these functions and goals. In other words, they exist because of what they do and so can be said to exist for the sake of doing that. They evolve by an *ex post facto* process of multigenerational sorting that shows no traces of intentional design. Still, they can be said to be biologically teleological in an empirically respectable sense that contrasts with large-scale inner-driven teleology (Wright 1976; Brandon 1981; Neander 1991; Lennox 1993; Depew 2015). Darwin would probably have been pleased.

7.3.2 Conceptual Clarification Unleashes Empirical Discoveries by Using Models: The Modern Evolutionary Synthesis's Secret of Success

It is true that whether or not the processes I have described are actually shown to exist in nature depends on empirical research. Ever since the

---

[6] Ernst Mayr, a founder of the Modern Synthesis, insisted that sudden genetic mutationism retains elements of species essentialism (Mayr 1980).

1940s it has been the ambition of the Modern Evolutionary Synthesis to undertake just this sort of research and by this means to integrate biology's fields into a general theory of evolution based on population genetic models. Without conceptual clarification and redefinition of concepts such as natural selection, mutation, genetic drift, species, adaptation, function, and others in the light of these models, however, the questions that evolutionary scientists ask might neither be perspicuously framed nor persuasively answered. True, other conceptual frameworks might have done this too (Bowler 2013). As it happened, however, successful prosecution of detailed research into questions posed in the manner of the Modern Synthesis unleashed a 50-year torrent of fruitful investigations whose validation supported the continued explanatory fecundity of the models that inspired them. If this had not happened, Darwinism would have continued to be plagued by pseudo-problems that arise not from its core concept of gradual natural selection, but from inadequate interpretations of it. One of the reasons intelligent design creationists are so frustrating to professional evolutionary biologists and teachers of biology is that they still insist on these old objections when almost a century of productive research on the basis of population genetic models has rendered them empty.

Darwin's idea of evolution by gradual natural selection might be thought of as a Kuhnian paradigm that first arose in Darwin's fertile mind and continues unchanged to this day. It is a concept, we might say, of which there have been various conceptions. Still, to appreciate the power and explanatory potential of Darwin's insight requires a more fine-grained approach to the history of science, an approach in which the Darwinian tradition in evolutionary biology has been held together by a series of research programs, each of which constructs and deploys a family of models (Depew & Weber 1995). There is nothing in the nature of things that dictates that population-genetic Darwinism, or even Darwinism more generally, will not become what Imré Lakatos called a "degenerating research program" (Lakatos 1970). Indeed, calls for replacing, expanding, or revising the Modern Synthesis in the light of new discoveries grow louder with each passing year (Pigliucci & Müller 2010; Huneman & Walsh 2018). But neither is there anything that forecloses the possibility that the Darwinian research tradition will knit itself together again even if the Modern Synthesis comes up against its limitations. I wouldn't bet against the possibility that new models of complex systems will uncover as yet untapped explanatory potential in the concept of natural selection (Depew & Weber 2013).

Once we frame the history of evolutionary biology in this way, we cannot help noticing that in Herschel and Darwin's day philosophy of science was infused with assumptions taken from Newtonian models of systems dynamics. Anything that didn't measure up was not considered a full-fledged science. Under that presumption, finding and applying universal, mechanistic, and deterministic laws of nature was taken to be the aim of scientific inquiry. With its built-in appeal to chance and historical contingency, evolution by natural selection fared badly under this dispensation. The probability revolution that so affected twentieth-century physics and chemistry did much better when applied to biology. It suggested models of systems dynamics that were, and are, very much at ease with chance and contingency (Gayon 1998). It should not surprise us, accordingly, that just as deterministic models affected what counted as good science in the nineteenth century, so too the statistical and probability revolution has reshaped contemporary philosophy of science.

On this view, whether some relatively discrete part of the real world exemplifies a particular conceptual model depends on probabilistically measured degrees of fit (van Frassen 1980). If the statistical fit between a model and some range of phenomena is poor, we should not conclude that this falsifies it. Instead, we should conclude only that for a certain range of phenomena this model is inapplicable to the phenomena in question and should look for a model with a better fit with the relevant range of data and phenomena (Beatty 1982).

Classical physics itself can be retrospectively reframed in these terms. Newton's laws of motion constitute a class of models that define how a Newtonian system will behave (Giere 1979). Whether a planetary system, or for that matter an Adam-Smithian economy, is a Newtonian system so defined is a matter of using statistical data and mathematical probability theory to empirically assess degree of fit. Suppose particles traveling at or near the speed of light don't behave in the way bodies in a planetary system do, as Einstein's experimentalist colleagues showed. If Newton's laws were taken to be universal conditions of the possibility of the world itself (as Immanuel Kant, for one, construed them), physics would have to go back to the drawing board. But on the philosophy of science inspired by the second or probabilistic scientific revolution,

Newton's laws will merely be declared inapplicable to phenomena operating over very large distances at very high speeds. At the level of middle-sized objects, they work just fine.

The idea that science meets the world through models has been so congenial to genetic Darwinism's admission of evolutionary contingency that some philosophers of biology have used it to defend the Modern Evolutionary Synthesis as high-grade, cutting-edge, not barely passable, science (Lloyd 1988). In spite of how they are usually taught to students, for example, Mendel's laws of inheritance are not universal laws of nature, but products of evolution that apply only to stretches of phylogeny, and – even then – far from perfectly (Beatty 1982, 1995). We should not infer from this that Mendel's laws have been falsified. Instead, we should conclude that not every genetic system is a Mendelian system.

The moral of the story is this. Scientific models are constructs. Constructs are concepts. On this view, concepts, which on a Baconian philosophy of science do little more than get in the way of empirical observation and induction, are in reality necessary conditions for making and validating the empirical discoveries that enable scientific inquiry to make advances.

7.4 Competing Conceptual Frameworks and Scientific Controversies: Genetic Darwinism, the Molecular Revolution, and Another Key Role for Concepts

The makers of the Modern Synthesis enthusiastically greeted the molecular revolution in genetics that began with Crick and Watson's 1953 model of the structure and function of the DNA molecule. The reason is not hard to find. This discovery confirmed Darwinism's assumption that variations arise without respect to their subsequent utility. It also revealed the mechanism by which this happens: spontaneous mutation in DNA's chemical bases. It was as if Darwin himself had left a place in his theory into which DNA could be inserted when it finally appeared (Mayr 1980).

This story assumes that the molecular gene and the gene of population genetics are equivalent. Actually, they aren't (Moss 2003). But the difference can be minimized by reversing the way the founders of the Modern Synthesis construed the relation between genes and phenotypes. Rather than existing for the sake of the phenotypes that the genes help reliably reproduce, and *a fortiori* for the sake of the life-activity of the organisms that exhibit these phenotypes, adapted traits can be thought of as evolving in order to maximize the built-in self-reproductive prowess of DNA.

The most notorious but not the only expression of "genocentrism," as it is called, is Richard Dawkins' "selfish gene hypothesis" (Dawkins 1986). On Dawkins' view, the gene is a stretch of DNA that codes for proteins. A gene survives repeated meiotic division because its pheno-typic expressions successfully interact with environmental factors to protect and enhance the self-replicating power of the genes themselves. To make this argument, Dawkins inflated the ontological reality of genes in order to magnify their independence and causal power. Corres-pondingly, he demoted the ontological integrity of organisms. Not only are they collections of independently evolved adaptations, but, as Daw-kins remarked in one of his least circumspect formulations, they have no more unity than "clouds in the sky or dust storms in the desert" (Dawkins 1986, pp. 34–35).

The selfish gene hypothesis illustrates how abstract philosophical notions about what kinds of beings actually make up the world – ideas, as Plato believed? Substances that persist through change, as Aristotle had it? Democritus's atoms? The inherently changing processes of Hera-clitus? – are enlisted to lend support to one or another interpretation of core concepts such as natural selection.

Conceptual frameworks can be recognized by the metaphors and similes their advocates use. Dawkins's metaphor of genes as "selfish" expresses a disaggregated, atomistic ontology of genes and denies sub-stantial integrity to organisms. Such exercises in "seeing as" have their uses, especially in struggles within a field to displace one research program with another. The selfish gene gambit makes perspicuous phe-nomena that can seem anomalous in standard versions of the Modern Synthesis. One of these is why so much of an organism's DNA appeared to consist of "junk DNA," that is, as functionless repeats.[7] Why not? After all, self-replication is what comes naturally to DNA. More signifi-cantly, the selfish gene hypothesis also seemed to resolve one of

---

[7] The notion that most DNA is functionless "junk" is no longer as compelling as it seemed only a decade ago. See Biémont 2010; Project Encode www.encodeproject.org.

Darwinism's most troubling anomalies: How can cooperation evolve in a competitive world? Genes, Dawkins replies, can collaborate with each other to produce phenotypes that serve the replicative interests of each. Phenotypes can be cooperative because genes are selfish. To see this all one has to do is to extend the causal influence of genes on a new generation from parents to a wider set of relatives. "Inclusive fitness," as this proposal is called, leads to "kin selection," which evolves cooperative traits in proportion to genetic relatedness. Whether a species exhibits this phenomenon is an empirical question to be answered by matching phenomena with models, as we saw in the previous section.

Biotechnicians gravitate toward genocentrism. The very nature of their work inclines them to see organisms as assemblies of gene-instructed parts that are independently optimized by natural selection and are ripe for manipulation or replacement. Accordingly, armed with talk about "genes for" this or that trait, molecular genetics has taken a prominent role in developing techniques of genetic manipulation, such as, most recently, CRISPR-Cas9. Organocentric and ecologically minded biologists, philosophers of biology, and bioethicists may protest, but so great is the prestige of genocentric Darwinism in contemporary culture that social scientists have confidently taken on board its atomized approach to adaptation in interpreting human behavior. In research programs such as Human Sociobiology and Evolutionary Psychology, "genes for" this or that aspect of human behavior, from predictable aggression against other groups not closely genetically related, to the supposed natural promiscuity of males, are hypothesized. Conceptual framing is hard to miss in the battles in which a research program seeks to become dominant within a discipline, but even harder to miss in efforts to extend these programs to other fields.

Still, genocentrism has methodological problems. For one thing, the fact that a phenotype fails to appear when a DNA sequence is knocked out or silenced does not prove that a particular sequence historically became fixed in a population in order to express that phenotype. Accordingly, it may not be an adaptation by the lights of the Modern Synthesis. More relevant to our present concerns is that concepts become excessive whenever commitment to a conceptual framework substitutes for evidence that is not currently at hand but, we are always reassured, is sure to turn up soon. The cure for question-begging of this sort is science's robust process of criticism, review, and commentary (Depew 2013).

These practices are especially necessary because leaning on philosophy, if only by choice of metaphors, to do the work of empirical research can easily escalate into unempirical metaphysical claims, such as when Dawkins implies that materialism and atheism follow from and are required to see the explanatory power of the genocentric interpretation of natural selection (Dawkins 2006). Appealing to potentially viciously circular tactics like this to support a scientific hypothesis may be a sign that a research program is coming to the end of its productive life.

Science is as replete with unintended ironies as any other human endeavor. Crude empiricism often makes it difficult for us to see that contestation between conceptual frameworks is not always a sign of weakness. Sometimes it is a sign that a new approach that better explains and integrates empirical evidence is on the rise. In recent decades, for example, the well-funded progress of molecular genetics has shifted the attention of researchers away from how structural genes code for proteins – the problem that preoccupied the generation of Crick and Watson – to how regulatory sectors of the genome turn protein production on and off during ontogeny. Many more "developmental resources" than DNA are required for this to happen, including various kinds of RNA and heritable epigenetic chemical side chains attached to DNA that are responsive to environmental contingencies (Uller 2013; Cavalli & Heard 2019; on "developmental resources," Griffiths & Gray 1995). Moreover, since RNA can slice and dice the genetic material at different sites in response to developmental needs, it is increasingly difficult to identify a particular chunk of DNA as "the gene for x," or even as a gene at all (Burian & Kampourakis 2013; Kampourakis 2017). As a result, the self-replicative power of DNA that figures so prominently in the selfish gene hypothesis is becoming as mythical as Crick and Watson's image of a "genetic program" running recursively, irreversibly, and autonomically through a fixed sequence from DNA to RNA to protein. It is ironic that the genocentric turn in social sciences is turning out to be as infected with an out-of-date version of evolutionary theory as Spencerian Social Darwinism once was.

In response to the revelation that DNA is only one of many developmental resources, some philosophers of biology and philosophizing biologists have argued that the interaction between organisms and environments is so dynamic that it requires framing organisms as processes rather than either substances or aggregates of atom-like units.

"The reality of metabolism," write John Dupré and Daniel Nicholson, "forces us to recognize that despite their apparent fixity and solidity organisms are not material things but fluid processes" (Dupré & Nicholson 2018, p. 13). "Organisms don't *have* life cycles," they say. "They *are* life cycles." In a nod toward the original process philosopher, Heraclitus, who said that you never put your foot in the same river twice, Nicholson and Dupré have titled a collection of essays on this theme *Everything Flows* (Nicholson & Dupre 2018). We are witnessing in recommendations of this sort a recoding of the biological knowledge arising from developmental genetics in a conceptual framework that promises to sort better with fast developing empirical evidence than either the original Modern Synthesis or its genocentric successor.

7.5 Conclusions: Toward a New Pedagogy of Science

I have been arguing that concepts play at least three key roles in scientific discovery: as metaphors, as models, and as conceptual frameworks that figure in controversies about the scope and limits of research programs. The bench scientist or the classroom science teacher may find it difficult to believe that metaphors, models, and conceptual frameworks are internal to scientific inquiry, and not mere commentary aimed at gaining support from public audiences and external stakeholders. This response is understandable, particularly when it is acknowledged, as I have, that conceptual frameworks can sometimes lose touch with empirical research, even if at other times they consolidate it. Nonetheless, I have tried to show that "science in action" depends on its conceptual aspects as much as on experiment and close observation, and far from making science unreliable turns it into genuine knowledge.

Teaching students about science, and *a fortiori* teaching their teachers about it, is of great importance in modern and modernizing societies. The influence of Baconian empiricism has long made itself felt in science pedagogy. Its legacy is omnipresent in the stress placed on learning facts, facts, and more facts. Specialists who study how students actually learn, however, have discovered that students do not come to the study of science as empty vessels waiting to be filled up. They bring to the scene of inquiry their own implicit theories of how the world works. The good science teacher will begin from there and lead students to engage in critiquing their own assumptions and in finding new ones that explain and interpret phenomena more successfully (Vosniadou 2013; Kampourakis & Reiss 2018).

In this connection, a certain amount of history of science should be integrated into science classes, even in early grades. It should not be confined to hagiographic stories about Marie Curie's self-martyrdom in her studies of radiation or to musings on the genius of a Newton, a Darwin, or an Einstein that ignore the fact that all of them were deeply familiar with and contributed to the state of play in their fields. Their innovations were interventions that challenged embedded assumptions, often with help from metaphors, models, and conceptual frameworks. Given the essential roles these play in scientific advancement, it is important to teach science teachers about how intimately connected are shifts in programs of scientific inquiry and shifts in philosophy of science.

Luckily, we need not wait for the future to bring us such treasures. Research of the sort can already be found in journals such as *Science & Education, Evolution, Education, and Outreach*, and others.

References

Beatty, J. (1982). What's Wrong with the Received View of Evolutionary Biology? In P. D. Asquith & R. N. Giere (eds.), *PSA 1980*, vol. 2, pp. 34–55. East Lansing, MI: Philosophy of Science Association.

Beatty, J. (1995). The Evolutionary Contingency Thesis. In G. Wolters & J. Lennox (eds.), *Concepts, Theories, and Rationality in the Biological Sciences*, pp. 45–81. Pittsburgh: University of Pittsburgh Press.

Biémont, C. (2010). A Brief History of the Status of Transposable Elements: From Junk DNA to Major Players in Evolution. www.genetics.org/content/186/4/1085

Black, M. (1962). *Models and Metaphors*. Ithaca, NY: Cornell University Press.

Bogen, J. & Woodward, J. (1988). Saving the Phenomena. *Philosophical Review* 97: 303–352.

Bowler, P. (1983). *The Eclipse of Darwinism: Anti-Darwinian Evolution Theories in the Decades Around 1900*. Baltimore, MD: Johns Hopkins University Press.

Bowler, P. (2013). *Darwin Deleted: Imagining a World without Darwin*. Chicago: University of Chicago Press.

Brandon, R. (1981). Biological Teleology: Questions and Explanations. *Studies in the History and Philosophy of Science* 12: 91–105.

Burian R. & Kampourakis, K. (2013). Against "Genes For": Could an Inclusive Concept of Genetic Material Replace Gene Concepts? In K. Kampourakis (ed.), *The Philosophy of Biology: A Companion for Educators*, pp. 597–628. Dordrecht: Springer Verlag.

Cavalli, G. & Heard, E. (2019). Advances in Epigenetics Link Genetics to the Environment and Disease. *Nature* 571: 489–499.

Darwin C. (1859). *On the Origin of Species*. E. Mayr (ed.). Cambridge, MA: Harvard University Press. Facsimile of 1st ed.

Darwin, C. (1861). *On the Origin of Species*. London: John Murray. 3rd ed.

Darwin, C. (1862). *On the Various Contrivances by which Orchids Are Fertilized by Insects*. London: John Murray.

Darwin, C. (1875a). *The Movements and Habits of Climbing Plants*. London: John Murray.

Darwin, C. (1875b). *Insectivorous Plants*. London: John Murray.

Darwin, C. (1881). *On the Formation of Vegetable Mould Through the Action of Worms, with Observations on Their Habits*. London: John Murray.

Darwin, C. (1958). *The Autobiography of Charles Darwin*. N. Barlow (ed.). London: Collins.

Darwin, C. (1975). *Charles Darwin's Natural Selection*. R. C. Stauffer, (ed.). Cambridge: Cambridge University Press.

Dawkins, R. (1986). *The Selfish Gene*. Oxford: Oxford University Press. 2nd ed. (1st ed., 1976).

Dawkins, R. (2006). *The God Delusion*. London: Bantam Books.

Depew, D. (2013). The Rhetoric of Evolutionary Theory. *Biological Theory* 7(4): 380–389.

Depew, D. (2015). Accident, Adaptation, and Teleology in Empedocles, Aristotle, and Darwin. In P. Sloan, K. Eggleson, & G. McKenny, *Darwin in the Twenty-First Century: Nature, Man, and God*, pp. 116–143. Notre Dame, IN: University of Notre Dame Press.

Depew, D. & Weber, B. (1995). *Darwinism Evolving*. Cambridge, MA: MIT Press.

Depew, D. & Weber, B. (2013). Challenging Darwinism: Expanding, Extending, or Replacing the Modern Evolutionary Synthesis. In M. Ruse (ed.), *The Cambridge Encyclopedia of Darwin*, pp. 405–411. Cambridge: Cambridge University Press.

Depew, D. & Weber, B. H. (2017). Developmental Biology, Natural Selection, and the Conceptual Boundaries of the Modern Evolutionary Synthesis. *Zygon* 52: 468–490.

Dobzhansky, T. (1937). *Genetics and the Origin of Species*. New York: Columbia University Press.

Dupré, J. & Nicholson, D. (2018). A Manifesto for a Processual Philosophy of Biology. In D. Nicholson & J. Dupre (eds.), *Everything Flows: Towards a Processual Philosophy of Biology*, pp. 3–45. Oxford: Oxford University Press.

Fisher, R. A. (1918). The Correlation between Relatives on the Supposition of Mendelian Inheritance. *Transactions of the Royal Society of Edinburgh* 52: 399–433.

Fisher, R. A. (1930). *The Genetical Theory of Natural Selection*. Oxford: Oxford University Press.

Forest, D. (2018). Darwin est-il resistable? Mill sur le Dessein et la selection naturelle. In F. Merlin & P. Huneman (eds.), *Philosophie, histoire, biologie: mélanges offerts á Jean Gayon*, pp. 183–194. Paris: Editions Materiologiques.

Gayon, J. (1997). The "Paramount Power" of Selection: From Darwin to Kauffman. In J. Della Chiara (ed.), *Structure and Norms in Science*, pp. 265–282. Dordrecht: Kluwer Academic Publishers.

Gayon, J. (1998). *Darwinism's Struggle for Survival*. Cambridge: Cambridge University Press.

Gigerenzer, G., Swijtink, Z., Porter, T., & Daston, L. (1990). *The Empire of Chance: How Probability Changed Science and Everyday Life*. Cambridge: Cambridge University Press.

Giere, R. (1979). *Understanding Scientific Reasoning*. New York: Holt Reinhart.

Gould, S. J. (1977). *Ontogeny and Phylogeny*. Cambridge, MA: Harvard University Press.

Griffiths, P. & Gray, R. (1995). Developmental Systems and Evolutionary Explanation. *The Journal of Philosophy* 91: 277–304.

Hesse, M. (1966). *Models and Analogies in Science*. Notre Dame, IN: Notre Dame University Press. Revised ed.

Herschel, J. W. F. (1831). *Preliminary Discourse on the Study of Natural Philosophy*. London: Longmans.

Hodge, M. J. (1982). Darwin and the Laws of the Animate Part of the Terrestrial System, 1835–1837: On the Lyellian Origins of His Zoonomial Explanatory Program. *Studies in the History and Philosophy of Biology* 6: 1–106.

Hofmann, J. & Weber, B. (2003). The Fact of Evolution: Implications for Science Education. *Science & Education* 12: 729–760.

Hull, D. (1978). A Matter of Individuality. *Philosophy of Science* 45: 335–360.

Huneman, P. & Walsh, D. (eds.) (2018). *Challenges to the Modern Synthesis* Oxford: Oxford University Press.

Kampourakis, K. (2017). *Making Sense of Genes*. Cambridge: Cambridge University Press.

Kampourakis, K. & Reiss, M. J. (2018). *Teaching Biology in Schools: Global Research, Issues and Trends*. New York: Routledge.

Kuhn, T. (1962). *The Structure of Scientific Revolutions*. Chicago: University of Chicago Press.

Lakatos, I. (1970). Falsification and the Methodology of Scientific Research Programmes. In I. Lakatos & A. Musgrove (eds.), *Criticism and the Growth of Knowledge*, pp. 91–195. Cambridge: Cambridge University Press.

Lakoff, G. & Johnson, M. (1980). *Metaphors We Live By*. Chicago: University of Chicago Press.

Latour, B. (1987). *Science in Action*. Cambridge, MA: Harvard University Press.

Laudan, L. (1977). *Progress and Its Problems*. Berkeley and Los Angeles: University of California Press.

Lennox, J. (1993). Darwin *Was* a Teleologist. Biology and Philosophy 8: 409–421.

Lloyd, E. (1988). *The Structure and Confirmation of Evolutionary Theory*. New York: Greenwood Press. Reprinted Princeton University Press, 1994.

Mayr, E. (1980). Prologue. In E. Mayr & W. Provine (eds.), *The Evolutionary Synthesis*, pp. 1–48. Cambridge, MA: Harvard University Press.

Mills, S. & Beatty, J. (1979). The Propensity Interpretation of Fitness. *Philosophy of Science* 46: 263–286.

Moss, L. (2003). *What Genes Can't Do*. Cambridge, MA: MIT Press.

Nicholson, D. & Dupré, J. (eds.) (2018). *Everything Flows: Toward a Processual Philosophy of Biology*. Oxford: Oxford University Press.

Pigliucci, M. (2007). Do We Need an Extended Evolutionary Synthesis? *Evolution* 61: 2743–2749.

Neander, K. (1991). Functions as Selected Effects. *Philosophy of Science* 58: 168–184.

Pigliucci, M & Müller, G. (eds.) (2010). *Evolution: The Extended Synthesis*. Cambridge, MA: MIT Press.

Ruse, M. (1975). Darwin's Debt to Philosophy: An Examination of the Influence of the Philosophical Ideas of John F. W. Herschel and William Whewell on the Development of Charles Darwin's Theory of Evolution. *Studies in the History and Philosophy of Science* 6: 159–81.

Snyder, L. (2006). *Reforming Philosophy: A Victorian Debate on Science and Society*. Chicago: University of Chicago Press.

Sober, E. (1993 [1984]). *The Nature of Selection: Evolutionary Theory in Philosophical Focus*. Chicago: University of Chicago Press.

Uller, T. (2013). Nongenetic Inheritance and Evolution. In Kampourakis, K (ed.), *The Philosophy of Biology: A Companion For Educators*, pp. 267–287. Dordrecht: Springer Verlag,

van Frassen, B. 1980. *The Scientific Image*. Oxford: Oxford University Press.

Vosniadou, S. (ed.) (2013). *International Handbook of Research on Conceptual Change*. 2nd ed. New York and London: Routledge.

Walsh, D. (2015). *Organisms, Agency, and Evolution*. Cambridge: University of Cambridge Press.

Watson, J. (1968). *The Double Helix*. New York: Simon and Schuster.

Weismann A. (1889). *Essays Upon Heredity*. Oxford: Clarendon Press.

Whewell, W. (1840). *Philosophy of the Inductive Sciences*. London: John W. Parker. Reprint of 2nd ed., New York, Johnson Reprint Co., 1967.

Wright, L. (1976). *Teleological Explanations*. Berkeley and Los Angeles: University of California Press.

Wright, S. (1932). The Roles of Mutation, Inbreeding, Crossbreeding, and Selection in Evolution. *Proceedings of the Sixth Annual Congress of Genetics* 1: 356–366.

Zammito, J. (2004). *A Nice Derangement of Epistemes: Post-Positivism in the Study of Science from Quine to Latour*. Chicago: University of Chicago Press.

8   How Can Conceptual Analysis Contribute to Scientific Practice?

The Case of Cultural Evolution

TIM LEWENS

8.1 Conceptual Analysis

Mainstream philosophers often ask questions about concepts that are of fundamental importance to our dealings with each other and with the world. They might ask what is knowledge, or justice, or art. Philosophers of science have asked similar questions about fundamental concepts that characterize science in general: What, for example, is explanation, or probability, or a law of nature? This kind of approach typically aims to understand key concepts in science better, but it does not always aim to inform or assist scientific practice. Moreover, this approach sometimes presupposes – unwisely, according to its critics – that scientists jointly employ some reasonably unified concept of explanation, probability, or a law of nature.

This chapter exemplifies an alternative approach to the philosophical analysis of scientific concepts that is (i) localized, in the sense that it focuses on a particular set of notions used in just one specialized community of scientific practice; (ii) pluralist, in the sense that it acknowledges considerable and justified divergence in how these concepts are understood and employed and (iii) ameliorative, in the sense that it aims to contribute to scientific practice. The focus in this chapter is on how specialists in the field of cultural evolution – primarily evolutionary theorists, biological anthropologists, and experts in animal behavior – understand and employ key concepts such as "culture," "social learning," and "cumulative cultural change." Ultimately the chapter aims to demonstrate the value of a philosophical approach in helping to avoid

I am grateful to the editors and also to Andrew Buskell, Cecilia Heyes, and Kevin Laland for exceptionally valuable comments on an earlier version of this chapter.

unnecessary disputes across disciplinary boundaries. The chapter also highlights how specific definitional practices can both facilitate focused inquiry while also blinkering researchers to potentially fruitful explanations. But before we can achieve any of these payoffs we must give a brief overview of the field of cultural evolution itself.

8.2 Cultural Evolution: The Basics

Ever since the 1930s, many researchers within the biological mainstream have held that the study of evolution should aim to understand changes in gene frequencies in populations. Sewall Wright, for example, made an elegant case for the gene-focused approach in 1942:

> we need a common measure for such diverse factors as mutation, crossbreeding, natural selection and isolation. At first sight these seem to be incommensurables but if we fix attention on their effects on populations, rather than on their own natures, the situation is simplified. Such a measure may be found in the effects on *gene frequency* in each series of alleles. (Wright 1942, p. 225)

In other words, while diverse evolutionary processes are at work in nature, they can all be tracked and compared in their effects via a common genetic accounting system.

Wright explained why he was so confident that this system would capture all important forms of evolutionary change:

> We shall restrict consideration to changes in the system of genes and aggregates of genes (chromosomes). There are relatively rare and obscure hereditary changes which must be attributed to other cell components but our knowledge of these does not warrant the elaboration of a statistical theory. (Wright 1942, p. 223)

Wright did not deny that other components of cells had the ability to change in ways that would provoke inherited differences in offspring. His claim was simply that such events were so unusual that there was no point trying to treat them using the type of statistical framework developed within population genetics.

Some researchers working on various forms of epigenetic inheritance would now disagree with Wright's verdict, and there is a lively debate underway regarding the adequacy of Wright's claim that genetic differences are the only ones worth attending to (Jabonka & Lamb 2005; Laland et al. 2015; Miska & Ferguson-Smith 2016; Charlesworth et al. 2017). We will not be concerned with this debate in this chapter. There is, however, one domain where calls for a broader approach to inheritance have been comparatively uncontroversial. In humans, there are a great many inherited differences that have nothing to do with differences in genes, and are instead based on what individuals learn from others. Population genetics uses statistical apparatus to ask how we can expect the genetic composition of populations to change under the influence of a variety of factors – mutation, migration, fecundity, longevity, and so forth – that affect the lives of individuals. Similarly, cultural evolutionary theory also aims at a statistical treatment of how populations can be expected to change as individuals are affected by their exposure to various forms of teaching and other types of social influence.

In short, cultural evolutionists argue that our understanding of change in human populations cannot be complete unless we move beyond the traditional evolutionary focus on genetic inheritance, and supplement it with an examination of the impact of various forms of learning. There is also abundant research on cultural evolution in a range of other species whose members can learn from each other. These include primates, whales and dolphins, birds and fish (e.g., Avital & Jablonka 2000; Whiten 2000; Rendell & Whitehead 2001; Laland & Janik 2006; Laland & Galef 2009).

These cultural evolutionary theories have focused on the general evolutionary origins of capacities to transmit information via various forms of learning and also on the specific changes that storehouses of skill, habit, belief, and so forth have undergone since the advent of these learning capacities (Creanza et al. 2017). An important strand of recent work has asked why so few of these cultural species – humans are perhaps the only completely uncontroversial example – seem to undergo cultural evolution of a "cumulative" nature, whereby the learning achievements of one generation are built on and elaborated in subsequent generations (e.g., Boyd & Richerson 1996; Lewis & Laland 2012; Dean et al. 2013; Laland 2017; Mesoudi & Thornton 2018).

This chapter makes a case for the contribution of philosophy to these cultural evolutionary projects. It aims to show the following:

- Cultural evolutionary theories often have different explanatory goals. Getting clear on what these goals are can help to dispel confusion about what cultural evolutionary theories can achieve and can help to reduce potential conflict between natural and social scientific approaches to change in human populations.

- Cultural evolutionary theorists tend to use the notion of culture in very specific and highly pragmatic ways. There are, in other words, broader and narrower ways to define what we might mean by "culture," with no single best definition that is defensible across contexts.
- Cultural evolutionary theorists' definitions of cultural "accumulation" are also offered with pragmatic goals in mind. There are risks, however, that the notion of social learning that underlies some of these definitions can blinker research. In particular, by foregrounding what humans and animals learn by interacting with each other, the dominant notions of social learning can downplay the potential for features of natural and technical environments to act as reservoirs of indirectly transmitted know-how.

In other words, scientists' conceptual choices can invite confusion when they are encountered by other disciplines, but they can also narrow research within disciplines.

8.3 Background: The Goals of Cultural Evolutionary Theory

Pioneering work on cultural evolution in the statistical style briefly outlined previously was first undertaken by Cavalli-Sforza and Feldman (e.g., 1973, 1981), and Boyd & Richerson (e.g., 1985). It is possible to discern two different general accounts of what these cultural evolutionary theories are intended to achieve (Lewens 2015).

Very roughly speaking, the first motivation offered for cultural evolutionary theory rests on the general idea that natural selection is a process that explains adaptation – that is, it explains the emergence of good fit between a trait and its environment (Campbell 1974). Darwin argued that natural selection acting on natural variation could explain the advent of such elegant structures as the vertebrate eye. Some cultural evolutionary theorists offer a parallel to Darwin's argument. Darwin proposed that a population of organisms in which there is inheritance and competition can, when that process is iterated repeatedly, evolve traits that are well suited to the organisms' environments. Many cultural evolutionists argue that if there is also inheritance and competition in the cultural realm – perhaps when ideas are passed from one believer to another, whereby only those ideas with appeal survive; perhaps when techniques of manufacture are passed from one craftsperson to another, whereby only those judged to be effective survive – then selection can explain the appearance over time of useful conceptual or technological traits, too.

Mesoudi, Whiten, & Laland (2006, p. 329), for example, began their case for a "unified science of cultural evolution" by arguing that "culture exhibits key Darwinian evolutionary properties." They mean that it instantiates the properties required for evolution by selection. In their view, these are "variation, competition, inheritance, and the accumulation of successive cultural modifications over time" (Mesoudi et al. 2004, p. 1). Richerson and Boyd – perhaps the most influential of all contemporary researchers in cultural evolution – also remarked that, "The logic of natural selection applies to culturally transmitted variation every bit as much as it applies to genetic variation" (Richerson & Boyd 2005, p. 76). Joseph Henrich followed them in linking culture to the emergence of a new mechanism enabling cumulative adaptation, albeit without such overt stress on the conditions of variation, competition, and inheritance:

> Probably over a million years ago, members of our evolutionary lineage began learning from each other in such a way that culture became cumulative. That is, hunting practices, tool-making skills, tracking knowledge and edible-plant knowledge began to improve and aggregate – by learning from others – so that one generation could build on and hone the skills and know-how gleaned from the previous generation. (Henrich 2016, p. 3)

The first general motivation for building cultural evolutionary theories draws on a comparatively ambitious theoretical claim about the importance of natural selection understood in the abstract. The second general motivation for these theories instead relies on a more matter-of-fact conjecture: If our understanding of populational change in cognitively sophisticated species is to be thorough, we must understand how various forms of learning can explain change and stasis in these populations over time.

For example, some of Marcus Feldman's earliest work on cultural evolution, undertaken in collaboration with Cavalli-Sforza, was dedicated to undermining overly simple (and highly misleading) inferences from claims about populational distribution of IQ scores to claims about genetic causation (see Cavalli-Sforza & Feldman 1973; Feldman & Lewontin 1975). Their models pointed to a number of important complications and confounding effects that could be introduced by forms of cultural transmission: "Given the existence of individual plasticity in response to the environment, correlations between biological relatives are expected even if there is no genetic variation whatsoever" (Cavalli-Sforza & Feldman 1973, p. 633). This aspect of their work is not premised on the idea that culture provides an alternative source of adaptation compared with natural selection acting on genetic variation, although it is compatible with that approach.

The same is the case for much of Boyd and Richerson's work. For example, they raised the question of why the capacity for culture – which they consider to be intimately linked to the ability to learn via interactions with others – should have emerged at all (e.g., Richerson & Boyd 2005). That question is worth asking, for humans appear to have especially acute social learning capacities, and one might also think that these capacities have some costs in terms of exposing populations to the possible transfer of misinformation.

To answer their question, Boyd and Richerson must integrate traditional approaches to evolution underpinned by the inheritance of genes – which many of their models represent as responsible for inheritable variation in our learning capacities – with approaches that allow us to understand how individual fitness will be affected because of the action of these learning dispositions in structured populations. In a semi-popular overview of their whole approach, they noted that, "The heart of this book is an account of how the population-level consequences of imitation and teaching work" (Boyd & Richerson 2005, p. 6). It is noteworthy that they did *not* say that the heart of their book is an account of how culture instantiates Mesoudi et al.'s "key Darwinian evolutionary properties."

The same stance has been adopted in more technical works. Henrich & Boyd (2002, p. 87), for example, asserted that, "Formal models of cultural evolution analyse how cognitive processes combine with social interaction to generate the distributions and dynamics of 'representations.'" Again, this aspect of cultural evolutionary theorizing has no general commitment to an image of cultural change as a process that parallels the action of natural selection in populations of organisms. Instead, it merely aims for the quantitative modeling of changing patterns of cognitive representations over time, in populations whose individual members can learn from each other. This means that there are many forms of cultural evolutionary theory that have no commitment to the idea that cultural change is a close analog of evolutionary change, and they have no commitment to the general idea that the primary goal of a cultural evolutionary theory should be the explanation of forms of cultural adaptation via cultural natural selection. In turn, this means that skeptics from outside the natural sciences, who doubt that there are close similarities between adaptation in the biological and cultural domains, need not pick fights with all forms of cultural evolutionary theory.

8.4 Genes and Culture

We have just seen, in preliminary fashion, some of the ways in which we can help to calm potential disputes over cultural evolutionary theories by paying philosophical attention to the goals of these theories. We will see in this section that some of these disputes can also be avoided if we are careful to pay attention to how the relationship between "genetic" and "cultural" inheritance is conceived within cultural evolution.

One of the most fruitful areas of research within the field of cultural evolution has been in terms of the action – and interaction – of genes and culture, where these are conceived as two distinct "channels" for inheritance. Perhaps the most thoroughly researched instance of this "gene-culture coevolution" – which I have discussed in further detail elsewhere (Lewens 2015, 2017) – is the evolution of lactose tolerance in humans. There is good evidence for a co-evolutionary story of the following broad form. On the one hand, humans have learned how to exploit dairy animals, thereby constructing an environment that offers a ready source of calories to those who are able to digest lactose after infancy. In other words, this cultural innovation establishes a selective niche in which genes that extend the production of the enzyme lactase into adulthood (i.e., genes that enable lactose tolerance via what is called "lactase persistence") are favored. On the other hand, those populations in which lactose tolerance is present are especially likely to adopt dairying practices. Both processes work together – and reinforce each other – to explain the historical coincidence of dairying and lactose tolerance. Hence this well-confirmed story shows that if we want to explain why genes favoring lactose tolerance have become more prevalent in some human populations than others, we need to pay due attention to what humans learn from each other (Feldman & Cavalli-Sforza 1989; Holden & Mace 1997).

The very idea of gene-culture coevolution – often referred to as "dual-inheritance theory" – has sometimes attracted skepticism from researchers with backgrounds in social science. Perhaps the clearest example comes from the anthropologist Maurice Bloch, who has suggested that the general approach issuing from cultural evolutionary theory wrongly conceives of culture and genetics as distinct forces that influence each other, when instead they should be thought of as a "a unified process" (Bloch 2012, p. 52). The social anthropologist Christina Toren has made a similar proposal for a "unified model of human being," which allows us to "discard the biology/culture distinction" (Toren 2018, p. 172).

Bloch has written that, "Social and natural scientists have come to hate each other. They cannot understand each other's purpose. They consider each other's methods either sloppy or dangerous. They are repulsed by each other's style and mode of presentation. They even dress differently" (Bloch 2012, p. 1). We will now see how philosophy of science can help to clarify whether social anthropology and cultural evolutionary theory – especially in the form of dual-inheritance theory – are in genuine conflict with respect to their stances on the relationship between culture and genetics.

In their widely cited work on the evolution of lactase persistence, Holden and Mace write that, "lactase persistence is a genetic trait, whereas pastoralism and milk-drinking are cultural traits" (Holden & Mace 1997, p. 604). This type of comment may appear to commit Holden and Mace (natural scientists) to the sort of distinction between cultural traits on the one hand, and genetic traits on the other, that Bloch and Toren (social scientists) reject. In truth, the two groups are less divided than one might think.

Like any trait, lactase persistence must emerge over ontogenetic time. Indeed, the very definition of lactase persistence is the continued ability to produce lactase beyond weaning. Lactase persistence is not guaranteed by the presence of genes that normally play a causal role in bringing it about. For example, research indicates that the persistent production of lactase can be affected by various forms of gut trauma, which can in turn be brought on via social influence on diet. Stress, when it affects individuals who are heterozygous for genetic variants that normally result in lactase persistence, can result in their becoming lactose intolerant (Swallow 2003). The ability to care for dairy animals – and even the ability to drink milk – must also develop in maturing individuals, and that development, too, is contingent not only on what others may be trying to teach the focal individual, but also on that individual's physiological capacities. Hence the development of pastoralism and milk-drinking are, in part, dependent on the genetic makeup of the person in question. In these developmental senses there are no traits that are purely genetic, or purely cultural.

Holden and Mace would not deny this familiar consensus regarding the interactions between genes and the socially constituted environments in which they have their effects. Moreover, this is just the consensus that Toren relies upon when she argues against any simple invocation of distinctions between biology and culture, between nature and nurture, and so forth. So what might Holden and Mac plausibly have meant when they wrote that "lactase persistence is a genetic trait, whereas pastoralism and milk-drinking are cultural traits"?

Lactase persistence – even if its ontogenetic causation relies on both genes and culture – is the sort of trait that is inherited "vertically." Lactase persistence is not the kind of trait that parents can acquire from their siblings, or that children can acquire from their friends. More generally it is not the sort of trait one can acquire from unrelated others. Instead, parents with genes conferring lactase persistence are likely to have offspring with lactase persistence, too. Dairying, on the other hand – even though its ontogeny relies on the contributions of genes – is the kind of trait that can be acquired "horizontally" or "obliquely," via the observation of others to whom one is not related, and via practice of the skill. This means that lactase persistence tends to spread comparatively slowly, via cycles of natural selection acting on successive generations of individual organisms. Dairying, on the other hand, can potentially spread far more quickly. It is possible, therefore, for the social environment to change rapidly in ways that promote the subsequent natural selection of lactase persistence. It is consequently reasonable to distinguish, as Holden and Mace did for the purpose of their explanatory modeling, between two types of trait, without the superficial implications that one develops in a manner that is free of cultural influence, while the other develops in a manner that is free of genetic influence. In short, once we clarify what explanatory modeling is aiming at, we show the compatibility between frameworks that might appear fundamentally opposed to each other. In this case the compatibility is between an approach that stresses the fundamental unity of genetic and cultural processes on developmental grounds, and an approach that instead focuses on the different pathways that vertical and horizontal cycles of inheritance can trace through populations.

8.5 Individual and Social Learning

Let us recap. We have seen in outline what cultural evolutionary theory aims to achieve. We have seen some potential misunderstandings of the commitments of "dual inheritance" theory, and we have seen how to defend that approach from the accusation that it drives too strong a wedge between nature and culture.

In this section I would like to move on to assess how culture is defined by cultural evolutionary theorists. My goal is to highlight a comparatively harmless problem first, and a more substantial problem second. The comparatively harmless one derives from the fact that the notion of "culture" is defined according to the precise goals of cultural evolutionary theory. This opens up the possibility for multiple reasonable definitional choices, which in turn lead to a confusing proliferation of terminology within the cultural evolutionary research community. The more substantial problem is that these definitional choices can invite misunderstandings when cultural evolutionary theory is assessed by those coming from different disciplinary backgrounds, with different goals.

Recall that cultural evolutionary theory is frequently motivated by the goal of understanding how forms of learning can affect how populations of organisms change over time. This pragmatic explanatory concern is why culture itself is almost always defined within this research community as a form of "information" that individual organisms acquire via interaction with others.

In their work on culture in cetaceans (i.e., whales and dolphins), for example, Rendell & Whitehead (2001) adapted Boyd & Richerson's (1996) highly influential definition of "culture" as: "information or behaviour acquired from conspecifics through some form of social learning." (Rendell & Whitehead 2001, p. 310). Laland & Janik (2006), similarly, argued for a "broad" definition of culture as: "all group-typical behaviour patterns, shared by members of animal communities, that are to some degree reliant on socially learned and transmitted information." In other words, "culture" here is defined in terms of notions such as "social learning" and "socially transmitted information." In turn, we must ask what these terms mean if we wish to understand the concept of culture.

There has been a proliferation of terminology in articles on social learning in animals that can make these debates hard to keep track of. For example, when Rendell & Whitehead (2001) clarified how they were using the term "social learning," they stated that they would use it to cover everything that Whiten & Ham (1992) labeled using the different term "social processes." Confusingly, when we turn to Whiten and Ham, we find that they distinguished between "social learning" and "social processes," and they used "social learning" to name a specific subvariety of "social process." On their view, "social learning" proper involves one animal learning from another, whereas "social process" is a broader category that also includes forms of "social influence." Cases of "exposure," for example, whereby merely by being with an animal A, animal B is exposed to a similar learning environment, count as "social influence" for Whiten and Ham, without counting as "social learning." In other words, Rendell and Whitehead cited Whiten and Ham with approval, while using the term "social learning" in a different manner. Although these differences can make it hard to follow some debates within the social learning literature, they are largely cosmetic: Whiten & Ham (1992) and Rendell & Whitehead (2001) all agree with the basic claim that it is the broader category of socially influenced learning that should be used to define culture.

There are, however, more substantive problems that arise from definitions of "social learning" and from the definitions of culture to which they give rise (Lewens 2017). In an important survey of work on social learning, Heyes (1994, p. 207) opened with the claim that:

> The term social learning refers to learning that is influenced by observation of, or interaction with, another animal (typically a conspecific) or its products … The complementary set is commonly known as "individual learning."

She then pointed out that "this usage is potentially confusing," since "in all cases it is ultimately individuals who learn."

There are two types of conceptual problem associated with definitions of social learning. Heyes here focuses on the first. It is always individuals who learn, hence if we want to draw useful distinctions we need terms that reflect the nature of the learning process, rather than the nature of the learner. Her suggestion was that we drop the term "individual learning," and instead use the term "asocial learning" to refer to learning "that does not involve social interaction."

There is a second problem, which Heyes paid less attention to in this early article (but not in later work: see Heyes 2018). She indicated that "individual learning" is meant to be the "complement" of social learning: In other words, the two categories are not meant to overlap. But now consider Joseph Henrich's much more recent account of these terms:

> *social learning* refers to any time an individual's learning is influenced by others, and it includes many different kinds of psychological processes. *Individual learning* refers to situations in which individuals learn by observing or interacting directly with their environment, and can range from calculating the best time to hunt by observing when certain prey emerge, to engaging in trial-and-error learning with different digging tools. (Henrich 2016, p. 12)

Thus defined, "social learning" and "individual learning" can overlap. That is because even if an individual is not observing or interacting directly with other organisms – even if it is experimenting alone with what it finds in its immediate biological or physical environment – that environment may have been altered by others in ways that influence what the focal individual learns.

Henrich was well aware of this, for he immediately noted that:

> the least sophisticated forms of social learning occur simply as a by-product of being around others, and engaging in individual learning. For example, if I hang around you and you use rocks to crack open nuts, then I'm more likely to figure out on my own that rocks can be used to crack open nuts. (Henrich 2016, pp. 12–13)

In other words, individual learning can give rise to a form of social learning, when it goes on in a pre-structured environment (see also Sterelny 2012).

This overlap between individual and social learning is quite widely acknowledged. Hobaiter et al. (2014), for example, noticed that the wild chimps they were studying had acquired a new behavior. Previously these chimps had made "sponges" by chewing up leaves, which they used to soak up water for drinking. Some then began to make sponges from moss instead. One chimp – called KW – acquired the moss-sponging behavior because she found, and re-used, an old moss sponge that another chimp had discarded. Hobaiter et al. did not claim that KW saw moss-sponging being performed by another chimp. From KW's perspective, the discarded sponge was simply an environmental feature, which she made use of through individual learning. At the same time, the actions of an earlier member of the group clearly facilitated KW's acquisition of the new behavior. This explains why Hobaiter et al. them-selves described their study as part of "a growing literature that refutes a strong distinction between individual and social learning" (Hobaiter et al. 2014; see also Heyes 2012; Lewens 2012, 2015).

Heyes's definition of "social learning" is a broad one. On her account, social learning may involve observation of another animal, but it can also involve observation of that animal's "products." In other words, Heyes's definition gives the result that Hobaiter et al.'s chimps are engaging in social learning. They are influenced by observing others' sponges, even if they do not observe other chimps. Recall, also, that Heyes recommended that we use the term "asocial learning" as the "complement" of social learning. This raises the question of how often learning is ever strictly asocial: Such learning would need to be free of any influence from conspecifics, and in a species such as our own it seems there will be almost no cases where the environments that affect our learning are untouched by others.

Deciding on when learning is "asocial" turns out to depend on precisely what we are to count as the influence on learning of another animal's "products." Consider, for example, that animals modify their environ-ments in countless ways via the construction of nests, the building of dams, the dispersal of seeds, the creation of trampled tracks, the scattering of discarded elements of foods, the production of offspring, the transmis-sion of parasites, and so forth (Laland & Brown 2018). We can potentially include a vast array of learning interactions with such modified features of the environment as cases of "social learning," and hence we will tend to include a vast number of species as endowed with forms of "culture."

Let us summarize this section so far. I have focused until now on what we might think of as the comparatively harmless problem of defining culture. There are some potentially confusing instances where different theorists use different terms to label the same processes and phenomena. We saw this with the case of "social learning" and its narrow and wider uses. We also saw how the concept of culture can vary because of scope for flexibility in what is to count as learning from the "products" of other animals.

These conclusions have further implications, which bring us toward a more substantial set of problems that arise from flexibility in how scientists define their terms. We saw that cultural evolutionists' notion of culture arises from a specific research problem – namely one focused on the populational effects of learning. That is why the very idea of culture is usually defined within these research communities by reference to social learning. This allows investigators in cultural evolutionary theory to build various explanatory models that examine the effects of different forms of learning. That does not mean, however, that when we look to different research contexts, all phenomena that deserve to be called "cultural" must involve forms of learning (Scott-Phillips et al. 2018). If we survey different human populations, we will find that posture, gait, and musculature vary. In some regions the economic landscape channels most people into agricultural work, and perhaps into walking or cycling as dominant modes of transport. In other areas dominant economic opportunities involve sedentary labor, and the built environment encourages the use of cars. Other bodily traits vary from population to population, for in some regions foods with high quantities of sugars are widely produced and consumed, whereas in other regions diet consists more in grains, or perhaps fish.

These would be classed by many as uncontroversial cases of *cultural* variation: Social interactions in one generation maintain dominant features of built environments, economic opportunities, and nutritional opportunities, which in turn are causally responsible for maintaining local forms of bodily difference. That said, it would be a strain to describe all of these forms of apparently "cultural" differences as the result of learned traditions, even of individual learning in structured environments. Instead, different forms of gait, posture, adiposity, and so forth emerge as plastic physiological reactions to collectively maintained features of local environments. Learning is not the only mechanism by which local communities can transmit idiosyncratic traditions across generations via plastic developmental reactions. In other words, by reflecting on quite mainstream instances of human "cultural" variation, we explain why there are some research traditions that would view the approach to "culture" taken by cultural evolutionists as too narrow.

The foregoing thoughts also help to explain why, for some, the very idea of culture is one that is of little use (Toren 2018). We have seen that the breadth of the category of "animal products" makes the category of "social learning" potentially very capacious. We have also seen that "culture" need not be sustained by learning at all. Ultimately, we might end up concluding that every species has a form of culture in some sense, for even when we are considering plants and bacteria we find that one generation leaves a structured array of developmental resources to which the next responds developmentally, in plastic fashion. However, we will now see that while we thereby explain why *some* theorists have been skeptical of the notion of "culture," this does not mean that the notion should be dropped.

8.6 Ecological Inheritance and Cultural Accumulation

When a concept – culture, in this case – is potentially highly elastic, it is important to find ways of ensuring that specific research questions are posed in clear ways. One way of achieving that is to build explanatory models focused on very localized phenomena. We have already seen, for example, that we can ask questions about the changes in dairying that have led to the proliferation of genes favoring lactose tolerance. The answers we give to specific questions such as these are not remotely threatened by any background concerns we may have about "social learning" being a notion whose general scope of application is not always clear, nor are they threatened by worries about what we might mean in pronouncing any individual species as capable of sustaining "culture." In other words, scientists working within a given disciplinary context can make headway by simply ignoring general definitional worries, focusing instead on testing fairly narrow explanatory models.

We can also try to maintain a theoretical focus on a comparatively broad phenomenon, but one that is better defined than "culture" in general. Boyd & Richerson (1996), for example, have suggested that "only observational learning [i.e., imitation of one individual by another] allows *cumulative* cultural change." In other words, we can narrow our focus from trying to ask what processes are necessary for culture, to asking what processes are necessary for cultural change that is cumulative.

It is one thing for different forms of song, say, to sweep through populations of birds, or whales, and something else for a population of animals to show iterated improvements over time, underpinned by learning from others, in some aspect of their behavioral repertoire, their use of tools, and so forth. The very elasticity of the notion of social learning suggests we should not attach too much significance to the very general question of whether any given species does indeed have the capacity for "culture," and it is now quite uncontroversial to credit species of many kinds with that capacity. It is also uncontroversial to credit humans with a heightened capacity for cumulative culture. In the domain of technology, in particular, we often find that inventions are built upon and refined across time and generations. What remains up for grabs, then, is the apparently substantive question of whether there are any non-human animals that have culture that is cumulative, and why humans seem so conspicuously good at sustaining cumulative forms of cultural inheritance (Lewis & Laland 2012; Laland 2017).

These questions cannot be answered without a reasonably tight definition of what "cumulative" culture might mean. In approaching this question, Mesoudi & Thornton (2018) followed Tomasello (1999) in pointing to the need for some kind of "ratchet" mechanism – understood as a "metaphor for the accumulation of increasingly effective modifications without reverting back to prior, less effective states" – to underpin cultural accumulation. They then moved on to suggest the following "minimum requirements for a population to exhibit CCE [cumulative cultural evolution]":

(i) a change in behaviour (or product of behaviour, such as an artefact), typically due to asocial learning, followed by (ii) the transfer via social learning of that novel or modified behaviour to other individuals or groups, where (iii) the learned behaviour causes an improvement in performance, which is a proxy of genetic and/or cultural fitness, with (iv) the previous three steps repeated in a manner that generates sequential improvement over time. (Mesoudi & Thornton 2018, p. 2)

It seems undeniable that cumulative cultural change requires social learning in some sense; however, that is in large part a consequence of the very broad definition usually attached to the term. We saw earlier that some forms of social learning – at least if we follow the sort of broad definition recommended by Heyes – need not involve imitation in the strict sense, and they need not even involve observation of one animal by another. Instead, they may simply involve one animal learning by interaction with the modified environmental or material products of another animal. This explains why the majority of recent work in this domain has asked what specific *forms* of social learning might be able to sustain cumulative adaptation.

Here there are risks that definitional choices may inadvertently exclude some worthwhile hypotheses from assessment. In this particular context, the risk is that by focusing on what type of interaction between individual animals explains cumulative adaptation, we overlook the ways in which animals may influence each other in highly indirect ways via their collective effects on their environments. We can illustrate these risks by considering the case of forest elephants. They forage in ways that result in the creation of networks of paths through their territories (Fishlock, Caldwell, & Lee 2016). Over time these networks become more elaborate as new territories are explored. The network of pathways is sustained across generations – that is to say, the pathways are lost neither when individuals die, nor with the death of the elephants that first trod them – via the collective activity of the elephants. The plastic nature of the forest itself – which retains the imprint of successive trampling, but which will eventually grow back once a pathway is neglected – acts as a form of external memory, or an ecological ratchet, for the elephants that follow the paths (see Clark & Chalmers 1998).

This increasingly elaborate network of pathways thus has some features that are cumulative, and it appears to satisfy Mesoudi and Thornton's definition for cumulative cultural evolution (see also Jesmer et al. 2018 for a similar case for cumulative cultural evolution in bighorn sheep). Some elephants open new pathways, and the growing network satisfies Heyes's liberal definition of an entity sustained via social learning. That is because the habits of any one elephant that follows a pathway are clearly influenced by the products – i.e., by flattened paths through the forest – of earlier conspecifics. As the network becomes more elaborate in terms of food sources it opens up, we can also anticipate it will have positive consequences with respect to elephants' fitness.

So while we have an apparent case of cultural accumulation via social learning here, the conjectured ecological ratcheting effect does not meet the definition of "stimulus enhancement," normally understood as a form of social learning in which a learner's attention is drawn to the object or site of the demonstrator's activity. Nor does it count as "local enhancement," a process in which the learner is attracted to the individual demonstrator (Heyes 1994). Indeed, when one animal simply responds to its collectively modified environment there need be no "demonstrator" at all. More generally, the elephants' responses to their environment do not meet the conditions for any of the sub-varieties of social learning found in many standard taxonomies offered within the animal learning literature.

Needless to say, this form of ecological ratcheting does not seem capable by itself of sustaining the breathtaking forms of change that we see in human technological trajectories, such as the iterated refinements to the compass, the car, or the computer. Even so, it does remind us of the potential role for environmental modification as a contributory mechanism that can account for aspects of accumulation (Sperber 2001, p. 307; see also Laland 2017). This mechanism seems especially worth considering as a contributor to cumulative innovation in the case of our own species, because of the exceptionally rich nature of our constructed technological environments, and their capacity to guide action in fine-grained ways. As Sterelny (2012) has pointed out, when humans engage in trial and error learning, they do so in surroundings that feature an array of previously selected raw materials and preexisting devices, which provide advantageous platforms to a new generation of potential innovators.

8.7 Conclusions

This chapter has explored a series of questions about how scientists should define terms such as "culture," "social learning," and "cumulative cultural change." It has made the case that there is no one privileged way to pick out what we should mean when we use these terms. Instead, we have choices about how to use these terms that are dictated by our investigative interests. This explains why there is such scope for misunderstandings across disciplinary boundaries, and it even explains why there is scope for confusion within the context of a single discipline.

The costs of confusion are often comparatively minor and can be resolved simply by clarity and standardization in use of language with any given context. Moreover, it is often unnecessary for the practice of fruitful scientific enquiry to attempt to define general terms with precision: Instead, scientists can aim to use explanatory models to give answers to narrow questions formulated with specificity, such as questions about the causes of the spread of lactose tolerance. There are also cases where general questions – such as questions about the capacity for cumulative culture – require that clear definitions are given for key notions. Here, the potential costs can be more substantial. That is because our definitional choices impose constraints on the range of explanatory hypotheses that scientists are likely to entertain. Philosophical analysis, when closely coupled to scientific practice, can help to expose and mitigate these risks.

References

Avital, E. & Jablonka, E. (2000). *Animal Traditions: Behavioural Inheritance in Evolution*. Cambridge: Cambridge University Press.

Bloch, M. (2012). *Anthropology and the Cognitive Challenge*. Cambridge: Cambridge University Press.

Boyd, R. & Richerson, P. (1985). *Culture and the Evolutionary Process*. Chicago: University of Chicago Press.

Boyd, R. & Richerson, P. (1996). Why Culture Is Common, but Cultural Evolution Is Rare. *Proceedings of the British Academy* 88: 77–93.

Campbell, D. (1974). Evolutionary Epistemology. In P. A. Schilpp (ed.), *The Philosophy of Karl Popper*. Chicago: Open Court.

Cavalli-Sforza, L. & Feldman, M. (1973). Cultural versus Biological Inheritance: Phenotypic Transmission from Parents to Children. *American Journal of Human Genetics* 25(6): 618–637.

Cavalli-Sforza, L. & Feldman, M. (1976). Cultural and Biological Evolutionary Processes: Selection for a Trait under Complex Transmission. *Theoretical Population Biology* 9(2): 238–259.

Cavalli-Sforza, L. & Feldman, M. (1981). *Cultural Transmission and Evolution: A Quantitative Approach*. Princeton, NJ: Princeton University Press.

Charlesworth, D., Barton, N. H., & Charlesworth, B. (2017). The Sources of Adaptive Variation. *Proceedings of the Royal Society B: Biological Sciences* 284(1855): 20162864.

Clark, A. & Chalmers, D. (1998). The Extended Mind. *Analysis* 58(1): 7–19.

Clarke, E. & Heyes, C. (2017). The Swashbuckling Anthropologist: Henrich on the Secret of Our Success. *Biology and Philosophy* 32(2): 289–305.

Creanza, N., Kolodny, O., & Feldman, M. (2017). Cultural Evolutionary Theory: How Culture Evolves and Why It Matters. *PNAS* 114(30): 7782–7789.

Darwin, C. (1871). *The Descent of Man and Selection in Relation to Sex*. London: John Murray.

Dean, L., Vale, G., Laland, K., Flynn, E., & Kendal R. (2013). Human Cumulative Culture: A Comparative Perspective. *Biological Reviews* 89: 284–301.

Dupré, J. (2012). *Processes of Life*. Oxford: Oxford University Press.

Feldman, M. W. & Cavalli-Sforza, L. L. (1989). On the Theory of Evolution under Genetic and Cultural Transmission with Application to the Lactose Absorption Problem. In M. W. Feldman (ed.), *Mathematical Evolutionary Theory*, pp. 145–173. Oxford: Princeton University Press.

Feldman, M. W. & Lewontin, R. C. (1975). The Heritability Hang-Up. *Science* 190(4220): 1163–1168.

Fishlock, V., Caldwell, C., & Lee, P. (2016). Elephant Resource-Use Traditions. *Animal Cognition* 19(2): 429–433.

Henrich, J. (2016). *The Secret of Our Success*. Princeton, NJ: Princeton University Press.

Henrich, J. & Boyd, R. (1998). The Evolution of Conformist Transmission and the Emergence of Between-Group Differences. *Evolution and Human Behavior* 19(4): 215–241.

Henrich, J. & Boyd, R. (2002). On Modelling Culture and Cognition: Why Cultural Evolution Does Not Require Replication of Representations. *Culture and Cognition* 2(2): 87–112.

Heyes, C. (1994). Social Learning in Animals: Categories and Mechanisms. *Biological Reviews* 69(2): 207–31.

Heyes, C. (2012). What's Social about Social Learning? *Journal of Comparative Psychology* 126(2): 193–202.

Heyes, C. (2018). *Cognitive Gadgets*. Cambridge, MA: Harvard University Press.

Hobaiter, C., Poisot, T., Zuberbühler, K., Hoppitt, W., & Gruber, T. (2014). Social Network Analysis Shows Direct Evidence for Social Transmission of Tool Use in Wild Chimpanzees. *PLoS Biology* 12: e1001960.

Holden, C. & Mace, R. (1997). Phylogenetic Analysis of the Evolution of Lactose Digestion in Adults. *Human Biology* 69: 605–628.

Ingold, T. (2013). Prospect. In T. Ingold and G. Palsson (eds.), *Biosocial Becomings: Integrating Social and Biological Anthropology*, pp. 1–21. Cambridge: Cambridge University Press.

Jablonka, E. & Lamb, M. J. (2005). *Evolution in Four Dimensions: Genetic, Epigenetic, Behavioral, and Symbolic Variation in the History of Life*. Cambridge, MA: The MIT Press.

Jesmer, B. et al. (2018). Is Ungulate Migration Culturally Transmitted? Evidence of Social Learning from Translocated Animals. *Science* 361(6406): 1023–1025.

Laland, K. (2017). *Darwin's Unfinished Symphony*. Princeton, NJ: Princeton University Press.

Laland, K. & Brown, G. (2018). The Social Construction of Human Nature. In T. Lewens & E. Hannon (eds.), *Why We Disagree About Human Nature*. Oxford: Oxford University Press.

Laland, K. & Galef, B. (eds.) (2009). *The Question of Animal Culture*. Cambridge, MA: Harvard University Press.

Laland, K. & Janik, V. (2006). The Animal Cultures Debate. *TREE* 21(10): 542–547.

Laland, K. N., Odling-Smee, J. & M. Feldman (2001). Niche Construction, Biological Evolution, and Cultural Change. *Behavioral and Brain Sciences* 23(01): 131–46.

Laland, K. et al. (2015). The Extended Evolutionary Synthesis: Its Structure, Core Assumptions, and Predictions. *Proceedings of the Royal Society of London* 282: 20151019.

Lewens, T. (2007). *Darwin*. London and New York: Routledge.

Lewens, T. (2012). The Darwinian View of Culture. *Biology and Philosophy* 27(5): 745–753.

Lewens, T. (2015). *Cultural Evolution: Conceptual Challenges*. Oxford: Oxford University Press.

Lewens, T. (2017). Human Nature, Human Culture: The Case of Cultural Evolution. *Interface Focus* 7: 20170018.

Lewis, H. & Laland, K. (2012). Transmission Fidelity is the Key to the Build-Up of Cumulative Culture. *Philosophical Transactions of the Royal Society B* 367(1599): 2171–2180.

Mace, R. (2010). Update to Holden and Mace's "Phylogenetic Analysis of the Evolution of Lactose Digestion in Adults." *Human Biology* 81(5–6): 621–624.

Mesoudi, A. & Thornton, A. (2018). What Is Cumulative Cultural Evolution? *Proceedings of the Royal Society B* 285(1880): 20180712.

Mesoudi, A., Whiten, A., & Laland, K. (2004). Is Human Cultural Evolution Darwinian? Evidence Reviewed from the Perspective of The Origin of Species. *Evolution* 58(1): 1–11.

Mesoudi, A., Whiten, A., & Laland, K. (2006). Towards a Unified Science of Cultural Evolution. *Behavioral and Brain Sciences* 29(4): 329–47.

Miska, E. & Ferguson-Smith, A. (2016) Transgenerational Inheritance: Models and Mechanisms of Non-DNA Sequence-Based Inheritance. *Science* 354(6308): 59.

Odling-Smee, J., Laland, K., & M. Feldman (2003). *Niche Construction: The Neglected Process in Evolution*. Princeton, NJ: Princeton University Press.

Ram, Y., Liberman, U., & Feldman, M. (2018). Evolution of Vertical and Oblique Transmission under Fluctuating Selection. *PNAS* 115(6): E1174–E1183.

Rendell, L. & Whitehead, H. (2001). Culture in Whales and Dolphins. *Behavioral and Brain Sciences* 24(2): 309–324.

Richerson, P. & Boyd, R. (2005). *Not by Genes Alone: How Culture Transformed Human Evolution*. Chicago: University of Chicago Press.

Scott-Phillips, T., Blancke, S., & Heintz, C. (2018). Four Misunderstandings about Cultural Attraction. *Evolutionary Anthropology* 27(4): 162–173.

Sperber, D. (2001). Conceptual Tools for a Natural Science of Society and Culture (Radcliffe-Brown Lecture in Social Anthropology 1999). *Proceedings of the British Academy* 111: 297–317.

Sterelny, K. (2012). *The Evolved Apprentice: How Evolution Made Humans Unique*. Cambridge, MA: MIT Press.

Swallow, D. (2003). Genetics of Lactase Persistence and Lactose Intolerance. *Annual Review of Genetics* 37(1): 197–219.

Tomasello, M. (1999). *The Cultural Origins of Human Cognition*. Cambridge, MA: Harvard University Press.

Toren, C. (2018). Human Ontogenies as Historical Processes. In T. Lewens & E. Hannon (eds.), *Why We Disagree About Human Nature*. Oxford: Oxford University Press.

Whiten, A. & Ham, R. (1992). On the Nature and Evolution of Imitation in the Animal Kingdom: Reappraisal of a Century of Research. *Advances in the Study of Behaviour* 21: 239–277.

Whiten, A. (2000). Primate Culture and Social Learning. *Cognitive Science* 24(3): 477–508.

Wright, S. (1942). Statistical Genetics and Evolution. *Bulletin of the American Mathematical Society* 48: 223–246.

9   What Methods Do Life Scientists Use?

A Brief History with Philosophical Implications

ERIK L. PETERSON

> An organism is something which the scientific method cannot deal with; it is a hard, round, smooth nut, which experimental analysis can neither crack nor lever open at any point. As soon as a hole is made in it, it explodes like a Prince Rupert drop and vanishes away.
>
>   – Joseph Needham, *Sceptical Biologist* (1929)

9.1 Introduction

This chapter is about the methods used in the life sciences. In it, I present a case that there are a limited number of methods that life scientists employ. On its face, this conclusion seems neither exciting nor novel; that is, unless we consider that the life sciences have long been engaged in a hidden battle between two opposing views regarding the number of scientific methods that exist. I am arguing against *both* sides. The two standard positions that I am arguing against are: (1) that there exists a single, if idealized, Scientific Method (i.e., unity), and (2) that the sciences are individualistic and, therefore, scientists operate without a common method (i.e., anarchy). I am asserting that in contrast to unity or anarchy, there are *several but a limited number* of methods possible in the bio-sciences. By combining general methods and sub-methods, there could be more than a dozen possible valid means by which to conduct a life sciences inquiry. I offer a table sketching out these methods and submethods at the conclusion of the chapter. But before I launch into explaining these two standard views and why a middle way, endorsing several but a limited number of methods, seems preferable, I want to begin with an example of modern life science *in vivo*, as it were.

9.2 The Case of Amyloid-ß[1]

Rather than playing their usual videogames at work late on a Friday afternoon in 2007, Massachusetts General Hospital neuroscientists Robert Moir and Rudolph Tanzi surfed the internet and sipped beer. Tanzi had a thought about a connection between the human immune response and his work on Alzheimer's disease. It was just a hunch. But still, half-consumed beer in hand, he poked his head around Moir's door. "What do you know about innate immunity in the brain?" By chance, Moir, too, had been thinking about the biochemistry of the brain and Alzheimer's. Suddenly, a spark; a moment of insight.

It took a few months. They met repeatedly over drinks to talk about the consequences of inverting a long-established scientific fact. Years earlier, biochemists identified amyloid-ß as the chemical system likely to be implicated in Alzheimer's. It constituted a sticky plaque; those tangled plaques were the universal sign of Alzheimer's – a gooey cellophane wrap around the brains of patients. But Moir and Tanzi wondered if the mere presence of amyloid-ß did not cause Alzheimer's. And if that wasn't enough, they made an even bolder claim: Amyloid-ß acted in the *functional* immune system in the brain – an extra layer of protection against fungal, bacterial, and viral invaders. So, the disease developed as a once-functional system spun out of control during the process of aging like a spinning top decelerating, wobbling.

Moir and Tanzi demonstrated the quite functional antibacterial properties of amyloid-ß in the lab in 2010. Instead of lining up in support, however, established figures – Nobel laureates among them – grew more skeptical. Still, they persisted; Moir had supportive research partners attached to the Cure Alzheimer's Fund. The better part of a decade passed from the first beer-fueled spark to any truly positive result. Just as they first postulated, their tests showed amyloid-ß fibers in model mice and worms could indicate a *functioning* brain immune system. Too many amyloid-ß tangles, however, and disease might ensue. More

---

[1] This account is drawn primarily from: Jon Hamilton, February 18, 2018, "Scientists Explore Ties between Alzheimer's and Brain's Ancient Immune System," National Public Radio (USA), www.npr.org/sections/health-shots/2018/02/18/580475245/scientists-explore-ties-between-alzheimers-and-brains-ancient-immune-system; and Vijaya Kumar et al. 2016, "Amyloid-beta Peptide Protects against Microbial Infection in Mouse and Worm Models of Alzheimer's Disease," *Science Translational Medicine* 8(340): 1–15, http://doi.org/10.1126/scitranslmed.aaf1059.

work remained to be done, but it was a start. They had tested and confirmed their conjecture years earlier that amyloid-ß alone was not the culprit.

I introduce this story as a glimpse at how ordinary bioscience is conducted today. Notice what steps emerge: a spark of insight, a conversation, hunches, more conversations, laboratory tests on model organisms, and eventually, publications. Of course, there is a great deal *not* in the reported story: unnamed laboratory employees, computer models, other laboratories working in the same area, and details about chemicals and experimental animals bought and paid for, etc. Some of these details can be found by excavating online supplements to published work.[2] Still, behind even these aspects of studies hide other factors: perhaps most importantly, the years of mundane work on other aspects of biology by Moir, Tanzi, and their unnamed colleagues. Together these hidden aspects inspired hunches that led to grants to pay for the experiments to test and, after a very long time in the face of substantial skepticism, confirm those hunches.

9.3 One Scientific Method or Everyone's Methods?

The winding, uncertain path of discovery outlined in the amyloid-ß Alzheimer's case certainly does not seem to match the widespread stock image of the five-step Scientific Method taught in many primary, secondary, and even post-secondary classrooms and held by public figures. That rigid Scientific Method moves from (1) Observation to (2) Hypothesis, then (3) Experiment, (4) Verification, and finally (5) Theory (Rudolph 2005).[3] At the university level, students may learn that the Scientific Method contains additional steps and is not quite so rigid. But they likely still learn from their professors that this method, this logical sequence sometimes called "hypothetico-deductive," is *the* way toward knowledge: "The scientist makes observations, designs instruments,

---

[2] A full recapitulation can be found here: https://stm.sciencemag.org/content/suppl/2016/05/23/8.340.340ra72.DC1.

[3] Ironically, John Dewey – who captured a five-step *description* of how his non-science students *did* think, rather than the *prescriptive* Scientific Method outlining for science students how they *should* think – intended something far more flexible than the descendent ideas became (Dewey 1910).

conducts experiments, collects data, takes measurements, makes calculations, draws conclusions" (Feibleman 1972, p. 1).

Anecdotally, science professors more familiar with philosophy even introduce their students to an idea of "falsificationism" when they introduce method. According to the popular rendering of Karl Popper's notion of falsification, scientists present informed conjectures that are able to be falsified; it is that *ability to be falsified* that renders ideas possible of being scientific and, therefore, getting at true statements about the world. No matter how rigorous they appear, ideas cannot be scientific unless they have the *potential to be rendered false* by an empirical test. Scientists propose conjectures/hypotheses that both they and others test by various means and either allow to pass along, or, if tests give negative results, they discard. If such a test cannot be devised – if a concept is always verified only in principle, in other words – then the conjecture is not scientific; it's pseudoscience. Upper-level students may even hear these two concepts – the five-step Scientific Method and Popperian falsification – stitched together. The fifth (Theory) step in this view is a conjecture that has passed a sufficient number of possibly falsifying tests without being demonstrated to be actually false. Astrology couldn't be a science, in this rendering, either because it does not propose hypotheses to be tested or because any hypotheses would always be corroborated. Eliminating the ability to be falsified means eliminating the possibility of being a science.

Without a doubt, there is pedagogical and rhetorical utility in presenting a step-wise Scientific Method and tying it to a version of Popper's falsificationism. Such a picture underwrites the unity of science – that physicists and biologists and geologists and psychologists, and so on, are all chasing after the same truth about the world even if not in the same domain of phenomena (Wilson 1998). The conjunction insinuates that scientific knowledge produced by individual scientists and laboratories is trustworthy because science writ large rigorously follows methods in a way that astrology, for instance, does not.

Though there might be good social reasons to cling to this view, historians and philosophers of science have long recognized it does not fit actual practice (Lakatos & Musgrave 1970). Among their number stood Paul Feyerabend – a physicist-turned-philosopher – who charged that *no* singular unified method exists, in a book entitled *Against Method* and initially subtitled *Outline of an Anarchistic Theory of*

Knowledge. "Successful research does not obey general standards," he insisted (Feyerabend 1993, p. 1). The assumption that there is any set of step-wise procedures that scientists follow to ensure success, he maintained, is either naïve or a piece of propaganda. Feyerabend and his fellow travelers hoped to drive a wooden stake through the heart of an undying conceit (perhaps dating back to Plato, 22 centuries earlier) that the tie between one thing called "science" and another thing also called "science" was a set of shared logically integrated methods (Dupré 1993). The kind of work scientists present at conferences and write in papers subjected to peer review in journals does reinforce power dynamics, hierarchies, and group identities (Bernstein 2000). Yet there is nothing in these practices, nor in the more particular practices each laboratory conducts, that could qualify as shared method. The day-to-day practice of science instead tends to scatter research communities and discourages any propensity to unify around a particular method, according to Feyerabend. Centrifugal forces dominate.

These, then, are the two extremes sketched out by philosophers, educators, and scientists themselves. At one pole, we find a carefully defined and rigorously applied unified Scientific Method enmeshed with comforting claims about the trustworthiness of science and its distinctness from pseudoscience. At the other, we find Feyerabend's anarchy – nearly anything goes when it comes to method since science is more about policing boundaries and power networks than following specific procedures (Galison 1996). Perhaps, this side of the argument says, science can be trustworthy sometimes – only as viewed backwards through the lens of history, maybe. But the pronouncements issuing from individual laboratories? Given one reading of Feyerabend's anarchistic theory of knowledge, there is no rigorously applied logical method to ensure individual scientists would be approaching truth through any set of behaviors in a laboratory.

If we refer back to the Moir and Tanzi example, we can easily see that it undercuts the simple Scientific Method taught in schools. Perhaps teachers should stop exclusively presenting that model in classrooms (Bonner 2005). But that example might also caution us against a strong version of Feyerabend's anarchy. Surely there must be *some* shared method within the life sciences community. Otherwise, how could one group evaluate the veracity of the work pursued in another lab? How would awards be settled, grants denied, papers published, professorships filled? Philosophical arguments about relativism aside, the strong version of Feyerabend's anarchy could be problematic if we don't see that in modern science – just like we don't see a unified stepwise Scientific Method.

Is there a way between these extremes? I think there is.

9.4 Searching for Methods in the Life Sciences

It seems reasonable that the number of scientific methods available to scientists is somewhere between one and n, with n being the number of active scientists or laboratories in a discipline (i.e., Feyerabend's claim). But how many methods are there, and what are they exactly? Unfortunately, the most prominent historians and philosophers of science who have dealt with this question directly have been interested in physics and have used examples from physics exclusively to address problems surrounding method (e.g., Laudan 1984; Chalmers 1990; Wootton 2015). In their defense, it is not easy to find any explicit discussion of separate biological methods in history of biology books (Mayr 2004, p. 2). Most older histories of biology were squarely *intellectual* histories: less interested in the *how* of the process of discovery in biology than the *what* and the *who* (e.g., Nordenskiöld 1935).

Recent histories of biology have paid more attention to method. Endersby (2007), for example, highlighted select model organisms and put forward the thesis that "tool" use – in this case, the model organisms *Pisum sativum* to *Zea mays* to *Arabidopsis thaliana*, for instance – shaped biological concepts. In other words, the tool used by the biologist influenced the possibility of discovery and the justification of that discovery.

This focus on tools and tool use is relatively common in histories of physics. One prominent example of this argument uses examples in astronomy: Galileo's and Kepler's telescopes shaped their support for heliocentrism; Tycho Brahe's lack of a telescope precluded his acceptance of heliocentrism (Huff 2010). Tool use itself was the central part of the geocentric-to-heliocentric revolution, according to this view. So perhaps we need only to trace the use of various scalpels and X-ray diffraction machines in biology to discover the evolution of methods.

That argument seems promising, but there is a problem with it. The life sciences correlate tool to the telescope in the physical sciences would be the microscope, of course. If we extended the tool-first argument to the life sciences, we should see a before-microscope to after-microscope transition. But there is no before-microscope to after-microscope story to match the before-telescope to after-telescope story. The telescope helped Galileo and Kepler revolutionize astronomy in the early seventeenth century. The microscope did not enable Leeuwenhoek or Malpighi to revolutionize biology in an analogous way in the later seventeenth century. There's an easy historical proof for this claim. Heliocentrism had replaced geocentrism among professional astronomers by the early 1700s, a century after Galileo's telescope at the latest (Hirshfield 2001). By contrast, humoral and blending inheritance concepts every bit as ancient as geocentrism hung around in medicine and physiology for centuries after Leeuwenhoek's microscope. The modern concepts that replaced humoral and blending inheritance concepts – Koch and Pasteur's germ theory and Mendelian genetics – appeared nearly simultaneously at the turn of the twentieth century. In other words, heliocentrism was adopted *three centuries* before germ theory and genetics, though microscopes and telescopes were born at nearly the same time.

This raises a question. If the microscope did not revolutionize biology in the way that the telescope revolutionized astronomy (which opened the door for Isaac Newton to revolutionize physics in Leeuwenhoek's day), then is that because there have not been real revolutions in biology? Charles Darwin's work is often held up as a strong reason to believe there has been at least one revolution in the life sciences equivalent to the geocentrism-to-heliocentrism one in astronomy. Perhaps Darwin's work was the revolution in biological *methods* equivalent to the Galilean-Keplerian revolution. I disagree. While Darwin's work codified a revolution in biological *concepts*, he did not present an equivalent revolution in *method*. I will get to that argument. But before that argument, it would be better to get a handle on how life scientists organize inquiry in general.

9.5 The Life Sciences Were Baconian First

Since Aristotle, philosophers have sorted the methods behind investigations of the natural world into two general approaches: induction and deduction. For the next 20 centuries after Aristotle, philosophers across

Asia and Europe sharpened and extended the depictions of these logical methods. But by the early seventeenth century, Francis Bacon and René Descartes, among others, argued for a reformation of these methods and the Aristotelian natural philosophy built with them. Famously, Bacon prioritized *induction-plus-extensive classification*; Descartes favored *deduction from self-evident first principles*. Both philosophers contributed to redirect the flow of natural philosophy into new streams: (1) an emphasis on experiments and instruments and (2) explanations in terms of corpuscles or atoms arranged as parts of a machine (van Berkel 2000; Gaukroger 2006, p. 249).

The biosciences stood slightly apart from these trends for a while. Yet there was one crucial distinction between Bacon's induction and Descartes' deduction that impacted biology immediately and more deeply than other natural sciences. Baconians did not permit *hypotheses*. A scientific hypothesis is "a preconceived idea as to what an investigation will disclose, maintained on the basis of a certain measure of doubt …. [B]oth doubt and some inclination toward belief" (Feibleman 1972, p. 91). Baconian induction, by contrast, emptied itself of such preconceived notions. Bacon derided them as "idols" perverting scientific observation. True knowledge of the world had to be built up from observed particulars. Relationships between the observations emerged naturally as the natural historian carefully collected and sorted them (Laudan 1981, p. 22).

By the eighteenth century, zoology and botany emerged as Baconian fields. Gone were those symbolic connotations at the core of natural histories of the past (Sloan 1972). Carl Linnaeus became the figurehead for this de-symbolized, Baconian classification of living creatures. Though by no means simple, the method of natural historians such as Linnaeus was at least straightforward: collect organisms, preserve them, find similarities between them, and classify them according to those similarities. Debates emerged over the particular features that indicated a relationship between organisms. Linnaeus's followers seemed convinced that the correct method of categorization would eventually appear to an experienced natural historian, just as Bacon suggested. Opponents, such as the renowned French naturalist Georges-Louis Buffon, repeatedly insisted that Linnaeus's methods left too much *hypothesis*, too many preconceived notions in his taxonomy. They didn't reflect the actual order of the world, in other words. Only by examining reproductive behaviors, organs, and processes, insisted Buffon, could one locate the causes that really mattered in the world of living things. But the point is that even in this case, theirs' was an in-house fight between Baconians; the debate was over the purity of their inductions, not between two different methods (Sloan 1976).

9.6 Modifying the Baconian Method, 1: "Intervening" & Machine Analogies

Anatomists and physiologists were as inductive as natural historians. This insight should be a little surprising since Andreus Vesalius and William Harvey feature prominently among the Scientific Revolution-aries (e.g., Westfall 1971; Wootton 2015; Burns 2016). But examine any of the major works of the physiologists and physicians, even those equipped with microscopes, and you find their methods are Baconian. They appealed to observations and decried the siren's call of hypotheses. Witness the responses of physiologists whenever the technique of vivi-section was challenged, for instance. Vivisection made excellent sense to a Baconian physiologist; *in vivo* ("in life") was the only sure way to witness natural processes. Without actually witnessing a heart pump, a lung inflate, a vein ripple, a salivary gland squeeze, how was a physiolo-gist to know how the body worked?

The practice of vivisection, however, paradoxically began to modify the strict inductive methods practiced by natural historians. Physiolo-gists recognized that organisms when perturbed attempted to regain their status quo through multiple, often redundant means. Intervening – in this case, disturbing an organism and observing its response – became a chief and mostly innocuous alteration to the Baconian method prac-ticed by life scientists. Of course, inorganic machines such as steam pumps returned to their status quo as well, as engineers were at pains to point out and describe through the early nineteenth century. In fact, the recognition of the functional similarities between machines and organ-isms prompted long-running mechanist–vitalist debates. And, while these debates were surely conceptual, they were also methodological.

The mechanist–vitalist debates were methodological because mech-anists built their arguments on analogies. To understand the living organism, scientists could observe a machine analog and study that. They could bootstrap their way to understanding very complex processes through combinations of simpler ones – more complex machines were just collections of simpler ones. Analogies had long been part of the tool kit of life scientists, of course. Galen of Pergamon studied pigs and apes to understand human anatomy centuries earlier. But the pervasiveness of the machine analogy became an eighteenth-century phenomenon in the life sciences and an alteration of strict Baconian induction (Riskin 2016). To be sure, careful observation and categorization still reigned. But the metallic machine analog was materially different than the carbon-based organism. This slight methodological shift often belied a deeper one, as we will see later.

9.7 Modifying the Baconian Method, 2: Dead Organisms *In Vitro*

In vitro ("in glass") examination of smaller and stranger pieces of the living organism further bent the method of the life sciences away from the Baconian ideal of screening out preconceptions. I argued earlier that the microscope alone did not cause a revolution in the life sciences concomitant with that precipitated by the telescope in astronomy. In part, that is because Leeuwenhoek and his microscope-enhanced colleagues used their devices as extenders of natural-historical induction, not as a replacement for it. Robert Hooke's 1665 *Micrographia* is possibly the best example of such an approach, though physicist Christiaan Huygens also practiced Baconian induction using his own microscope. By the first few decades of the nineteenth century, peering at rotifers and daphnia through microscopes was both serious science and a hobby for the growing middle classes (Croft 2006).

Microscopy altered induction but not radically. As microscopists chased down smaller and more elusive elements of the organism, however, they often had to stop an organism mid-process – kill it, in other words. This would be a slightly more substantial modification to *in vivo* induction. While a dead flea looks quite a bit like a living one, the flea's hematophagic digestion (blood eating) can only be inferred after its death. And that alteration in Baconianism changed even more dramatically as biologists adopted new tools to make microscopy easier. In the 1830s, Jan Evangelista Purkyně and Gabriel Valentin fashioned a microtome that could slice tissues much thinner than possible by a human hand holding a knife. Dividing segments of organisms via the microtome, staining relevant parts with dyes, observing the stained slices *in vitro*, and mentally reassembling them to describe the processes of a living organism – these steps were in service of describing in greater detail the regular order of the living world and properly organizing it. They still proceeded without explicit hypotheses. *In vitro* observation may have been a different submethod than *in vivo*, but it was still induction.

9.8 Theoretical Entities Require a Different Method[4]

Galileo advocated viewing the world as a vast mechanism, yet even that philosophy was not completely incompatible with Bacon's inductive method. Galileo's mechanism extended from actually sensed regularities. When eighteenth-century physiologists analogized from mechanisms to living organisms, they merely extended Galileo's extension of Baconian induction. Invisible entities, however, must be incompatible with Bacon. The seventeenth-century corpuscular philosophy endorsed by Descartes, Robert Boyle, Pierre Gassendi, and (in his optical and alchemical theories at least) Isaac Newton, among others, relied upon entities that explained phenomena but could not themselves be witnessed. These important scientists derived their method from Descartes; that method undergirded much of physics and chemistry beginning in the eighteenth century. The life sciences followed suit, but gradually and only in the wake of an early nineteenth-century quasi-mystical *Naturphilosophie* movement that seemingly had little to do with advancing science.

The corpuscular philosophy hypothesized the existence of tiny entities with few properties that could combine into objects and phenomena with many properties. Because particles resembling corpuscles could not be observed directly for centuries, corpuscularians required a different method to support their work.[5] Instead of gleaning general information about the world by collecting and organizing particulars, as Bacon suggested, corpuscularians hypothesized first and used

---

[4] In this section, I lean heavily on Laudan (1981), especially pp. 20–26 and pp. 181–91.

[5] Not to mention that corpuscles do not actually exist so far as we know, given that molecules, atoms, and subatomic particles do not conform to the depictions of corpuscles on offer by these major figures from the 1600s–1700s.

observations deductively to test their hypotheses. Descartes, as is well-known, cast wholesale doubt on the entire enterprise of induction, given possible errors in human perception. Secure knowledge, he insisted, could only be gained by postulating and then devising mental tests to adjudicate the postulates. Boyle concurred and, as is also well-known, created devices such as air pumps and operated them in front of audiences to promote acceptance of particles and vacuums the audience could not actually observe. Moreover, despite protests that he did not "feign hypotheses," Newton parroted Gassendi's theory of light composed of corpuscles in the *Opticks* (Newton 1999, p. 943). Their significant shift in method was away from describing general principles given observations and toward creating experiments that tested their conjectures – conjectures that employed theoretical entities. To borrow Laudan's (1981, p. 24) terminology, it was a methodological shift from "experience as theory-generator" to "experience as theory-tester."

Biology's turn toward the hypothetico-deductive method that would support unobservable entities followed generations later in the nineteenth century. That transition came as a result of several new factors including cell theory and the notion of descent with modification from a common ancestor. Evidence for both emerged in the 1700s or before – Robert Hooke identified and named the cells he witnessed through a microscope in the 1600s. But serious consideration for common descent from an original unobservably tiny primal organism emerged out of German Romantic philosopher Lorenz Oken's *Naturphilosophie* (1803).[6] In the 1840s, German physiologist Theodor Schwann and lawyer-turned-botanist Matthias Schleiden looked through their microscopes at cells as well. But now they saw structures that could function as Oken's primal units. Rudolf Virchow cobbled these observations into an entire "cell theory": tissues were merely collections of cells, disease was a function of cells, and cells had their origin in other cells rather than emerging out of some undifferentiated mass.

Doubtless some other physiologist had engaged in hypothetico-deductive theory-testing rather than Baconian theory-generation before him, but in Virchow's "cell theory" biology had definitively crossed a methodological line. Each of his major claims were hypotheses to be

---

[6] Erasmus Darwin, grandfather of Charles R. Darwin, supported the idea simultaneously in epic poems such as *Zoonomia* (1803).

tested by experiment. Each claim was verified, to be sure. My point is that by the mid-nineteenth century, "cell theory" launched a second distinct methodological path of conjecture and experiment in the life sciences mirroring that method already dominating the physico-chemical sciences. It was an exceedingly successful method. By 1896, E. B. Wilson could proclaim without hesitation that "the key to every biological problem must finally be sought in the cell" (Wilson 1925).

9.9 What Did Darwin Do?

Because historians and philosophers of science as well as biologists hold up Charles Darwin as the revolutionary *par excellence* in the life sciences, it seems worthwhile to address briefly how he situated himself in this methodological debate. In the mid-1800s, just after Darwin returned from his long circumnavigation onboard the *HMS Beagle*, well-known polymaths William Whewell and John Stuart Mill locked horns over the meaning of induction. Scholars have argued for years about these philosophers' influence on Darwin. Darwin likely knew Whewell personally and read some Mill (Ruse 1975; Thagard 1977). Both Whewell and Mill claimed that induction was the proper way to proceed in science. By "induction," Whewell meant something slightly different than Mill, however.

Whewell called his method of reasoning the "consilience of inductions." And he illustrated what he meant by the phrase in several volumes, including *The History of the Inductive Sciences* (1837) and *The Philosophy of the Inductive Sciences, Founded Upon Their History* (1840), effectively inventing the field of History and Philosophy of Science along the way. According to Whewell, the grand history of physics and chemistry (he excluded the biosciences from his analysis) revealed a two-step process that captured the true essence of Baconian induction. From Whewell's perspective, Bacon never intended to claim that observation alone was the method of science. Instead, the method of science demanded a back-and-forth weaving from new observations to broader concepts, back-and-forth hooking low-level observations to higher-level generalizations already held by the community of sage-like master scientists. Strangely similar to later philosophers of science, Whewell admitted that *a priori* commitments limited the kinds of questions scientists asked and the kinds of observations they believed were worth doing, but that those commitments could be positive restraints, not perversions.

Mill denied that what Whewell described truly counted as *induction*. Holding views prior to making observations was precisely what Bacon called "idols." Nevertheless, Mill recognized that there could be real value to Cartesian hypothetico-deduction. That might be the only available method when "direct methods of observation and experiment" were out of reach to the scientist – when the phenomena or process required reference to theoretical entities, for instance (Mill 1882, p. 325). But that sort of method should only be used when two positions were roughly equivalent and evidence could not cleanly adjudicate between them. Unfortunately, according to Mill, there were a large number of phenomena with such a plethora of inputs that "there is much difficulty in laying down with due certainty the inductive foundation necessary to support the deductive method." Biology, for instance, presented one of the toughest cases, thought Mill. It was often difficult or impossible to disentangle and isolate variables without killing the organism. Mill interpreted the death-of-organism as a larger deviation from Baconian induction than I suggested previously. In fact, when discussing the problem of moving away from *in vivo* situations, Mill broke into poetry:

– following life, in creatures we dissect,
We lose it, in the moment we detect. (Mill 1882, p. 327)

For Mill, Whewell's modification of Baconian induction opened the door to problematic speculation. Combined with that other modification to Baconian induction, *in vitro* observation of dead organisms, Mill worried that physiology in particular was heading down an insecure path.

Still, Whewell and Mill thought of themselves as inductivists. Both agreed that science only proceeds by slow accumulations of observations. Both feared that investigators were liable to be dragged astray by their own preconceived, *a priori* conjectures (Forster 2011). And both reinforced in Charles Darwin the importance of careful induction minus distracting hypotheses.

Even late in life, Darwin wore his commitment to the Baconian method of induction on his sleeve. In his *Autobiography*, Darwin reflected upon his own methods in the long lead up to the publication of *The Origin of Species*:

> I worked on true Baconian principles, and *without any theory* collected
> facts on a wholesale scale, more especially with respect to domesticated
> productions, by printed inquiries, by conversation with skillful breeders
> and gardeners, and by extensive reading. When I see the list of books of
> all kinds which I read and abstracted, including the whole series of
> Journals and Transactions, I am surprised at my industry . . . . [F]ifteen
> months after I had begun my *systematic inquiry*, I happened to read for
> amusement Malthus on Population [*sic*], and being *well prepared to
> appreciate* the struggle for existence which everywhere goes on from
> *long-continued observation* of the habits of animals and plants.
> (Darwin 1902, p. 40 [emphasis added])

Here was Darwin's perspective on his method. Hypotheses were to be
avoided. Inquiry was to be slow and systematic. Organizing principles
would fall out of experience, rather than being the method by which
hypotheses about the world might be tested. Some commentators doubt
the sincerity of Darwin's statements here (e.g., Gould 1980). But considering that Darwin penned these words late in life, we can rightly
imagine that *he believed* he was still acting in accordance with the
inductive methods he learned as a young naturalist, whether or not he
actually followed them.

Even an exception illustrates the larger rule. While his *The Origin of
Species* (1859) accomplished many things, Darwin did not provide a
strong mechanism for inheritance and variation. That gap clearly worried him. He attempted to address it in the first major book of his
evolutionary trilogy – for which *The Origin of Species* was merely an
abstract – *Variation in Animals & Plants Under Domestication* (Darwin
1868). Just as he had done many years earlier, Darwin inductively collected accounts from other naturalists, farmers, breeders, and the like.
From those observations a theory emerged: Organisms are intricately
adapted to their environments in the grand struggle for existence. But if
organisms fit environments so very well – if selection was so pervasive –
where did variation come from? Why would an organism ever alter *away*
from that fitted-ness? Obviously, life circumstances must knock the
organism into or out of perfect fitness somehow. Variations had to
be acquired and then inherited, in other words. But exactly what in the
organism acquired those variations? As Descartes, Gassendi, and Boyle
had done centuries earlier, Darwin postulated theoretical entities.

He called them "gemmules." And by that term he seemed to mean special cells or particles in the organism that took on an imprint from the environment or from other life circumstances and then passed those on to offspring (Darwin 1868). To get to gemmules, Darwin observed and theorized from experiences, including his own show-pigeon breeding, exactly as he had for natural selection. And dedicated to Baconian induction as he was, Darwin contrived no major experiments to *test* his pangenesis theory.

Instead it was Francis Galton, Darwin's cousin, who sensed the first rays of the new methodological sunrise that was already dawning by the 1870s. Darwin made generalizations. Galton recognized them for what they were: conjectures of theoretical entities ready for experimental test. After consulting with his cousin, Galton obtained some rabbits of one color, injected some of their blood into rabbits of a completely different color, and allowed the recipients to mate. No surprise; baby rabbits looked like their parents without a spot of color from the blood donor rabbits. Galton did more experiments exchanging more blood, finally whole transfusions. Aside from piles of rabbits, the experiments also created a legitimate test of Darwin's conjecture. And the experiments did not support Darwin's conjecture (Galton 1871). Darwin, seemingly unaware of the dawning of the new method, complained about Galton's experiments. He wanted it both ways: He wanted to be a Baconian – like Newton he did not "feign hypotheses" – but he did not want have to give up his hypothetical entities.

9.10 Biology's New Theoretical Entities and New Hypothetico-Deductive Methods

Maybe Darwin was the last great natural historian. With the advent of germ theory and genetics at the end of the nineteenth and beginning of the twentieth centuries, entities too small or too diffuse to assess empirically became the cornerstones of the best bioscience. Methods Darwin extolled were derided by physicists, who had long been comfortable with unobservable entities, as so much "stamp collecting." That's not to suggest that germs or genes have no empirical validity, nor does it mean that they remain unseen in our day. It's just that they operated as entities in those explanatory environments long before they could *actually* be observed. Within the hypothetico-deductive method, these entities functioned *as if* they were actually there before anyone actually saw them.

Thankfully, sometimes the scientists involved revealed the secret that these entities were unobserved but useful to support the conjectures being tested. During his Nobel acceptance address, for instance, geneticist Thomas Hunt Morgan revealed:

> What is the nature of the elements of heredity that Mendel postulated as purely theoretical units? What are genes? ... There is no consensus of opinion amongst geneticists as to what the genes are – whether they are real or purely fictitious – because at the level at which the genetic experiments lie, it does not make the slightest difference whether the gene is a hypothetical unit, or whether the gene is a material particle. (Morgan 1934)

More than a half century after Darwin speculated about gemmules, geneticists conjectured about genes. Only, unlike Darwin, modern geneticists such as Morgan utilized the hypothesis-then-test method. Morgan, trained originally as an embryologist during the last fleeting hours during which unfettered induction was still a laboratory ideal, may have been the last prominent laboratory scientist to actually feel the transition in biology's dominant method (Allen 1978).

A wonderful example of the completeness of this transition appeared in the journal *American Biology Teacher* around a decade ago. Biologist José Bonner suggested that methods in biology were not one-size-fits-all. Without apparent knowledge of history of method in science, Bonner (2005) nevertheless summarized the debate I rehearsed earlier. He identified a "Method A" that started with conjectures to be tested – just as Descartes and Morgan did – and a "Method B" that – as Bacon, Mill, and Darwin would have suggested – induced from observations without preconceived hypotheses. Interestingly, Bonner chose the 1960s work of Marshall Nirenberg on RNA-coding to illustrate how the best science could be made through induction (i.e., "Method B[acon]"). Biologist Anton Lawson (2010) disagreed with Bonner. Only one method, the hypothetico-deductive "Method A," was the Scientific Method. No self-respecting life scientist – and students least of all – could actually countenance Method B. Everyone *must* have guiding preconceptions to conduct a proper experiment. Even Nirenberg was, in truth, formulating hypotheses, according to Lawson; Nirenberg's hypotheses were simply implicit. He then had followed the by-then standard method: test hypotheses by experiences, rather than generate them from experiences. Lawson insisted the Nobel nomination signified Nirenberg's proper use of the one and true "Method A."[7]

The moral of this quotidian dispute is clear. Scientists, as Bonner admitted, really do induce from observations to generalities in a way reminiscent of natural history a great deal of the time (e.g., Brown & Lahr 2019). But the Baconian – and Darwinian – ideal to empty oneself of hypotheses and merely observe-categorize-observe-categorize seems quaint at best, possibly even unscientific. The advent of invisible processes and entities being taken seriously in the biosciences ushered in the age of hypothesis testing by the turn of the twentieth century. Even if we have the ability to witness these theoretical entities now – though they are certainly not what Darwin envisioned with his "gemmules," nor Pasteur with his "germs" – the practice of making conjectures and testing them came to dominate the biosciences just as it did the physico-chemical sciences centuries earlier.

9.11 A Categorization of Methods

Likely some readers made it this far in spite of their protests that I created a false dichotomy between the inductive method and the deductive one.[8] Don't we do *both* in the sciences – don't we read and observe like Darwin did, and then create our conjectures to test like Morgan did? Certainly we do. As the Whewell–Mill debate effectively demonstrated, the inductive and deductive methods of Bacon and Descartes, respectively, are at the poles of a spectrum. That spectrum is our general method of knowledge acquisition. Inasmuch as we make use of entities in our theories that we cannot actually witness, we make use of Cartesian theory-testing methods. Inasmuch as we move from

---

[7] The problem with students, according to Lawson, is that they are *too open*. They must be trained to conjecture – to narrow down possible explanations – and only then to create tests to see if their hypotheses hold. I think Bacon, Whewell, and the rest would also have chided Lawson's students but for a different reason. They would have criticized students not for being *open*, but for being too lazy or too rushed and, therefore, too likely to move from only a few scant observations to general statements, leaving aside – as Darwin put it – *systematic* inquiry.

[8] I should give credit/share blame for the dichotomy to/with Laudan (1981), since that is where I encountered the clearest rehearsal of it.

TABLE 9.1 *Methods and submethods operating in the biosciences*

General method of knowledge acquisition:	inductive <—> deductive
Experimental/observational setting:	*in vivo* (field) — *in vitro* (laboratory) — *in silico* (software)
Interference of subject:	non-interference — interference
Focal distance:	high (wholes) <—> low (parts)

individual instances to generalizations, we still make use of Baconian theory-generation methods.

But the question posed by my title requires something more precise. There are more than just these two very general methods of knowledge acquisition, inductive and deductive, operating in the life sciences. I have hinted throughout this chapter that there are submethods as well. I now want to address those explicitly and offer a tentative categorization of them (Table 9.1). A more detailed explanation follows.

Scientists long recognized a distinction between *in vivo* and *in vitro* submethods. In this age of ubiquitous computers and digital models preceding physical experiments, if not replacing them entirely, we now need to add *in silico* (Autoimmunity Research Foundation 2015). As I noted previously, researchers recognized in the nineteenth century, if not earlier, that killing an organism to study it nudged the methods away from pure Baconian induction. Vitalists, of course, went further, claiming that this intervention corrupted the observations entirely. Moving organisms from outdoor habitats into indoor, climate-controlled laboratories to study "in glass" likewise altered observations, though it was harder to say precisely how. Modern computerized models are rapidly becoming the first and sometimes the only stage of scientific investigations. It is easy to see why. They limit stochastic behavior of organisms and environments even further than *in vitro* settings. *In silico* studies even allow for imaginative scenarios and variables never actually witnessed to be fully gamed out – an almost inconceivable departure from Bacon's and Darwin's views of biological method. I would submit that *in vivo*, *in vitro*, and *in silico* approaches are "setting" submethods under the general inductive to deductive umbrella. They do not map precisely onto these general methods – I can imagine a Baconian inquiry using an *in vitro* setting, for instance.

Whether an investigator interferes with/disturbs the experimental subject also presents an interesting submethodical variation. As we saw earlier, scientists interfered with organisms in even quite severe ways to observe how the organism returned to a previous state. "Interference" was still squarely inductive in most instances: The point was to see what the organism did to heal or move. But it did differ from an earlier hands-off approach. In principle, a researcher could switch back and forth between observing without disturbing the subject organism and then intervening to study the response of an organism. Some of the most accomplished experimental embryologists of the nineteenth century, such as Karl Ernst von Baer, seem to have approached their work this way (Coleman 1971).

The "focal distance" at which the investigator sets her/his experiments impacts the study as well. By "focal distance," I mean the tendency to explain *wholes* (a high focal distance) versus *parts* (a low focal distance). Russell (1972, 1982) extensively studied this division in the life sciences of the eighteenth and nineteenth centuries and came away convinced that the division between what he called "unity theories" and "particulate theories" was the single most important one in the whole history of biology. Not surprisingly, given the history I outlined previously, Russell found that those with a high focal distance (i.e., unity theorists) congregated earlier on the timeline; the number of particulate theory advocates increased over the nineteenth century, though some unity theorists could still be found in the twentieth century (Goldstein 1963). Even here, I should offer the same caveats about the focal distance being a spectrum, not a dyad (though Russell himself remained convinced the relationship was mutually exclusive). "Organicists," such as biochemist Joseph Needham and epigeneticist C. H. Waddington, insisted they could move freely between these focal distances (Peterson 2017).

Table 9.1 offers an abstract categorization of these methods and submethods. I make no claims to definitiveness. Nor do I regard these categories as mutually exclusive. The only somewhat firmer relationship I posit between them is that they correlate chronologically, such that the methods to the right side are more pronounced today than those to the left, which dominated historically. Even a quick glance at the table helps reinforce my initial assertion that, though these categories are certainly connected to a degree, *several but a limited number* of methods and submethods combine to operate in the life sciences.

9.12 Conclusion: The Case of Amyloid-ß, Revisited

Moir and Tanzi reported a spark of insight – a generalization from particulars – and amyloid-ß is hardly a hypothetical entity.[9] So is modern biology less divorced from Baconian induction than I insinuated, or have the sciences simply perfected the method? Do we, as a rule, now patiently induce from observations without prejudice and then build experiments to test our induced generalizations deductively? I would answer "no" to these questions.

The social context of science has changed radically in the centuries since Bacon or even Darwin. They were independently wealthy English gentlemen; today's scientists work for a paycheck. Discovering great philosophical truths about the world is rarely the goal of life science investigations today. CRISPR, the most recent great discovery, illustrates the transition well. While certainly an important step that will likely merit a Nobel nomination, it is an unquestionably technical rather than conceptual innovation. In this social environment, the ultimate goals of the life scientist are technique-oriented: to secure funding, to pursue publication in highly ranked journals, and to patent research processes such as CRISPR or entities such as Oncomouse. Whole laboratories and trans-institutional research networks now orient in this direction (Latour & Woolgar 1979). Moir and Tanzi's story may have some elements of old induction in them. But for the most part, they exhibit "deductive-in vitro/in silico-interference-low focal distance" methods. This particular combination of methods and submethods dominates present day science to such a degree that some might regard it as *the* Scientific Method.

Yet, as this brief historical flyover suggests, life scientists have used more than one method to pursue what they consider a satisfactory solution to the puzzle set by an organism or phenomenon. Philosopher Paul Feyerabend may have posited anarchy to undercut our faith in the unitary hypothetico-deductive Scientific Method. He was right about the lack of a singular Method, but he overstated his case. Researchers use multiple methods, yet the choices of method are limited. To be sure, Feyerabend understood that the choice of methical orientation is not

---

[9] Though there remains something quite theory-laden about the way genes are often portrayed in scientific accounts outside of the domain of molecular biology (Falk 2000; Kampourakis 2017).

free; there is a thumb on the scales. Given the structure of European and American research models over the last two centuries – with more funding and prestige flowing toward those researchers who produce publications, grant applications, and commercializeable innovations quickly – the tendency has been to move faster than Darwin's induction process allowed (Kay 1993; Krimsky 2003).

A generation ago, biologist Barry Commoner decried this trend in "In Defense of Biology" (1961). He foresaw that satisfactory solutions to contemporary biological puzzles would regularly arrive in the language of theoretical entities to the exclusion of all else. The old inductive biology, he believed, still had worth, though it was being rapidly displaced. Were he alive today, Commoner would no doubt see the craze for genome-wide association studies (GWASs) as symptomatic of this trend. Stengers' (2018) "Slow Science" seems to be another version of Commoner's jeremiad. What Commoner and advocates for Slow Science are calling for is not new, but a reprioritization of methods that once ruled the practice of the life sciences. Their hope may be misplaced. Still, given how many of the most significant discoveries in the history of the life sciences came about prior to the sort of theory-testing methods currently prioritized, there can be no harm in presenting the plurality of available methods actually used by life scientists (*contra* Lawson 2010).

References

Allen, G. (1978). *Life Science in the Twentieth Century*. Cambridge: Cambridge University Press.

Autoimmunity Research Foundation. 2015. Differences between in vitro, in vivo, and in silico studies. *The Marshall Protocol Knowledge Base*. https://mpkb.org/home/patients/assessing_literature/in_vitro_studies (accessed May 22, 2019)

Bernstein, B. (2000). *Pedagogy, Symbolic Control and Identity*. Oxford: Rowman & Littlefield.

Bonner, J. J. (2005). Which Scientific Method Should We Teach and When? *American Biology Teacher* 67(5): 262–264.

Brown, M. & Lahr, D. (2019). "'Micro Snails' We Scraped From Sidewalk Cracks Help Unlock Details of Ancient Earth's Biological Evolution," *The Conversation* (Feb. 28). https://theconversation.com/micro-snails-we-scraped-from-sidewalk-cracks-help-unlock-details-of-ancient-earths-biological-evolution-112362

Burns, W. E. (2016). *The Scientific Revolution in Global Perspective*. New York: Oxford University Press.

Chalmers, A. F. (1990). *Science and Its Fabrication*. Milton Keynes, UK: Open University Press.

Coleman, W. (1971). *Biology in the Nineteenth Century: Problems of Form, Function, and Transformation*. New York: J. Wiley & Sons.

Commoner, B. (1961). In Defense of Biology. *Science* 133(3466): 1745–1748.

Croft, W. J. (2006). *Under the Microscope: A Brief History of Microscopy*. Singapore: World Scientific Publishing Co.

Darwin, C. (1868). *The Variation of Animals and Plants Under Domestication*. 2 vols. London: John Murray.

Darwin, C. (1902). *Charles Darwin: His Life Told in an Autobiographical Chapter, and in a Selected Series of His Published Letters*, ed. F. Darwin. London: John Murray.

Dewey, J. (1910). *How We Think*. Boston: D. C. Heath & Co.

Dupré, J. (1993). *The Disorder of Things: The Metaphysical Foundations of the Disunity of Science*. Cambridge, MA: Harvard University Press.

Endersby, J. (2007). *A Guinea Pig's History of Biology*. Cambridge, MA: Harvard University Press.

Falk, R. (2000). The Gene: A Concept in Tension. In P. Beurton, R. Falk, & H.-J. Rheinberger (eds.), *The Concept of the Gene in Development and Evolution*, pp. 317–348. Cambridge: Cambridge University Press.

Feibleman, J. K. (1972). *Scientific Method: The Hypothetico-Experimental Laboratory Procedure of the Physical Sciences*. The Hague, Netherlands: Martinus Nijhoff.

Feyerabend, P. (1993 [1975]). *Against Method*. 3rd ed. New York: Verso.

Forster, M. (2011). The Debate Between Whewell and Mill on the Nature of Scientific Induction. *Handbook of the History of Logic* 10. doi: 1016/B978-0-444-52936-7.50003-3

Galison, P. (1996). Introduction: The Context of Disunity. In P. Galison & D. J. Stump (eds.), *The Disunity of Science*, pp. 1–33. Stanford, CA: Stanford University Press.

Galton, F. (1871). Experiments in Pangenesis, by Breeding from Rabbits of a Pure Variety, into Whose Circulation Blood Taken from Other Varieties Had Previously Been Largely Transfused. *Proceedings of the Royal Society of London* 19: 123–29.

Gaukroger, S. (2006). *The Emergence of a Scientific Culture: Science and the Shaping of Modernity, 1210–1685*. New York: Oxford University Press.

Goldstein, K. (1963 [1939]). *The Organism*. Boston: Beacon Press.

Gould, S. J. (1980). Darwin's Middle Road. In *The Panda's Thumb: More Reflections in Natural History*. New York: Norton.

Hirshfeld, A. (2001). *Parallax: The Race to Measure the Cosmos*. New York: Henry Holt.

Huff, T. (2010). *Intellectual Curiosity and the Scientific Revolution: A Global Perspective*. Cambridge: Cambridge University Press.

Kampourakis, K. (2017). *Making Sense of Genes*. Cambridge: Cambridge University Press.

Kay, L. (1993). *The Molecular Vision of Life: Caltech, the Rockefeller Foundation, and the Rise of the New Biology*. New York: Oxford University Press.

Kohler, R. E. (1994). *Lords of the Fly: Drosophila Genetics and the Experimental Life.* Chicago: University of Chicago Press.

Krimsky, S. (2003). *Science in the Private Interest: Has the Lure of Profits Corrupted Biomedical Research?* New York: Rowman & Littlefield Publishers, Inc.

Lakatos, I. & Musgrave, A. (eds.) (1970). *Criticism and the Growth of Knowledge.* Cambridge, UK: Cambridge University Press.

Latour, B. & Woolgar, S. (1979). *Laboratory Life: The Construction of Scientific Facts.* Los Angeles: Sage Publications.

Laudan, L. (1981). *Science and Hypothesis: Historical Essays on Scientific Methodology.* Dordrecht, Netherlands: D. Reidel Publishing Co.

Laudan, L. (1984). *Science and Values: The Aims of Science and Their Role in Scientific Debate.* Berkeley, CA: University of California Press.

Lawson, A. E. (2010). How Many Scientific Methods Exist? *The American Biology Teacher.* 72(6): 334–336.

Mayr, E. (2004). *What Makes Biology Unique? Considerations on the Autonomy of a Scientific Discipline.* Cambridge: Cambridge University Press.

Mill, J. S. (1882 [1843]). *System of Logic Ratiocinative and Inductive.* 8th ed. New York: Harper & Bros. http://www.gutenberg.org/files/27942/27942-h/27942-h.html

Morgan, T. H. (1934). "The Relation of Genetics to Physiology and Medicine," Nobel Lecture. NobelPrize.org. Nobel Media AB. https://www.nobelprize.org/prizes/medicine/1933/morgan/lecture (accessed May 27, 2019).

Newton, I. (1999 [1726]). *Philosophiae Naturalis Principia Mathematica.* 3rd ed. I. B. Cohen & A Whitman, trans. Berkeley, CA: University of California Press.

Nordenskiöld, E. (1935 [1928]). *The History of Biology.* 2nd ed. L. B. Eyre, trans. New York: Knopf.

Peterson, E. (2017). *The Life Organic: The Theoretical Biology Club and the Roots of Epigenetics.* Pittsburgh, PA: University of Pittsburgh Press.

Riskin, J. (2016). *The Restless Clock: A History of the Centuries-Long Arguments Over What Makes Living Things Tick.* Chicago: University of Chicago Press.

Rudolph, J. (2005). Epistemology for the Masses: The Origins of "The Scientific Method" in American Schools. *History of Education Quarterly* 45(3): 341–376.

Ruse, M. (1975). Darwin's Debt to Philosophy: An Examination of the Influence of the Philosophical Ideas of John F. W. Herschel and William Whewell on the Development of Charles Darwin's Theory of Evolution. *Studies in the History and Philosophy of Science* 6(2): 159–181.

Russell, E. S. (1972 [1930]). *The Interpretation of Development and Heredity.* Oxford: Clarendon.

Russell, E. S. (1982 [1916]). *Form and Function: A Contribution to the History of Animal Morphology.* Reprint ed. Chicago: University of Chicago Press.

Sloan, P. R. (1972). John Locke, John Ray, and the Problem of the Natural System. *Journal of the History of Biology* 5: 1–53.

Sloan, P. R. (1976). The Buffon–Linnaeus Controversy. *Isis* 67: 356–375.

Stengers, I. (2018). *Another Science Is Possible: A Manifesto for Slow Science.* S. Muecke, trans. Cambridge: Polity Press.

Thagard, P. (1977). Darwin and Whewell. *Studies in the History and Philosophy of Science* 8: 353–356.

van Berkel, K. (2000). Descartes' Debt to Beeckman. In S. Gaukroger, J. Schuster, & J. Sutton (eds.), *Descartes' Natural Philosophy*, pp. 46–59. New York: Routledge.

Westfall, R. S. (1971). *The Construction of Modern Science: Mechanisms and Mechanics*. New York: John Wiley and Sons.

Whewell, W. (1847). *The Philosophy of the Inductive Sciences, Founded Upon Their History*. 2 vols, 2nd ed. London: John W. Parker.

Whewell, W. (1984 [1857]). *Selections from History of the Inductive Sciences from the Earliest to the Present Time*. 3rd ed. In Y. Elkhana (ed.), *Selected Writings on the History of Science*. Chicago: University of Chicago Press.

Wilson, E. B. (1925 [1896]). *The Cell in Development and Heredity*. 3rd ed. New York: Macmillan.

Wilson, E. O. (1998). *Consilience: The Unity of Knowledge*. New York: Vintage.

Wootton, D. (2015). *The Invention of Science: A New History of the Scientific Revolution*. London: Penguin.

10 Is It Possible to Scientifically Reconstruct the History of Life on Earth?

The Biological Sciences and Deep Time

CAROL E. CLELAND

10.1 Introduction

The data-based study of long past events and processes is common throughout the sciences. Some examples are the astrophysical hypotheses that the universe began with a cosmic explosion ("big bang"), which is supported by measurements of the cosmic microwave background radiation pervading the modern universe; the hypothesis that the end-Cretaceous mass extinction was caused by a meteorite impact, which is supported by an iridium anomaly and large quantities of shocked quartz in K-Pg (Cretaceous-Paleogene) boundary sediments; and the hypothesis that all life on Earth shares a common ancestor, which is supported by analyses of shared segments of ribosomal RNA found in contemporary organisms. My interest in the methodology of the historical sciences and how it differs from that of stereotypical or "classical" (as I later dubbed it) experimental science was first piqued in the 1990s by the writings of so-called "scientific [more accurately, biblical] creationists." Scientific creationists and their successors, members of the "Intelligent Design Network,"[1] extol classical experimental research (the testing of hypotheses under controlled laboratory conditions) as the paradigm of good science, contending that historical scientific research is inferior because it uses "a form of abductive reasoning that produces competing historical hypotheses, that lead to an inference to the best current explanation rather than to an explanation that is logically compelled

---

[1] The "Intelligent Design Network" is a politically active organization dedicated to undermining the teaching of Darwinian evolution in US public schools. They differ from "scientific creationists" primarily in eschewing the word "God" in favor of "intelligent designer," thereby hoping to avoid the charge of violating the First Amendment of the US Constitution (on the separation of religion and state).

by experimental confirmation."[2] Proponents of intelligent design are not alone, however, in denigrating the work of historical scientists. Articulating a view held by a surprising number of experimentalists, Henry Gee, at the time a senior editor of *Nature*, declared that no science can be historical because conjectures about the past cannot be tested by means of controlled laboratory experiments (Gee 1999).

As a philosopher of science, these arguments against the scientific status of historical science struck me as very naïve. As Section 10.2 discusses, they mistakenly treat classical experimentation – the testing of hypotheses under controlled laboratory conditions – as the standard for all scientific research. Moreover, philosophers long ago established that the widespread view that classical experimentation is capable of *logically compelling* the acceptance and rejection of hypotheses (and theories) is false. This raises the question of how the practices of historical scientists actually differ from those of experimental scientists and, most importantly, whether these differences justify the claim that the former is scientifically inferior to the latter. This is the topic of Section 10.3. Section 10.3 presents an analysis (see Cleland 2001, 2002, 2011, 2013 for more detail) of the differences in practice between historical science and classical experimental science that is grounded in a pervasive physical time asymmetry of causation,[3] as opposed to a purely logical relation between evidence and hypothesis. Viewed in light of this "asymmetry of overdetermination" (as it is known to philosophers) neither practice can be judged more scientific (rational or evidence-based) than the other; each practice is designed to exploit the information that nature puts at its disposal. Sections 10.4–10.6 flesh out the causal analysis, presented in Section 10.3, of the differing methodologies of experimental and historical science in the context of well-known cases from the biological science.

---

[2] www.intelligentdesignnetwork.org/Statement_of_Objectives_Feb_12_07.pdf

[3] Derek Turner (2007) labeled this time asymmetry of causation "metaphysical," but as explained in detail in Cleland (2011), it is *physical*. Indeed, it is well-known to physicists in the form of the second law of thermodynamics (statistically interpreted) and the radiation asymmetry, which affects all wave phenomena. Nevertheless, to my continuing frustration, some philosophers of science continue to parrot Turner's mistaken claim about it being metaphysical (above or beyond physics, or based on speculative or abstract reasoning that is unsubstantiated by empirical evidence). It isn't.

10.2 Experimental Science and the Scientific Method

Experimental physics provides the traditional model for the fabled "scientific method." Yet the testing of hypotheses and theories by means of controlled experiments represents only a small part of what goes on in experimental physics. Some experiments are exploratory, as opposed to hypothesis or theory driven, amounting to a search for new phenomena in need of theoretical explanation (Hacking 1983; Steinle 1997; 2002). Two famous illustrations are Francesco Grimaldi's seventeenth-century experiments with light, which led to the discovery of diffraction, and nineteenth-century measurements of atomic spectra, which led to the puzzling discovery that they are quantized, setting the stage for the development of quantum mechanics in the early twentieth century. Other roles for experiment include demonstrating a new experimental technique and measuring quantities that are of physical interest, such as Avogadro's number and the mass of an electron; see Allan Franklin (1990) for a discussion of these and other roles for experiment in physics. In short, the testing of hypotheses under controlled conditions is only one among many types of experiment performed in physics. Nevertheless, most critics of historical science view physics as the proverbial queen of sciences – the model for all the sciences – and treat classical experimentation as its distinguishing method.

The alleged superiority of classical experimentation, extolled by afficionados of experimental physics as "*the* scientific method," rests upon the supposition that experiments conducted under controlled conditions have the power to *logically compel* (in a manner analogous to mathematical proof and disproof) the acceptance or rejection of a scientific hypothesis. Philosophers long ago established that this claim is false. The notion that successful predictions can logically compel the truth of scientific hypotheses runs up against the infamous *problem of induction* (see Henderson 2019 for more detail). Famously articulated by the eighteenth-century philosopher David Hume, this is the problem of inferring the truth of a universal empirical generalization – a statement of the form "**All** As are Bs" – from a finite body of successful observations (where every A *thus far observed* has been a B, but there are nonetheless As and Bs that have not been observed): How can one eliminate the possibility that some of the unobserved instances of A and B violate the generalization concerned? As a salient illustration, consider the empirical claim "All swans are white," which prior to 1770 was supported by an enormous body of evidence, acquired from all over world (Europe, Africa, Asia, and the Americas). Indeed, this claim was widely used by logicians and philosophers as a reliable illustration of a universal empirical generalization. Somewhat amusingly, after Captain Cook's discovery of Australia, with its numerous black swans, it was replaced by "All ravens are black," which is still in use today; philosopher Carl Hempel's famous "raven paradox" provides a salient illustration (Hempel 1945).

Like "All swans are white," most of the hypotheses tested by means of classical experimentation have the form of law-like (universal, exceptionless) generalizations. That is, they designate *all* cases (past, present, and future) of a particular *type* of phenomenon. Consider, for example, the simplistic hypothesis "All copper expands when heated." It applies to *every* case in which copper is heated, whether deliberately in an experiment or incidentally by, for example, naturally occurring processes. This hypothesis boldly conjectures that there are no cases in which copper is heated and fails to expand. But even supposing that it has successfully withstood a large number of experimental tests, no scientist or team of scientists can possibly heat *every* piece of copper that has, is, or will exist in our universe to confirm that copper *always* expands when heated. Put another way, successful predictions *under-*determine the hypotheses that they support in the sense that no body of evidence collected by humans, however large, can ensure the truth of a timeless empirical generalization. In truth, claiming that a body of successful predictions can do so violates logic; it commits the formal fallacy of affirming the consequent.

The logic underlying falsificationism – the claim, usually attributed to Karl Popper (e.g., 1963, p. 33–39), that although scientific hypotheses and theories cannot be proven they nevertheless can be disproven – is *prima facie* more compelling. After all, the discovery of a single black swan surely disproves the hypothesis that all swans are white. But what if the swan's feathers were dyed black, or what appears to be a swan is a previously undescribed species of bird closely resembling swans? Here, again, a disanalogy between mathematical and scientific reasoning emerges: In mathematics, everything relevant to the truth or falsity of a conjecture, such as the Pythagorean Theorem, is included in the pertinent (Euclidean) set of axioms. Scientific hypotheses, in contrast, cannot be investigated empirically independently of ancillary conditions and processes present in a test situation. Some of these conditions and processes may be poorly understood or even unknown, and hence not properly accommodated by means of controls in the experimental setup. A prediction might fail not because the target hypothesis is false but because some unforeseen condition (e.g., an unknown black, swan-like species of bird) is present or an unanticipated process interfered in the causal chain between the test condition (e.g., heating of a piece of copper) and the predicted result (expansion of the copper). Such a result is known as a false negative; the prediction fails even though the hypothesis is true. Along the same lines, a prediction might succeed (despite a hypothesis being false) because some factor other than the test condition produced the anticipated result; such a result is known as a false positive. False positives and false negatives are common in experimental investigations, which explains why experimentalists spend so much time trying to avoid them by controlling for potentially interfering conditions and processes. How plausible a potentially interfering condition or process seems depends, however, upon the state of our knowledge. There are no guarantees, logical or empirical, that an experimental program of controls has not overlooked a critical causal factor present in the test situation.

In sum, classical experimentation cannot disprove scientific hypotheses because (in sharp contrast to mathematical deductions) one can never rule out false negatives. Moreover, in addition to facing the classical form of the problem of induction, experimentalists must also deal with the threat of false positives: Are all of the ostensibly successful predictions truly successful? It follows that the central claim upon which most attacks on the scientific status of historical science rests – that classical experimental research is superior because, unlike historical research, its findings *logically compel* the truth or falsity of scientific hypotheses – is a myth. This raises the question: How does the method of the historical sciences really differ from that of classical experimental science, and do these differences undermine the scientific status of the former vis-à-vis the latter?

10.3 Historical Science, Experimental Science, and the Asymmetry of Overdetermination

The logical form of many of the hypotheses investigated in the historical sciences differs from those investigated in classical experimental science. Classical experimentalists are concerned with testing timeless, law-like, generalizations, for example Boyle's gas law and Mendel's laws of inheritance. In contrast, (what I have dubbed) "prototypical" historical research – the most common type of research in the historical sciences – is concerned with investigating *particular* past events such as how Earth's unusually large moon formed and what caused the end-Cretaceous mass extinction. It is thus not too surprising that there are differences in how researchers in these different areas of science go about "testing" their hypotheses.[4] What is surprising is that these differences are founded upon causal, as opposed to logical, differences in the relationship between hypothesis and evidence. More specifically, the way in which our universe is causally structured in time is asymmetric: *Localized* events in the present, which is all that any scientist working within the confines of a laboratory or field site has access to, evidentially *over*determine their past causes and evidentially *under*determine their future effects. As discussed later, this time asymmetry of causation – known to philosophers as the "asymmetry of overdetermination" (Lewis 1991) – profoundly affects the manner and degree of

---

[4] As discussed in Cleland (2002, 2011), when a somewhat extensive and diverse body of historical evidence is available from field studies, historical scientists may behave like classical experimentalists and advance generalizations about the phenomena concerned. Such generalizations are typically virtually impossible to "test" in a laboratory by means of classical (controlled) experimentation. Nonetheless, there are exceptions, especially in biology. A good illustration is research on the evolution of bacteria and viruses, which reproduce extremely quickly, as models for evolutionary processes (mutation, drift, selection, etc.) considered more generally. The focus of this chapter, however, is on prototypical historical research and how it differs from classical experimental research. Prototypical historical research focuses on the investigation of *particular* past events, such as the end-Cretaceous mass extinction, which is thought to differ from the end-Permian mass extinction, and (clearly) differs from the currently ongoing Anthropocene mass extinction. Interestingly, some researchers have tried to identify a general cause for major extinctions on Earth, most famously Raup's & Sepkoski's (1984) "nemesis star" (a dim companion star or large planet, with a very asymmetric orbit, that periodically swings in closer to the sun, disrupting asteroids and comets, sending them careening around the inner solar system, and then moves far away from the sun, beyond the Oort cloud). Most paleontologists, however, believe that mass extinctions are not the product of a single kind of cause. The leading contender for the end-Permian mass extinction (of around 250 million years ago) is extreme (Siberian) flood volcanism, whereas the leading hypothesis for the end-Cretaceous mass extinction is meteorite impact and for the current Anthropocene extinction, human activity. In any case, it should be clear that, despite its generality (which, notably, is limited to one solar system, and most likely one planet, Earth) the nemesis star hypothesis cannot be tested via classical experimentation in a laboratory setting.

support offered by empirical evidence for scientific hypotheses and theories.

As an illustration of the overdetermination of the past by the localized present, consider the many, highly diverse traces left at a murder scene: DNA, blood spatters, finger prints, shoe prints, strands of hair, a bloody glove, etc. A detective doesn't require all of this material – e.g., every segment of foreign DNA found on the victim's body and suspect's weapon – in order to identify the perpetrator. Any small sample of foreign DNA, found in appropriate places at the crime scene, provides equally good evidence, especially in combination with a subset of other evidence available; one doesn't need every spot of blood, shoe print and finger print, for instance. It is in this sense (the availability of multiple, potentially incriminating, bodies of evidence, only one of which is required to identify the perpetrator) that traces of the past found in the present may be said to (evidentially) "*over*determine" a prior event, such as a murder.

Like forensic scientists, historical scientists generate competing hypotheses (in essence, suspects) for a puzzling body of traces found in the present and search for a "smoking gun," that is, a trace (or traces) that distinguishes one or more of these hypotheses as providing a better explanation than the others for the *total* body of evidence thus far available. Ideally, this procedure culminates in a single best explanation. Depending upon the character of the evidence available, however, the process of settling on a best explanation may be prolonged; it may be contingent upon the discovery of new lines of evidence and perhaps even the formulation of previously unconceived rival hypotheses. The important point, for purposes of this discussion, is that the asymmetry of overdetermination underwrites the methodology of the historical sciences: Because the present typically contains many diverse traces of past events, it is highly likely (but not completely certain) that telling evidence exists for a long past event such as the end-Cretaceous mass extinction. The challenge is discovering this evidence in the messy, uncontrollable world of nature. As Section 10.4 discusses, the contentious history of the scientific debate over what caused the end-Cretaceous mass extinction provides a salient illustration of how this process unfolds.

Some philosophers (e.g., Sober 1988; Turner 2007) contend that information-destroying processes significantly undermine the potential for reconstructing the past on the basis of traces found in the present. As I have argued (Cleland 2011, 2013), however, they overestimate the capacity of information-degrading processes to completely destroy information and underestimate the potential for recovering information from attenuated traces. While it is true that information about the past becomes increasingly degraded over time, it doesn't follow that it is unrecoverable. A striking illustration is the unanticipated discovery of melanosomes within the fossilized feathers of well-preserved dinosaur specimens, which has allowed paleontologists to reconstruct the colors of some dinosaurs; see, for example, Smithwick and colleagues' (2017) discussion of the evidence for a racoon-like, feathered mask around the eyes of *Sinosauropteryx*. The point is one can never rule out finding a smoking gun for a hypothesis about a long past event. Recognizing a trace for what it represents may take time, requiring the use of sophisticated tools and methods. But, as Adrian Currie (2018) discusses, historical scientists have a variety of resources for dealing with the problem of degraded traces.

As an illustration of the *under*determination of the future by the localized present, consider the difficulty of predicting the course of a disease, such as stage 3 colon cancer. Approximately 64 percent of victims survive but it is impossible to predict whether a particular individual, such as your uncle, will be among them. In contrast, it is fairly easy for a forensic medical examiner to assess whether someone's death was caused by colon cancer. The point is, although the present contains records (traces) of the past, it does not contain records of the future. This underscores the logical problems (discussed in Section 10.2) with claims that classical experimentation can prove or disprove conjectured generalizations. The test condition(s) specified by a hypothesized generalization is rarely (if ever) the *total* cause of an experimental result. An experimental scientist can never be sure that all the causal factors involved in producing the result have been satisfactorily accounted for through the manipulation of experimental controls. In order to rule out false positives and false negatives, she needs to know that the hypothesized test condition combined with other causal factors present in the experimental situation in the way anticipated. But experimentalists never have such assurances because they can never be certain that they have controlled for every potentially relevant causal factor (contributing and confounding) involved in the production of an experimental result. It is in this sense that bodies of evidence acquired in the present, however extensive, underdetermine (are, logically speaking, insufficient to prove or disprove) the occurrence of a predicted effect.

The remainder of this chapter, Sections 10.4–10.6, explore, in the context of actual cases, how researchers in the biological sciences exploit the asymmetry of overdetermination in their evidential reasoning. Sections 10.4 and 10.5 focus on historical research, explaining how researchers exploit the overdetermination of past events by traces found in the present in reconstructing particular episodes from the history of life on Earth, namely the cause of the end-Cretaceous mass extinction and the existence of a common ancestor for all known life on Earth. Section 10.6 contrasts historical research with experimental research in the biological sciences through a case study, namely Tom Cech's discovery of ribozymes (catalytic RNA molecules), exploring how experimentalists attempt to circumvent the underdetermination of predicted results by experimental controls and the dilemma they face in deciding whether a failed prediction represents a false negative or a true negative.

10.4 The End-Cretaceous Mass Extinction

The development of scientific thought about what caused the end-Cretaceous mass extinction, which is estimated to have extinguished 75 percent of all (eukaryotic) species on Earth around 66 million years ago, provides a salient illustration of historical reasoning in the biological sciences. Prior to the 1970s, scientists entertained a large number of rival hypotheses for the end-Cretaceous extinctions, including evolutionary competition (most famously, by clever mammals against dim-witted dinosaurs!), climate change, supernova explosion, meteorite impact, and volcanism. The discovery of a mysterious iridium anomaly in K-Pg boundary sediments[5] functioned as a smoking gun for the volcanism and meteorite impact hypotheses over their competitors; these were the only geophysical processes thought to be capable of causing an

---

[5] The K-Pg (previously known as K-T) boundary is a thin band of sedimentary rock, radiometrically dated to approximately 66 million years ago, which is found around the world and marks the end of the fossil records of distinctively Cretaceous flora and fauna, such as the non-avian dinosaurs and ammonites.

iridium anomaly. By the late 1990s most members of the geological community (including paleontologists) concurred that the impact of an enormous (Mt. Everest sized) meteor, centered on what is now the town of Chicxulub, on the Yucatan peninsula, triggered the end-Cretaceous extinctions.[6] As discussed in Cleland (2001, 2002, 2011), the evidence for the impact of a massive bolide – iridium anomaly and large quantities of impact debris (shocked minerals, tektites, and glassy spherules) found in K-Pg boundary sediments from around the world – is overwhelming. In addition, major changes in the fossil records of an extensive variety of small organisms – e.g., foraminifera (single celled protists with shells) and small bivalves (especially ammonites) – as well as plant material (especially pollen and fern spores) have been found across K-Pg boundary sediments from around the world, suggesting that a mass extinction event coincided with the bolide impact. Following an extensive review of the available evidence by an international panel of geoscientists (Schulte et al. 2010), the Chicxulub meteorite became the official explanation, enshrined in college textbooks, for the end-Cretaceous mass extinction.

A small but growing number of paleontologists disagree, however. Unlike early proponents of the volcanism hypothesis – see Powell (1998, ch. 6) for a discussion – these researchers don't reject the claim that a gigantic meteorite slammed into Earth around the time of the end-Cretaceous mass extinction; the geological evidence for the impact is too extensive. Instead, they contend that massive volcanic eruptions in the Deccan plateau region of what is now India were at least in part (if not wholly) responsible for the extinctions. The Deccan Traps eruptions were immense, covering an area about half the size of modern India, and undoubtedly released extensive quantities of toxic and climate altering gases. From a theoretical perspective, they seem just as capable of causing a mass extinction as the impact of a gigantic meteorite.

Three different versions of the contemporary "volcanism" hypothesis for the end-Cretaceous mass extinction are currently under investigation: (1) the Deccan traps eruptions alone caused the mass extinction before the impact of the Chicxulub meteorite; (2) the Deccan Traps eruptions decimated late Cretaceous flora and fauna and the Chicxulub meteorite finished them off; and (3) the meteorite impact severely

---

[6] See Powell 1998 for an excellent review of the history of this debate.

compromised late Cretaceous flora and fauna, which were already weakened by earlier Deccan volcanism, and a surge in Deccan volcanism after the impact finished them off. The first represents a revival of the old volcanism hypothesis, which held that the Deccan eruptions alone caused the end-Cretaceous mass extinction. Versions #2 and #3 represent new hypotheses about the cause of the end-Cretaceous extinctions. No longer treated separately as competitors, meteorite impact and volcanism are conjectured to have jointly caused the mass extinction. The latter are thus best characterized as combined meteorite-volcanism hypotheses.

Proponents of the original version (#1) of the volcanism hypothesis appeal to (i) recent, high resolution, geological reconstructions of the timeline for the Deccan eruptions (vis-à-vis the Chicxulub impact) and (ii) more fine-grained studies of the fossil records of late Cretaceous flora and fauna in the time interval between the beginning of the Deccan eruptions and the impact of the Chiculub meteorite (Punekar, Mateo, & Keller 2014; Sakamoto, Benon, & Venditti 2016; Schoene et al. 2019). In support of the conjecture that the Deccan eruptions caused the mass extinction, some of these researchers (Schoene et al. 2019) cite new geological studies indicating that the Deccan eruptions intensified during the thousand years preceding the impact of the Chicxulub bolide and argue that this interval of geological time exhibits notable changes in the fossil records of late Cretaceous flora and fauna.

In this context, it is worth discussing the fossil record of the (non-avian) dinosaurs, which is sometimes cited (especially in the popular press) as supporting the volcanism hypothesis: No fossilized dinosaur bones have been found within 3 meters of the K-Pg boundary, suggesting, to some people, that they were already extinct by the time the impact occurred. It is important to appreciate that the so-called "3-meter problem" does not provide very strong evidence in support of the volcanism hypothesis for the end-Cretaceous extinctions. As Powell (1998, ch. 8) explained, fossilization requires very special geochemical conditions and hence the fossil record of organisms is never complete. In addition, large organisms, such as dinosaurs, are much less common than tiny organisms, and as a result their fossils are very rare in comparison to those of the latter. Even supposing that they were thriving at the time of the impact, dinosaur fossils would be few and far between and hence more difficult to find than those of small bivalves, foraminifera, pollen, and fern spores, which are found right up to the lower edge of the K-Pg boundary. Lack of dinosaur remains close to the K-Pg boundary could simply be because they have yet to be found. For this reason, paleontologists studying mass extinctions focus most of their research on tiny organisms. This situation is exacerbated by the fact that well-preserved geological outcrops of the K-Pg boundary are rare; most have been removed by erosion (water and wind) and others remain deeply buried, awaiting the attentions of future researchers. In short, in the case of the fossil record of the dinosaurs, absence of evidence is not evidence of absence. The argument that the 3-meter gap indicates that the dinosaurs were already extinct by the time of the impact is not very compelling.

On the other hand, if paleontologists found dinosaur fossils in close association with K-Pg boundary sediments, it would significantly undermine the conjecture that the dinosaurs (and other end-Cretaceous flora and fauna) were extinct by the time of the Chicxulub bolide impact. This helps to explain excitement over the discovery of what seems to be a killing field laid down within hours after the meteor slammed into the Yucatan peninsula (DePalma et al. 2019). The deposits, found at a small field site in North Dakota, consist of an enormous variety of fossilized remains (fish, insects, mammals, etc., and plants, including trees) ripped apart and jumbled together in an area that experienced catastrophic, tsunami-like, flooding. The debris field includes extensive deposits of meteorite impact markers, some of which (glassy spherules) are actually embedded in tree resin and the gills of fish, which presumably inhaled them in desperate efforts to breathe. Radiometric analysis indicates that the deposits are the same age as the Chicxulub crater. In short, DePalma and colleagues claim to have discovered an actual record of what happened on the day the Cretaceous ended. Most importantly, for the debate over whether the Deccan eruptions were solely responsible for the end-Cretaceous extinctions, the expedition leader, Robert DePalma, *claims* to have found fossil remains of a Triceratops and Hadrosaur, as well as miscellaneous dinosaur feathers, mixed in with the debris. While this discovery was reported in a popular article in *The New Yorker* (Preston 2019) it has not (to the consternation of many paleontologists) yet appeared in a peer reviewed science journal; DePalma and colleagues' 2019 PNAS article did not mention dinosaur remains. If DePalma's claim about finding dinosaur remains can be substantiated to the satisfaction of the paleontological community, then support for the meteorite-impact hypothesis will be greatly enhanced vis-à-vis version #1 of the volcanism hypothesis.

DePalma's alleged findings of dinosaur remains would not, however, rule out the newer meteorite-volcanism hypotheses. According to the first of these hypotheses (Section 10.2), the Deccan eruptions severely compromised the end-Cretaceous flora and fauna, and the Chicxulub bolide finished them off. Advocates of this hypothesis cite evidence from the fossil record of two distinct extinction pulses, one earlier than the impact and another shortly after the impact (Petersen, Dutton, & Lohmann 2016). It is possible that the dinosaurs survived the first extinction pulse and succumbed to the second. The second version of the meteorite-volcanism hypothesis (Section 10.3) is based on evidence that the Deccan eruptions accelerated immediately after the Chicxulub bolide hit, perhaps as a result of the impact, and hence that the final blow to Cretaceous flora and fauna was delivered by massive flood volcanism (Sprain et al. 2019). If the latter hypothesis is correct, the dinosaurs may have survived the earlier Deccan eruptions and suc-cumbed to the combined one-two punch of a powerful bolide impact immediately followed by massive Deccan eruptions. More field studies on the tempo of the Deccan eruptions need to be done, however, to convince the geological community that either of the combined meteorite-volcanism hypotheses provide a more compelling explanation for the end-Cretaceous extinctions than the officially sanctioned meteorite-impact hypotheses.

The back-and-forth history of the debate over whether Deccan volcan-ism and/or the Chicxulub meteorite is responsible for the end-Cretaceous mass extinction illustrates important characteristics of historical scien-tific research: Conclusive proof and disproof is not any more possible in historical science than it is in classical experimental science. One can never rule out discovering that what initially seemed to be a smoking gun for a historical hypothesis (e.g., lack of dinosaur remains close to the K-Pg boundary and evidence that the Deccan volcanism was over by the time of the Chicxulub impact) is not as conclusive as originally supposed. Similarly, one can never rule out discovering new traces (e.g., evidence that the Deccan volcanism may have intensified after the Chicxulub impact and dinosaur remains found among debris associated with the impact) which, when combined with the total body of evidence available, provide compelling evidence in support of a formerly dismissed hypothesis (#1) or, alternatively, suggest new hypotheses (#2 or #3).

10.5 The Last Universal Common Ancestor (LUCA) and the Phylogenetic Tree of Life

As a somewhat different illustration of historical reasoning in the biological sciences, consider the widely accepted hypothesis that all life on Earth descends from a last universal common ancestor. Prior to the advent of Darwin's theory of evolution by natural selection, biologists were mesmerized by the astounding morphological diversity of life on Earth. This diversity, as illustrated by diatoms, jelly fish, mushrooms, mosses, red wood trees, butterflies, snakes, and elephants, is so extreme that they failed to recognize the significance of telling similarities suggestive of common ancestry. Darwin recognized the import of these similarities, however, and formulated a naturalistic theory that explained them in terms of common ancestry. Most importantly, his theory of evolution by natural selection also explained differences among organisms – differences previously extolled as evidence that each species was created separately by an intelligent designer – in terms of evolutionary adaptations to changing environments. That Darwin's theory could explain both similarities and differences among diverse groups of living and extinct (fossilized) organisms in this manner raised an intriguing possibility: Perhaps all life on Earth shares a common origin. In a private letter to his friend and colleague Joseph Hooker, Darwin entertained this possibility, speculating that life on Earth may have originated in a "warm little pond."

It wasn't until the twentieth century, however, that compelling *empirical* evidence for the existence of a Last Universal Common Ancestor (LUCA) emerged. Evidence for LUCA was not, as in the case of the debate over what caused the end-Cretaceous mass extinction, acquired through field studies. It was acquired through laboratory analyses of the biomolecules (especially, proteins and nucleic acids) of contemporary organisms. Nonetheless, as discussed later, the evidential reasoning involved is just as historical as that of paleontologists exploring what caused the end-Cretaceous mass extinction.

The structural, enzymatic, and hereditary material of all known life on Earth, from bacteria to redwood trees to human beings, is synthesized from the very same molecular building blocks (amino acids, nucleotide bases, phosphates, sugars, and lipids). Proteins, which supply the bulk of the enzymatic and structural material for Earth life, are synthesized from the same approximately 20 amino acids, despite the fact that over a hundred amino acids occur in nature. Moreover, biochemists have synthesized nonstandard proteins from suites of abiotic amino acids that would be functional in the right organismal environments. Yet no known organism on Earth uses them. Nucleic acids (DNA and RNA), which supply the hereditary material for life on Earth, display equally striking similarities, using the same nucleobases, genetic code, and coding scheme. Biochemists (e.g., Yang et al. 2011) have artificially synthesized nonstandard nucleic acids from alternative suites of building blocks (nucleobases) using alternative coding schemes (e.g., codons consisting of quadruplet vs. triplet sequences of bases) and assignments of amino acids to codons. This indicates that familiar life could have been different at the molecular level. Accordingly, the best current explanation for the similarities among the basic biomolecules of contemporary Earth life is that all life on Earth shares a last universal common ancestor from which it inherited these similarities.

Viewed from a historical perspective, the proteins and nucleic acids found in contemporary organisms represent molecular fossils, that is, remnants (traces) of breeding events passed down to us from our ancestors. That these biomolecules are so similar in molecular composition and architecture and, most importantly, could have been different in some of the ways mentioned previously, without affecting their biofunctionality, provides strong support for the LUCA hypothesis. Had laboratory investigations established that the proteins and nucleic acids of organisms as dissimilar as bacteria, mushrooms, elephants, and redwood trees do not closely resemble each other at the molecular level, the hypothesis that life on Earth shares a common ancestor would have been undermined.[7]

---

[7] Indeed, should scientists create novel microorganisms using nonstandard biomolecules with alternative genetic codes, and perhaps different coding schemes, and should these microorganisms escape into the natural environment and survive and reproduce, alien biologists arriving on our planet, long after humans have become extinct, will have compelling empirical evidence that not all life on Earth descends from a last universal common ancestor!

The famed phylogenetic tree of life, which is rooted in LUCA, provides a salient illustration of the impact of the asymmetry of overdetermination on the evidential reasoning of biologists. It is far easier to construct a tree of life for past life on Earth, by comparing and contrasting genomic sequences found in extant organisms, than it is to construct one for future life on Earth. The genomes of extant organisms do not contain traces of future breeding events. One can speculate about the descendants of humans and other animals under specific assumptions about how changes in climate will affect various terrestrial and oceanic environments, but one cannot be sure that these changes will occur in the manner supposed, or that they won't precipitate other changes that have yet to be taken into consideration. Which organisms will survive the current anthropocentric extinction event and how will they change? Will frogs and bees survive or disappear like the ammonites? Will human beings become extinct in the near future or evolve in such a way as to adapt to much warmer climatic conditions, or will melting of Greenland's vast continental ice sheets precipitate a new ice age, resulting in a different evolutionary trajectory for our species? It is not an accident that the phylogenetic tree of life taught to biology students extends deeply into Earth's past and ends abruptly in the present.

10.6 The Discovery of Ribozymes

The scientific research discussed in Sections 10.4 and 10.5 illustrate ways in which scientists exploit the asymmetry of overdetermination in reconstructing particular episodes from the history of life on Earth. The other side of the asymmetry of overdetermination (the underdetermination of the future by the localized present) sheds light on aspects of classical experimental science that are problematic insofar as they fail to conform to the much lauded (but as discussed in Section 10.2, logically problematic) "scientific method" of legend. More specifically, because the present does not contain records of future occurrences, scientists are faced with a serious problem when it comes to interpreting the evidential significance of successful and failed experimental tests of predictions. The test condition specified by a hypothesized generalization is rarely if ever the total cause of the predicted effect; as discussed in Section 10.2, there is always the threat of poorly understood or unknown interfering factors that could produce a misleading result. As a consequence, it is difficult to determine whether a successful prediction represents a false positive or a true positive, and most importantly, given the widespread veneration of Popperian falsificationism, whether a failed prediction represents a false negative or a true negative.

Tom Cech's serendipitous discovery of ribozymes (catalytic RNA molecules) provides a salient illustration of the difficulties faced by experimentalists in determining whether a failed prediction represents a false negative or a true negative; for more detail, see Cech 1989. Cech and his team were investigating the (then newly discovered) phenomenon of RNA cutting and splicing, which occurs when an RNA molecule is being assembled from a DNA template. This process involves removing ("cutting") pieces of precursor RNA and attaching ("splicing") the ends back together to form a functional RNA molecule. Cech assumed, in keeping with the so-called Central Dogma of Molecular Biology,[8] that a protein enzyme was catalyzing the splicing. His goal was to isolate it by adding protein enzymes to a purified sample of precursor RNA until they found one that spliced it. Cech's team artificially synthesized precursor RNA from genes for ribosomal RNA acquired from a ciliated protozoan. To their surprise they observed splicing occurring in the supposedly protein-free preparation. Instead of rejecting the hypothesis that splicing is catalyzed by a protein, however, they interpreted the results of their experiment as a false negative, and pursued various strategies – boiling the rRNA preparation in detergents at high temperatures and treating it with several nonspecific proteases, which break peptide bonds – for removing the protein that they thought must be interfering with their experiment.

As biologists now know, Cech's conjecture that a protein enzyme was responsible for the premature splicing reaction was false. In violation of the Central Dogma, the rRNA was splicing itself. In other words, Cech was dealing with a true negative. For purposes of this discussion, however, the important point is that it was perfectly reasonable for Cech and his team to treat the initial results of their experiment as a false negative. Up until the time of their research, the preponderance of evidence available indicated that nucleic acids could not function as catalysts,

---

[8] There are different versions of "the" Central Dogma of Molecular Biology. The one that Cech's discovery violated held that nucleic acids never function as enzymes; this version is attributed to Francis Crick (1958).

and indeed, this is typically the case. Most naturally occurring nucleic acids are not catalytically active. What Cech and his team discovered is an exception (and it is not the only one) to the generalization that nucleic acids never function as biological catalysts.[9] The important point for purposes of this chapter, however, is the challenge faced by Cech in evaluating whether the results of his experiment represented a false negative or a true negative. Cech's quandary was a consequence of the underdetermination of predicted (future) occurrences by experimental conditions; one can never eliminate the possibility that an unknown or poorly understood factor affected the result of an experiment. This helps to explain why experimentalists do not follow Popper's recommendation and immediately reject their hypotheses in the face of failed predictions. Like Cech, they conduct additional experiments – controlling for theoretically and empirically plausible interfering conditions that could be responsible for a false negative[10] – before concluding that they are dealing with a true negative.

10.7 Conclusion

In sum, the practices of historical scientists are not inferior to those of experimental scientists: Each practice is tacitly designed to exploit the information that nature makes available to it. The research practices of historical scientists are designed to exploit the overdetermination of the past by traces available in the localized present. The asymmetry of overdeterminations makes it highly likely that a "smoking gun" for the "correct" historical hypothesis (assuming, of course, that it is among those under consideration) exists somewhere in the messy,

---

[9] Viewed in this light, it is more accurate to think of experimental scientists as testing the *robustness* – rather than, as portrayed in traditional accounts of scientific methodology, falsity or probable truth – of their hypothesized generalizations in the face of contingency. The robustness of a generalization is its degree of insensitivity to varying conditions. The most robust generalizations persist in the face of changing circumstances, or as philosopher of biology Sandra Mitchell characterizes it, exhibit the greatest "stability over changes in context" (Mitchell 2002, p. 334). Developing an account of the methodology of experimental science along these lines is, however, beyond the scope of this discussion.

[10] What counts as a *plausible* interfering condition depends upon the current state of scientific knowledge about the phenomenon being investigated. Due to their commitment to the Central Dogma, the possibility of self-catalytic RNA was highly implausible for Cech and his team.

uncontrollable world of nature, just waiting to be discovered. Classical experimentalists, on the other hand, must contend with the other side of the asymmetry of overdetermination, the underdetermination of the future by the localized present. The test condition inferred from a hypothesized generalization is never the total cause of an experimental result; there is always the threat of interfering conditions producing misleading results. This helps to explain why experimental scientists never draw conclusions based on a single experimental result. Instead, they perform a series of experiments manipulating ("controlling for") the most plausible (given the current state of their knowledge) conditions that could be responsible for a false positive or a false negative (Cleland 2002).

Historical scientists also face the threat of false positives and false negatives. The way in which they deal with this threat is significantly different, however, from the way in which (classical) experimental scientists deal with it. This difference in methodology reflects their differing evidential predicaments vis-à-vis the asymmetry of overdetermination. Historical scientists cannot affect the past in the manner in which experimentalists affect the future, namely, by manipulating localized events and processes through experimentation.[11] Derek Turner (2007) contends that this represents a serious methodological defect of historical science. But as I argued (Cleland 2011, 2013), Turner downplays the extent to which diverse bodies of traces available in the present overdetermine their past causes, providing a rich source of information about the past for evaluating the threat of potentially misleading evidential traces.

Like experimentalists, historical scientists don't draw conclusions based on a single line of suggestive evidence; it could represent a false positive. Acceptance of the meteorite impact hypothesis for the end-Cretaceous mass extinction required much more than Walter Alvarez's

---

[11] Historical scientists do perform "proxy experiments" using computer models and simulations. But as discussed in Cleland and Brindell (2013), these "experiments" are significantly different from the physical experiments performed by experimental scientists; they cannot reveal poorly understood and unknown interfering conditions for the simple reason that these conditions are not (being poorly understood or unknown) incorporated into the model. In contrast, such conditions are present, albeit unbeknownst to a researcher, in an experimental situation, which significantly increases the likelihood that (as in Cech's and his team's discovery of ribozymes) they will eventually be recognized for what they represent.

famous discovery of an iridium anomaly in K-Pg boundary sediments near Gubbio, Italy. The anomaly might have represented a local phenomenon, a minor impact event, or poorly understood geological processes that (in a manner analogous to how gold placer deposits form) concentrated iridium from other places. Discovering an iridium anomaly in K-Pg boundary sediments found around the world, on the other hand, provided strong evidence that the impactor was enormous, and hence potentially catastrophic for life. The discovery of extensive deposits of impact debris, such as shocked minerals and glassy spherules, associated with the iridium anomaly in K-Pg boundary sediments from around the world, provided even stronger evidence of a meteorite impact capable of causing a mass extinction. But even this was not enough to support the hypothesis that the impact caused the mass extinction. Evidence of impact needed to be coupled more closely with evidence of a mass extinction. The discovery of extensive changes in the fossil records of a wide variety of organisms, immediately below and above K-Pg boundary sediments found around the world, provided very strong support for the hypothesis that the end-Cretaceous mass extinction was triggered by a meteorite impact.

Like experimental scientists, historical scientists also face the threat of false negatives. As discussed in Section 10.4, failure to find fossilized dinosaur bones within 3 meters of the lower edge of the K-Pg boundary is not, as some early proponents of the volcanism hypotheses claimed, compelling evidence that the end-Cretaceous mass extinction preceded the Chicxulub bolide impact; the absence of dinosaur remains could represent a false negative. This explains why much of the paleontological research into the end-Cretaceous mass extinction focused on small, highly abundant, biological entities (forams, ammonites, pollen, etc.), which being far more numerous than large organisms, such as dinosaurs, have a much greater likelihood of their remains being preserved in close proximity to K-Pg boundary sediments. On the other hand, Peter Ward's failure to find small ammonite fossils immediately below the K-Pg boundary at his favorite field site in Northern Spain did not, as he initially claimed, "falsify" the meteorite impact hypothesis; see Powell (1998, pp. 146–149) for more details. Ward subsequently discovered an abundance of ammonite fossils close to the lower edge of the K-Pg boundary a few miles to the north, in southern France. The ammonites in northern Spain appear to have been decimated before the

K-Pg impactor slammed into Earth by a local ecological problem that didn't extend very far northward. In short, the absence of ammonite fossils in northern Spain represented a false negative.

Finally, it is a mistake to conclude that the diverse and extensive body of evidence currently available for the meteorite impact hypothesis *proves* that the end-Cretaceous mass extinctions were triggered by the impact of a massive meteorite. For as discussed in Section 10.4, new, somewhat controversial, field studies suggest that Deccan flood volcanism may have played a major role in the end-Cretaceous mass extinction after all. The point is (as discussed in Sections 10.2 and 10.4) there is no conclusive proof in either historical or experimental science. What is important, for purposes of this discussion, however, is that the differing ways in which historical scientists and classical experimentalists deal with the ever present threat of potentially misleading empirical findings (false positives and false negatives) reflect the asymmetric character of their evidential situations vis-à-vis the asymmetry of overdetermination.

References

Cech, T. R. (1989). "Self-Splicing and Enzymatic Activity of an Intervening Sequence RNA from Tetrahymena" – Nobel Lecture, NobelPrize.org. *Nobel Media AB 2018.* www.nobelprize.org/prizes/chemistry/1989/cech/lecture

Cleland, C. E. (2001). Historical Science, Experimental Science, and the Scientific Method. *Geology* 29(1): 987–990.

Cleland, C. E. (2002). Methodological and Epistemic Differences between Historical Science and Experimental Science. *Philosophy of Science* 69(3): 474–496.

Cleland, C. E. (2011). Prediction and Explanation in Historical Natural Science. *British Journal of Philosophy of Science* 62(3): 551–582.

Cleland, C. E. (2013). "Common Cause Explanation and the Search for a Smoking Gun." In V. Baker (ed.), *125th Anniversary Volume of the Geological Society of American: Rethinking the Fabric of Geology*, pp. 1–9.

Cleland, C. E. & Brindell, S. (2013). Science and the Messy, Uncontrollable World of Nature. In M. Pigliucci & M. Boudry (eds.), *Reconsidering the Demarcation Problem: Philosophy of Pseudoscience*, pp. 183–202. Chicago and London: University of Chicago Press.

Crick, F. (1958). On Protein Synthesis. *Symp. Soc. Exp. Biol.* 12: 138–167.

Currie, A. (2018). *Rock, Bone, and Ruin: An Optimist's Guide to the Historical Sciences*. Cambridge, MA: The MIT Press.

DePalma, R. A., Smit, J., Burnham, D. A., Kuiper, K., Manning, P. L., Oleinik, A., Larson, P., Maurrasse, F. J., Vellekoop, J., Richards, M. A., Gurche, L., & Alvarez,

W. (2019). A Seismically Induced Onshore Surge Deposit at the KPg Boundary, North Dakota. *PNAS* 116(17): 8190–8199.

Franklin, A. (1990). *Experiment: Right or Wrong.* Cambridge: Cambridge University Press.

Gee, H. (1999). *In Search of Deep Time.* New York: The Free Press.

Hacking, I. (1983). *Representing and Intervening: Introductory Topics in the Philosophy of Natural Science.* Cambridge: Cambridge University Press.

Hempel, C. (1945). Studies in the Logic of Confirmaton. *Mind* 54(213): 1–26.

Henderson, L. (2019). "The Problem of Induction." In E. N. Zalta, ed., (spring 2019 edition), *The Stanford Encyclopedia of Philosophy.* https://plato.stanford.edu/archives/spr2019/entries/induction-problem

Lewis, D. (1991). Counterfactual Dependence and Time's Arrow. In F. Jackson, (ed.), *Conditionals,* pp. 46–75. Oxford: Oxford University Press,

Mitchell, S. D. (2002). Ceteris Paribus – An Inadequate Representation for Biological Contingency. *Erkenntnis* 57(3): 329–350.

Petersen, S. V., Dutton, A., & Lohmann, K. C. (2016). End-Cretaceous Extinction in Antarctica Linked to Both Deccan Volcanism and Meteorite Impact via Climate Change. *Nat. Commun* 7: 12079. doi: 10.1038/ncomms12079

Popper, K. (1963). *Conjectures and Refutations.* London: Routledge & Kegan Paul.

Powell, J. L. (1998). *Night Comes to the Cretaceous.* San Diego, New York, and London: Harcourt Brace & Company.

Preston, D. (2019). "The Day the Dinosaurs Died." *The New Yorker,* April issue. www.newyorker.com/magazine/2019/04/08/the-day-the-dinosaurs-died

Punekar, J., Mateo, P., & Keller, G. (2014). "Effects of Deccan Volcanism on Paleoenvironment and Planktic Foraminifera: A Global Survey." *The Geological Society of American,* Special Paper 505.

Raup, D. M. & Sepkoski, J. J. Jr. (1984). Periodicity of Extinctions in the Geologic Past. *PNAS* 81(3): 801–805.

Sakamoto, M., Benon, M. J., & Venditti, C. (2016). Dinosaurs in Decline Tens of Millions of Years before Their Final Extinction. *PNAS* 113(18): 5036–5040.

Schoene, B. Eddy, M. P., Samperton, K. M., Keller, C. B., Keller, G., Adatte, T., & Khadri, S. F. R. (2019). U-Pb Constraints on Pulsed Eruption of the Deccan Traps across the End-Cretaceous Mass Extinction. *Science* 363(6429): 862–866.

Schulte, P., Alegret, L., Arenillas, I., Arz J. A., Barton, P. J., Brown, P. R., Bralower, T. J., Christeson, G. L, Claeys, P., Cockell, C. S., Collins, G. S., Deutsch, A., Goldin, T. J. Goto K., Grajales-Nishimura, J. M., Grieve, R. A, Gulick S. P., Johnson, K. R., Kiessling, W., Koeberl, C., Kring, D. A., MacLeod, K. G., Matsui, T., Melosh, J., Montanari, A., Morgan, J. V., Neal, C. R., Nichols, D. J., Norris, R. D., Pierazzo, E., Ravizza, G., Rebolledo-Vieyra, M., Reimold, W. U., Robin E., Salge, T., Speijer, R. P., Sweet, A. R., Urrutia-Fucugauchi, J., Vajda, V., Whalen, M. T., & Willumsen, P. S. (2010). The Chicxulub Asteroid Impact and Mass Extinction at the Cretaceous-Paleogene Boundary. *Science* 327(5970): 1214–1218.

Smithwick, F. M., Nicholls, R., Cuthill, I. C., & Venther, J. (2017). Countershading and Stripes in the Theropod Dinosaur *Sinosauropteryx* Reveal Heterogeneous

Habitats in the Early Cretaceous Jehol Biota. *Current Biology* 27(21): 3337–3342. e2. doi: https://doi.org/10.1016/j.cub.2017.09.032

Sober, E. (1988). *Reconstructing the Past*. Cambridge, MA: The MIT Press.

Sprain, C. J., Renne, P. R., Anderkluysen, L., Pande, K., Self, S., & Mittal, T. (2019). The Eruptive Tempo of Deccan Volcanism in Relation to the Cretaceous-Paleogene Boundary. *Science* 363(6429): 866–870.

Steinle, F. (1997). New Fields: Exploratory Uses of Experiment. *Philosophy of Science* 64 (Supplement): S65–S74.

Steinle, F. (2002). Experiments in History and Philosophy of Science. *Perspectives on Science* 10(4): 408–432.

Turner, D. (2007). *Making Prehistory: Historical Science and the Scientific Realism Debate*. Cambridge: Cambridge University Press.

Yang, Z., F. Chen, J. B. Alvarado, & S. A. Benner (2011). Amplification, Mutation, and Sequencing of a Six Letter Synthetic Genetic System. *J. Am. Chem. Soc.* 133(38): 15105–15112.

11 What Is the Basis of Biological Classification?

The Search for Natural Systems

THOMAS A. C. REYDON

11.1 Dealing with Diversity Requires Classifying Diversity

Life on Earth is astonishingly diverse. Biologists have so far discovered, named, and described approximately 1.9 million species of organisms (Ruggiero et al. 2015). But on most counts, this is only a very small part of the actual diversity of life on our planet. According to one estimate, there currently exist around 8.7 million (give or take 1.3 million) species of eukaryotic organisms (Mora et al. 2011) – not even counting bacteria and other prokaryotes, which make up by far the largest part of the biomass. Other estimates put the total number of extant species anywhere between 2 million and 1 trillion, with a recent rough estimate being 1–6 billion species (Larsen et al. 2017). But life on Earth is not only extremely diverse when it comes to organismal diversity. Biologists also talk about biodiversity in terms of genetic diversity (e.g., the genetic variation within a particular species, such as the variation of alleles within the human genome, or the genetic variation that exists in a particular local population), ecosystem diversity (the variety of ecosystems found in a particular geographic region, or on the planet as a whole), and the variety of trophic groups in a particular region, as well as other types of diversity (Faith 2016).

So, how diverse *is* life on Earth? The short answer is that we don't know. A longer answer is that several unresolved issues complicate the search for good estimates of the diversity of life on Earth. One aspect of the problem is practical: Large parts of our planet have not yet been explored to a sufficient extent, and much more field work is still needed to "map out the biosphere" (Wheeler et al. 2012). And while the whole genomes of species are mapped out at an increasing speed, still only a diminishingly small number of species' genomes and their allelic diversity are known. Solving these practical matters will still require a lot of time and resources. But the task also involves a number of philosophical and conceptual questions that have proven very hard to solve.[1] These issues pertain to some of the central concepts and assumptions in the study of the diversity of life, including the concept of biodiversity itself. Finding at least approximate solutions to them is necessary for measures of diversity to be meaningful.

In this chapter, I focus on the philosophical and conceptual questions related to classificatory practices in biology. Arguably, these are among the most important issues when it comes to dealing with biodiversity. The question of what the diversity of life on Earth is, is primarily a question about the *diversity of kinds* of biological entities that exist. What we want to know is not so much how individual living beings differ from each other, but rather how many species there are in a particular taxon, how many kinds of genes and alleles there are in the genome of a particular species, how many kinds of ecosystems there are in a particular geographical region, and so on. In addition, we want to know how these kinds of biological entities are best characterized: What are the traits that the members of a kind typically exhibit, and in what ways do the members of one kind differ from the members of other kinds? This means that before we can even begin to attempt to estimate diversity, we must find suitable systems of classification for the various biological entities – genes, proteins, cells, organisms, biomes, communities, ecosystems – that we are interested in. Classification is a crucial aspect of biological science, because dealing with life's diversity requires classifying biological entities first.

But classifying biological entities is much less straightforward than it may seem. This chapter aims to explain why classification is difficult by providing a brief introduction to some of the main issues that philosophers of biology have long been highlighting in relation to this topic. The biologists who deal with life's diversity on a day-to-day basis are of course deeply familiar with the practical problems of classification. These include the identification of those traits that are most indicative of species membership, deciding if a newly discovered population should

---

[1] Good entry points into the philosophical issues surrounding the general topic of biodiversity are provided by Maclaurin & Sterelny (2008) and Faith (2016).

be counted as a new species or as a variety of a species that is already known, and so on. But such practical problems have a specific theoretical context and history. They are connected to a persistent lack of clarity regarding what aspects of nature biological classifications are supposed to represent, as well as to debates regarding the meaning of many core terms that feature in biological classifications (such as "species" and "gene"), and to how these concepts have changed as biology developed as a science.

Practical problems in classifying organisms partly arise because it is unclear which criteria we should use to group individual organisms into species and to delimit species from one another. The same holds for questions about genetic diversity, where biologists disagree on what segments of DNA count as genes, and how these are best grouped into gene families and so on. History and philosophy of science are important here, because they illuminate the historical background and the theoretical context of many problems that working scientists are confronted with in their daily practice. This is one reason why history and philosophy of science are important for working biologists, for biology educators, and for anyone who wants to understand how biology as a science works and what the nature is of the knowledge that it produces about the living world.

In what follows, I illustrate this by examining two cases of biological classification: the classification of organisms into species and other taxa, and the classification of genes into kinds. I focus on what a natural system of classification actually is; the assumption that there is a natural order in the living world; the conceptual difficulties surrounding the classificatory concepts of "species" and "gene"; and the theory-dependence of classifications. While I do not address these four issues in great detail, and have to pass over many important matters, this will help to understand why biological classification is not an easy task, and why it has been hotly debated by biologists as well as philosophers for at least as long as biology has existed as a science.

11.2 Natural and Artificial Systems of Classification

Ever since the beginning of the systematic study of living nature, two central questions have been what kinds of living beings there are on our planet, and on what basis we should group them together into species and higher taxa. These are two philosophically distinct, but intimately connected, issues. The former issue is a *metaphysical* one, i.e., a question about what the world is like. It concerns what the living world is made up of, or what kinds of organisms constitute the "furniture of the world," as philosopher of science Mario Bunge (1977) put it. The second issue is *epistemological* and *practical* in nature. The question here is which of the numerous possible ways of grouping organisms we should choose, given that we use classifications for various theoretical and practical purposes.

One might think that there is a single best way of grouping entities: the grouping that exactly represents the various kinds of organisms that exist "out there." However, this is not the case. To see this, consider a fundamental dichotomy that is made in philosophical discussions of scientific classification between *natural* systems of classification and *artificial* systems of classification. One could say that a natural system of classification represents "the natural order of things" (whatever that may be – I address this issue later), while an artificial system of classification groups things together in a way that suits whatever purposes we might have but does not represent the natural order. Classificatory systems such as the Standard Model in particle physics, the Periodic System in chemistry, and the Linnaean System in biology are supposed to be natural systems in this sense. They tell us something about what nature is supposed to be like, independently of us. This makes scientific classifications into hypotheses or theories about the state of affairs in the natural world (Reydon 2013, p. 192; 2014, p. 127), which tell us that the world consists of, among other things, elementary particles, atoms, and organisms, and that these entities come in numerous kinds. The grouping of fruits and vegetables in the supermarket, in contrast, is an artificial system that helps us quickly find the food items that we are looking for when we shop for groceries. We usually group tomatoes, bell peppers, and chili peppers with the vegetables and not with the fruits, even though botanically they are fruits, because we typically use them in salads and savory dishes rather than in desserts or sweets. Even though this way of classifying does not adequately represent "the natural order of things," it still is extremely well suited for specific practical purposes.

While artificial systems of classification may be useful for some purposes, in science the focus is usually on natural systems of classification.[2] This is not only because an aim of science is to find out what the world consists of, but also because science aims toward general knowledge. Not many people would be interested in studying, say, fruit flies if the knowledge that we obtain applies only to *that* particular sample of fruit flies in *this* particular laboratory at *that* particular time. Researchers are looking for knowledge that applies more broadly, such as to all organisms of the species *Drosophila melanogaster*. An important aim of science is to produce knowledge that, once obtained by studying a sample of a kind, can be extrapolated to all other members of the same kind. The practical/epistemological aspect of classification and the metaphysical aspect of classification connect in this aim of producing general knowledge.

Being able to produce general knowledge that applies to all or most members of a kind on the basis of the study of a limited number of members of the kind is a practical aim (scientists need to know that their predecessors actually worked on the same kind of organisms as they themselves are studying), as well as an epistemological aim (the aim is the production of well-supported and confirmed general knowledge about the world). But underlying the practical/epistemological aspect is a metaphysical aspect: For knowledge to be generalizable from a limited sample to all past, current, and future members of the kind, the kind must in some way or another be grounded in nature (Reydon 2014). That is, there must be a cause in nature that explains why inferences from a limited subset of a kind's members to all its members hold, and that cause is often assumed to be found in the natural state of affairs. The assumption here is that naturally occurring entities of the same kind are very similar to each other, for example because they share a set of innermost properties that explain typical observable properties and behaviors, or because they are produced by the same ongoing natural process, or because they are members of the same line of ancestry in which properties are passed on from one generation to the next, or because of some other factor in nature.

To be sure, this assumption of kinds being grounded in nature, and about the groupings in a natural system tracking natural factors, is quite vague and in need of clarification. Indeed, a long-standing tradition in

---

[2] For more extensive discussion and philosophical background, see Reydon 2013, p. 193ff.; 2014, p. 127ff.

philosophy more broadly is concerned with this issue under the keyword "natural kinds." It is impossible to go into detail about the various views that philosophers have advanced about what natural factors underlie natural kinds.[3] For the purposes of the present chapter it is sufficient to know that the guiding idea is that a *natural classification must track the natural factors that make the members of the kind similar in many of their properties*, such that the classification supports generalizations about the typical properties and behaviors of the members of a group. The metaphysical project, then, is to clarify what this supposed natural basis of classifications is in the various areas of science.

The question of what constitutes the basis for a natural system has proven difficult to answer. The history of biology shows that there are diverging ways to think about what a natural system is supposed to represent. Simply searching for a classification that exactly represents *the* natural order of things is misguided because there is no natural order of things in any straightforward sense – at least not in the living world. Living beings and other biological entities are similar and different from each other in so many ways that it appears impossible to identify one single natural way of ordering them. This is not an uncontroversial claim, however, and we can understand it better by examining the thinking of three major figures in the history of biology: Aristotle, Linnaeus, and Darwin. These three authors held considerably different views on how a natural system of classification of organisms should be constructed and what it is supposed to represent.

11.3 Aristotle and Linnaeus on the Natural Order of Things

In his *History of Animals*, Aristotle grouped animals on the basis of various inherent properties, such as whether they possessed a circulatory system or not ("blooded" or "bloodless"), how they moved (e.g., "terrestrial," "aquatic," "flying"), how they reproduced (e.g., "viviparous," "oviparous," "larviparous"), and so on. This reflected one of the central elements in Aristotle's metaphysics, namely the notion of form: On his view, every existing entity manifests a particular form that determines what it *is* – one could say that every thing is matter that

---

[3] Good entry points into the literature on natural kinds are Koslicki (2008); Hawley & Bird (2011); Reydon (2014); Kendig (2016); and Bird & Tobin (2018).

has assumed a particular form (for more details, see Ainsworth 2016). Accordingly, Aristotle's way of grouping animals highlighted the various forms of animals that existed – for example, "terrestrial and viviparous" – and in this way identified the shared nature of the animals that belonged to the same group, while at the same time explaining their similarities by means of this shared nature.

Even though Aristotle's way of grouping was intended to represent important aspects of nature (i.e., forms), and in this sense could be thought of as a natural system of grouping, it has been argued that it cannot be thought of as a classificatory system in a strict sense (Balme 1987, pp. 72–73; Walsh 2006, pp. 428–430). Most importantly, Aristotle's way of grouping did not allocate individual living beings to stable groups that each had their fixed place within an encompassing system. Rather, it was a tool for comparative analysis of organisms: It started with highlighting a group of organisms that were notably similar (e.g., birds) and proceeded to highlight the traits that distinguished the various forms that existed within this group from each other (flying vs. flightless birds, birds living in water vs. birds living on land, and so on). Aristotle called the group of similar organisms that constituted the starting point of the analysis the *genus*, but did not intend this to express a taxonomic level. A genus was simply a group of similar organisms within which subgroups could be identified according to stable differences.

But even though it was not a classificatory system in the strict sense, Aristotle's system can still serve as an excellent example of a system of grouping of organisms that is intended to represent *something* about nature. While Aristotle was interested in the various forms of organisms that exist, I do not think that he is best interpreted as assuming that these forms constitute a fixed and unchanging natural order. Aristotle famously held that the forms (natures) of things are inherent in the things themselves and that therefore examination of the entities that we find in the world can yield knowledge about their natures. This view does not rest on the assumption that there is a fixed set of forms that grounds the natural order of things and that individual entities instantiate. Rather, it involves the notion of form to analyze individual entities and conceives of forms in terms of the similarities and differences that happen to exist between entities.

This is different for Linnaeus, the father of the classificatory system that is by and large still in use in contemporary biology, who *did* believe in a natural order. Linnaeus' system allocated individual living beings to stable groups, each of which had their fixed place within an encompassing system. It consisted of a hierarchy of nested groups, with organisms being grouped into species, species into more inclusive genera, genera into more inclusive orders, and so on, using for example the shape and arrangement of reproductive organs as the foundation of his classification of plants. While Linnaeus himself only recognized five levels of groups in his classificatory system (species, genera, orders, classes, and kingdoms), contemporary biological systematics recognizes a far greater number of levels, including domains, phyla, subphyla, subclasses, families, subfamilies, subgenera, and more. Still, the basic idea of Linnaeus' classificatory system has not changed: In principle (though not necessarily in practice, as I discuss later) all organisms are exhaustively allocated to species, each organism belonging to one and only one species, and species are grouped into overarching groups that constitute more inclusive kinds of organisms.

In a similar way as the Periodic System gives us an inventory of the various chemical elements with brief characterizations of their main properties as well as an ordering of the elements into more encompassing groups (such as the alkaline earth metals, the chalcogens, and the halogens), the Linnaean system as it is used today and is presented in zoos, herbaria, and natural history museums, seems to provide us with a well-ordered – albeit very incomplete – inventory of the various kinds of organisms that exist (as well as those that are now extinct). The system tells us, for example, that fruit flies come in a large number of different kinds that are very similar to each other and distinct from other kinds of flies: While most biologists are familiar with *Drosophila melanogaster*, an important model organism, the genus *Drosophila* alone encompasses around 1,600 distinct species (O'Grady & DeSalle 2018) and the biological taxon of fruit flies (family Tephritidae) encompasses around 4,700 species of fruit flies ordered into almost 500 genera.

For Linnaeus and many of his contemporaries and successors, underlying this view of classification was the assumption that the natural order that was represented by the natural system was the plan of the Christian god's creation. This view was widespread throughout the eighteenth and nineteenth centuries (Grene & Depew 2004, pp. 213–214). Linnaeus expressed it as follows:

There are as many species as the Infinite Being produced diverse forms in the beginning ... Species are as many as there were diverse and constant forms produced by the Infinite Being; which forms according to the appointed laws of generation, produced more individuals but always like themselves. Therefore there are as many species as there are diverse forms or structures occurring today. (Linnaeus, quoted in Wilkins 2009, p. 41).

A century later, Louis Agassiz, a contemporary of Darwin and founder of the Museum of Comparative Zoology at Harvard University, expressed a similar view: "Natural History must in good time become the analysis of the thoughts of the Creator of the Universe, as manifested in the animal and vegetable kingdoms, as well as in the inorganic world" (Agassiz 2004, p. 137), and "all we can really do is, at best, to offer imperfect translations into human language of the profound thoughts, the innumerable relations, the unfathomable meaning of the plan actually manifested in the natural objects themselves" (Agassiz 2004, p. 187).

The problematic aspect of Linnaeus' and Agassiz' view is the presupposition that there *is* such a thing as a natural order in the first place. What grounds do we have for this assumption? Linnaeus' and Agassiz' view that the natural system of classification should represent the Christian god's plan of creation was rooted in religious belief and does not sit well with contemporary natural science.[4] Aristotle's view seems a little closer to our everyday experiences and, as such, better suited as the foundation of scientific classification. When looking around in our everyday surroundings we immediately see all sorts of similarities and differences between natural entities – if you have seen one red-bellied woodpecker (species *Melanerpes carolinus*), it seems that you have seen them all. It thus comes natural to group entities on the basis of the similarities and differences that they exhibit. But Aristotle's views encompass a quite speculative metaphysical view of what the world is like, including the idea that individual entities embody particular forms that constitute part of their inner nature and explain their typical

---

[4] Most importantly, the assumption that the world was created by a divine being that imposed a natural order onto the world conflicts with the principle of *methodological naturalism*, according to which any appeal to the actions of supernatural beings, supernatural causes, and so on, is unacceptable in the context of science. For more on methodological naturalism, in particular in relation to current debates on creationism and evolution, see Draper 2005; Boudry, Blanke, & Braeckman 2010; or Sober 2011.

properties and behaviors (Grene & Depew 2004, ch. 1; Walsh 2006). Adopting Aristotle's views thus entails having to buy into a particular metaphysics that is controversial and of which it is questionable whether it fits contemporary science at all. The claim that a natural system of classification represents the natural order of things implies the assumption that there is such a natural order – an assumption that neither Aristotle's nor Linnaeus' and Agassiz' views are able to support sufficiently.

11.4 Darwin's Alternative and the Question of Species

Darwin's views provided an alternative to the views of Aristotle, Linnaeus, and Agassiz. Writing about a century after Linnaeus, but publishing in the same year as Agassiz, Darwin criticized Linnaeus (and implicitly many biologists of both Linneaus' and Darwin's own time) for not being clear about what exactly the foundation of the natural system is supposed to be. In a famous passage in *The Origin of Species*, that summarizes Darwin's view so clearly that it warrants being quoted at length, he put it as follows:

> Naturalists try to arrange the species, genera, and families in each class, on what is called the Natural System. But what is meant by this system? Some authors look at it as merely a scheme for arranging together those living objects which are most alike ... But many naturalists think that something more is meant by the Natural System; they believe that it reveals the plan of the Creator; but unless it be specified whether order in time or space, or what else is meant by the plan of the Creator, it seems to me that nothing is thus added to our knowledge. Such expressions as that famous one of Linnaeus ... that the characters do not make the genus, but that the genus gives the characters, seem to imply that something more is included in our classifications, than mere resemblance. I believe that something more is included; and that propinquity of descent, – the only known cause of the similarity of organic beings, – is the bond, hidden as it is by various degrees of modification, which is partially revealed to us by our classifications. (Darwin 1859, pp. 413–414)

Here, Darwin credited Linnaeus for suggesting that a natural system of classification of living beings should involve more than the mere grouping of organisms on the basis of their resemblances and differences.[5] But, as Darwin rightly pointed out, merely referring to the Creator's plan without explicating what it encompasses won't do. Instead, Darwin suggested that common descent should be taken as the basis for the natural system: What makes living beings belong together in the same species, genus, or higher taxon is that they all are descended from the same ancestral population.

Darwin's view involves a rejection of the idea that there is a natural order of things – at least in the strong sense of Linnaeus. Darwin did not assume that the various organisms that existed were members of fixed kinds that constituted a stable natural order. On the contrary, he presented us with a picture of the living world in the shape of a continuous tree-like structure in which all living beings are connected by ancestry, starting from the very first life forms on Earth. The *Origin of Species* included precisely one illustration, which showed this tree-like structure. But as the genealogical nexus is continuous, the question that arises in relation to the classification of organisms is how this continuity is best broken up into discrete groups. All we have is a very large continuous family tree, which shows that living beings share close ancestral relations with some other living beings, more distant ancestral relations with others, and very distant ancestry with still others. A first problem for Darwin's view, then, is that the continuous family tree of life on Earth does not come with any pre-given cuts between ancestral and descendant species. Dividing up a continuum into discrete groups involves decisions on where one group ends and the next begins – decisions that can be made in multiple ways, yielding different ways of breaking up the continuum. Any decision on where an ancestor species ends and its descendant species begins will therefore involve judgments on the part of researchers regarding the question of whether any evolutionary novelties in the descendant group would merit recognizing it as a new species.

This difficulty is enhanced by the ongoing disagreement among biologists and philosophers of biology on the questions of how the basic

---

[5] This is the meaning of the passage, "the characters do not make the genus, but ... the genus gives the characters" (Darwin 1859, p. 413): Groupings of organisms such as genera are not *defined* by organismal traits, but traits can be used to *diagnose* groupings that are defined on a different basis ("something more").

groups in the classification of organisms – species – are to be delimited and what they are. Species occupy a central position in biology, as much of the knowledge that is produced in biological science is expressed in relation to species; as biologist Ernst Mayr (2004, p. 171) wrote, "[m]ost research in evolutionary biology, ecology, behavioral biology, and almost any other branch of biology deals with species." This centrality of the concept of species notwithstanding, biologists and philosophers of biology have been unable to reach agreement on what species are. Central – and unanswered – questions in what has come to be known as the *species problem* include:[6] On what basis do individual organisms belong to their species? (This is the *grouping problem*.) Are species real entities in nature (as many biologists assume), or are they merely instruments for devising a stable classification of the living world? What makes a particular grouping of organisms a *species*, rather than, say, a genus or a variety? (This is the *ranking problem*.) Do the groups that biologists recognize as species, and give binomial names, have anything in common that groups at other taxonomic levels lack? If not, then why do we consider the various recognized species of birds, insects, flowering plants, fungi, microbes, etc., to be *species*, i.e., groups that are all located at the same taxonomic level and thus should in some sense be the same kind of groups?

Darwin did not think that there was anything special about species. As he put it in the *Origin*:

> I look at the term species, as one arbitrarily given for the sake of convenience to a set of individuals closely resembling each other, and ... it does not essentially differ from the term variety, which is given to less distinct and more fluctuating forms. The term variety, again, in comparison with mere individual differences, is also applied arbitrarily, and for mere convenience sake. (Darwin 1859, p. 52)

This view largely avoids many of the questions highlighted earlier, as organisms can be grouped in numerous ways on the basis of their similarities, and groups are ranked as species on arbitrary grounds. On Darwin's account, grouping organisms into species and higher taxa would thus yield an artificial system of classification. However, many

---

[6] For entries into the debates, see Reydon 2004, 2005, 2013; Richards 2010; Reydon & Kunz (2019); and Zachos 2016.

biologists disagree with Darwin's view and hold that species groupings are real, or at least non-arbitrary. That is, there is one uniquely correct way of dividing up the genealogical continuum into species, and the challenge is to determine on what basis this should be done. Underlying this challenge lies the question, how we should conceive of species: What exactly *are* species and what holds organisms together in their species (Reydon 2013, pp. 209–210)?

A surprisingly large number of responses to this question – so-called species concepts, or rather definitions of the scientific term "species" – have been put forward throughout the history of biology, and at present it is far from clear which of these – if any – would be the correct answer. In the literature, around 30 distinct species concepts are being discussed (Mayden 1997; Wheeler & Meier 2000; Ereshefsky 2001, pp. 80–92; Wilkins 2009, section 11), but the actual number of definitions that have been advanced throughout the history of biology is an order of magnitude higher (for lists of more than 100 definitions, see Lherminer & Solignac 2000; Wilkins 2009). Definitions affect how organisms are grouped into species, such that a good estimate of biodiversity at the species level will not be possible until the biological community has agreed which definition should be used (Reydon & Kunz, 2019).

As Richards (2010, p. 5) wrote, "[t]his then is *the species problem*: there are multiple, inconsistent ways to divide biodiversity into species on the basis of multiple, conflicting species concepts, without any obvious way of resolving the conflict." Why this is the case can be clarified by considering the classification of molecular entities, namely genes.

11.5 The Classification of Genes and the Theory-Dependence of Classifications

A question that is to be answered before classification is: What are the entities that we are trying to classify in the first place? Genes, for example, are not simply found in nature as discrete material objects, but exist as sections of the genomes of organisms that have to be identified and delimited before they can be classified into kinds such as *Drosophila melanogaster teashirt* or human *VKORC1* genes. But which sections of a genome are identified as genes and where the boundaries of genes are placed depends on how exactly genes are conceived of.

The notion of the gene was introduced by the Danish biologist Wilhelm Johannsen in the early twentieth century. Johannsen intended the concept to refer to a fully instrumental unit – as he put it, the term "gene" should express "the simple picture ... that a property of the developing organism is or can be (co-)determined by "something" in the gametes. No hypothesis about the nature of this "something" should be formulated or supported" (Johannsen 1909, p. 124; my translation). Johannsen emphasized that the new term should be free of any hypotheses regarding the nature of genes. For Johannsen, a gene could be any factor that co-determined an organism's phenotype and could be passed on from generation to generation in the gametes. Here, two things can be seen. First, genes are *theoretical entities*. Genomes do not come in the form of pre-delimited genes lined up as beads on a string, but genes have to be individuated and their boundaries have to be delimited on the basis of theoretical considerations. Second, genes were from the start thought of as *functionally defined entities*: Whatever parts of the gametes co-determine the traits of organisms can be identified as a gene.

This functional view of what genes are has been retained as more clarity was achieved regarding their material basis. Two main research contexts should be distinguished here: classical and molecular genetics. In the context of classical genetics, genes were not simply seen as the causes or (co-)determinants of organismal traits but rather featured indirectly in the explanation of traits in the way that *differences* between two organisms' phenotypes were explained by *differences* in their genotype. As Waters (1994, p .172) explained, "[t]he basic dogma of classical genetics was that gene differences cause phenotypic differences .... [W]hat were studied were character differences, not characters, and what explained them were differences in genes, not the genes themselves." In classical genetics, then, genes were seen as functional units that were identified by means of differences in their effects.[7] In molecular biology, genes are instead conceived of as molecular entities involved in the production of particular kinds of macromolecules. As Waters (1994, p. 178) explained: "The fundamental concept underlying the application of 'gene' in molecular biology is that of *a gene for a linear sequence in a product at some stage of genetic expression* .... Genes are for linear sequences in products of genetic expression." In biological

[7] For more details, see Waters 1994, 2007, 2017 and Reydon 2009.

practice, then, individual parts of an organism's genome are both individuated *as genes* – that is, as members of the category of genes – and individuated as members of particular kinds of genes on the basis of their functions. Entirely non-functional stretches of a genome are not thought of as genes. Briefly put, genes are "things you can do with your genome," that is, "ways in which cells utilize available template resources to create the biomolecules that are needed in a specific place at a specific time" (Griffiths & Stotz 2006, p. 500; 2013, p. 75).

At present, several gene concepts are used side by side, a situation that gives rise to a difficulty similar to the case of organisms (albeit that the number of competing gene concepts is considerably less than the number of species concepts). The way organisms are classified into species depends on the species concept that is used, without there being clarity regarding which of the many species concepts is the best. Similarly, the way genes are classified depends on the gene concept that is used, without there being clarity regarding which concept is the best. A difference is that, in the case of genes, the entities that are to be classified themselves depend on the concept that is used.

Is there, then, a natural system of classification of genes? I want to suggest that there is not. What is clearly visible in both the case of species and the case of genes is that *classification is inherently theory-dependent*. While Aristotle's view was metaphysical and Linnaeus' and Agassiz' views theological, they all were theories about what the world is like, and their views of what a natural system of classification is depended on the theories they adopted. This is also true for Darwin's view (a scientific rather than a metaphysical or theological one) that all life on Earth is connected by relations of ancestry and descent. Darwin's view of the classification of organisms as based upon common descent is thoroughly rooted in the evolutionary theory that he presented in The *Origin of Species* and has been part and parcel of evolutionary thinking ever since.

Evolutionary theory, however, does not encompass any concrete principles regarding the individuation and classification of organisms and genes, other than that genes are part of genealogical copying lines in the same way as organisms stand in lines of ancestry and descent (Reydon 2009). Most available species concepts and all available gene concepts are compatible with evolutionary theory, and evolutionary theory does not provide us with any means to select a preferred species or gene concept. Different species and gene concepts can be useful in different research contexts. For example, in zoology a species concept based on interbreeding can be useful, while such a concept does not make sense in microbiology. Similarly, different gene concepts can be useful in different research contexts (a sequence-based concept in molecular biology vs. a concept related to phenotypic traits in population biology, for instance). In the same way as the classification of organisms is faced with the problem of dividing a genealogical continuum into discrete and delimited kinds, the classification of genes faces the problem of dividing up a genealogical continuum into discrete and well-delimited kinds. In the case of genes, though, there is also the problem of dividing up continuous DNA sequences into discrete entities that are recognized as genes and can be classified into kinds. When classifying genes, the problem of dividing up a continuum into discrete units arises twice.

11.6 Why Biological Classification Is Difficult

My aim in this chapter was to highlight some of the main philosophical and conceptual difficulties that arise in relation to biological classification. Within the confines of a short chapter, it is not possible to address all important issues, and I have only examined two cases of classification.[8] Still, in these two cases the most important difficulties are clearly manifest.

Perhaps the greatest difficulty is the assumption that the classifications are natural classifications, that is, ways of grouping entities that represent the natural order of things. But in biology it is very hard to answer what this supposed natural order actually is, and how our classifications could best represent it. In particular, there is no foundation for the assumption that there actually *is* a natural order of the living world that our classifications should represent. As Darwin showed, organisms do not come neatly prepackaged into species and higher taxa, but rather are parts of a genealogical continuum. Similarly, genes do not come neatly sorted into kinds, but constitute parts of a genealogical

---

[8] An important aspect of the classification of organisms that I have for instance been forced to skip are the philosophical problems in phylogenetic systematics. For more on these issues, see Reydon 2013 or Hamilton 2014.

continuum as well as of larger continuous molecules (i.e., chromosomes). Dividing up such continua into discrete units and kinds inherently involves theoretical considerations regarding, among other things, the meanings of concepts such as "gene" and "species," and the question of what aspects of the natural world our classifications should represent.

Classification, then, is crucially theory-dependent and the answer to the question of how a natural system of classification should be constructed depends on the theory within which one works. In the biological sciences, arguably the most important theory is evolutionary theory. But evolutionary theory underdetermines biological classification in several ways, as it does not tell us how to classify either organisms into species and higher taxa, or genes into kinds. This is where philosophy of science comes into play. The search for a natural system of classification cannot be disconnected from the theoretical viewpoint from which one works. But if this viewpoint does not provide researchers with clear foundations for the classification of the entities they study, the challenge is to clear up conceptual confusions and to find an interpretation of the theory that yields solid foundations. And this has always been a job *par excellence* for philosophers of science.

References

Agassiz, L. (2004 [1859]). *Essay on Classification.* Lurie, E., (ed.). Mineola, NY: Dover Publications.

Ainsworth, T. (2016). "Form vs. Matter." In Zalta, E. N. (ed.), *The Stanford Encyclopedia of Philosophy (Spring 2016 Edition).* https://plato.stanford.edu/archives/spr2016/entries/form-matter/

Balme, D. M. (1987). Aristotle's Use of Division and Differentiae. In A. Gotthelf & J. G. Lennox, (eds.), *Philosophical Issues in Aristotle's Biology*, pp. 69–89. Cambridge: Cambridge University Press.

Bird, A. & Tobin, E. (2018): "Natural Kinds." In: Zalta, E. N. (ed.), *The Stanford Encyclopedia of Philosophy (Spring 2018 Edition).* https://plato.stanford.edu/archives/spr2018/entries/natural-kinds/

Boudry, M., Blanke, S., & Braeckman, J. (2010): How Not to Attack Intelligent Design Creationism: Philosophical Misconceptions about Methodological Naturalism. *Foundations of Science* 15(3): 227–244.

Bunge, M. (1977). *Treatise on Basic Philosophy, Vol. 3 – Ontology I: The Furniture of the World.* Dordrecht: Reidel.

Draper, P. (2005). God, Science and Naturalism. In Wainwright, W. J. (ed.), *The Oxford Handbook of Philosophy and Religion*, pp. 272–303. Oxford: Oxford University Press.

Ereshefsky, M. (2001). *The Poverty of the Linnaean Hierarchy: A Philosophical Study of Biological Taxonomy*. Cambridge: Cambridge University Press.

Faith, D. P. (2016). "Biodiversity." In Zalta, E. N. (ed.), *The Stanford Encyclopedia of Philosophy (Summer 2016 Edition)*. https://plato.stanford.edu/archives/sum2016/entries/biodiversity/

Grene, M. & Depew, D. (2004). *The Philosophy of Biology: An Episodic History*, Cambridge: Cambridge University Press.

Griffiths, P. E. & Stotz, K. (2006). Genes in the Postgenomic Era. *Theoretical Medicine and Bioethics* 27(6): 499–521.

Griffiths, P. E. & Stotz, K. (2013). *Genetics and Philosophy: An Introduction*. Cambridge: Cambridge University Press.

Hamilton, A. (ed.) (2014). *The Evolution of Phylogenetic Systematics*, Berkeley, CA: University of California Press.

Hawley, K. & Bird, A. (2011). What Are Natural Kinds? *Philosophical Perspectives* 25: 205–221.

Johannsen, W. (1909). *Elemente der exakten Erblichkeitslehre (deutsche wesentlich erweiterte Ausgabe in fünfundzwanzig Vorlesungen)*. Jena: Verlag von Gustav Fischer.

Kendig, C. E. (ed.) (2016). *Natural Kinds and Classification in Scientific Practice*. London & New York: Routledge.

Koslicki, K. (2008). Natural Kinds and Natural Kind Terms. *Philosophy Compass* 3(4): 789–802.

Larsen, B. B., Miller, E. C., Rhodes, M.K., & Wiens, J. J. (2017). Inordinate Fondness Multiplied and Redistributed: The Number of Species on Earth and the New Pie of Life. *Quarterly Review of Biology* 92(3): 229–265.

Lherminer, P. & Solignac, M. (2000). L'espèce: Définitions d'auteurs', *Comptes Rendus de l'Académie des Sciences de Paris, Sciences de la Vie* 323: 153–165.

Maclaurin, J. & Sterelny, K. (2008). *What Is Biodiversity?* Chicago & London: University of Chicago Press.

Mayden, R. L. (1997). A Hierarchy of Species Concepts: The Denouement in the Saga of the Species Problem. In M. F. Claridge, H. A. Dawah, & M. R. Wilson (eds.), *Species: The Units of Biodiversity*, pp. 381–424. London: Chapman and Hall.

Mayr, E. (2004). *What Makes Biology Unique? Considerations on the Autonomy of a Scientific Discipline*. Cambridge: Cambridge University Press.

Mora, C., Tittensor, D. P., Adl, S., Simpson, A. G. B., & Worm B. (2011). How Many Species Are There on Earth and in the Ocean? *PLoS Biology* 9: e1001127.

O'Grady, P. M. & DeSalle, R. (2018). Phylogeny of the Genus *Drosophila*. *Genetics* 209(1): 1–12.

Reydon, T. A. C. (2004). Why Does the Species Problem Still Persist? *BioEssays* 26: 300–305.

Reydon, T. A. C. (2005). On the Nature of the Species Problem and the Four Meanings of "Species." *Studies in History and Philosophy of Biological and Biomedical Sciences* 36(1): 135–158.

Reydon, T. A. C. (2009). Gene Names as Proper Names of Individuals: An Assessment. *British Journal for the Philosophy of Science* 60(2): 409–432.

Reydon, T. A. C. (2013). Classifying Life, Reconstructing History and Teaching Diversity: Philosophical Issues in the Teaching of Biological Systematics and Biodiversity. *Science & Education* 22(2): 189–220.

Reydon, T. A. C. (2014). Metaphysical and Epistemological Approaches to Developing a Theory of Artifact Kinds. In M. P. M Franssen, P. Kroes, T. A. C. Reydon, & P. E. Vermaas (eds). *Artefact Kinds: Ontology and the Human-Made World*, pp. 125–144. Dordrecht: Springer.

Reydon, T. A. C. & Kunz, W. (2019): "Species as Natural Entities, Instrumental Units, and Ranked Taxa: New Perspectives on the Grouping and Ranking Problems." *Biological Journal of the Linnean Society* 126(4): 632–636.

Richards R. A. (2010). *The Species Problem: A Philosophical Analysis*. Cambridge: Cambridge University Press.

Ruggiero, M. A., Gordon, D. P., Orrell, T. M., et al. (2015). A Higher Level Classification of All Living Organisms. *PLoS ONE* 10: e0119248.

Sober, E. (2011). Why Methodological Naturalism? In G. Auletta, M. Leclerc, & R. A. Martínez (eds.), *Biological Evolution: Facts and Theories*, pp. 359–378. Rome: Gregorian & Biblical Press.

Walsh, D. (2006). Evolutionary Essentialism. *British Journal for the Philosophy of Science* 57(2): 425–448.

Waters, C. K. (1994). Genes Made Molecular. *Philosophy of Science* 61(2): 163–185.

Waters, C. K. (2007). Causes That Make a Difference. *Journal of Philosophy* 104(11): 551–579.

Waters, C. K. (2017). No General Structure. In M. H. Slater & Z. Yudell (eds.), *Metaphysics and the Philosophy of Science: New Essays*, pp. 81–107. New York: Oxford University Press.

Wheeler, Q. D., Knapp, S., Stevenson, D. W., et al. (2012). Mapping the Biosphere: Exploring Species to Understand the Origin, Organization and Sustainability of Biodiversity. *Systematics and Biodiversity* 10(1): 1–20.

Wheeler, Q. D. & R. Meier (eds.) (2000). *Species Concepts and Phylogenetic Theory*. New York: Columbia University Press.

Wilkins, J. S. (2009). *Defining Species: A Sourcebook from Antiquity to Today*. New York: Peter Lang.

Zachos, F. (2016). *Species Concepts in Biology: Historical Development, Theoretical Foundations, and Practical Relevance*. Cham: Springer.

12 What Is the Nature of Scientific Controversies in the Biological Sciences?

MICHAEL R. DIETRICH

12.1 Introduction

Biological knowledge is not created by individuals in isolation but through a process of review and response within scientific communities. Criticism then is a normal and necessary part of this process. Occasionally, however, lasting disagreements arise during this process that become scientific controversies. In modern biology, some of the most well-known controversies have been relative significance disputes, which are disagreements about the relative importance of features of a biological system. These do not admit all-or-nothing resolutions, but instead often start as strongly stated opposing positions only to find resolution in some middle ground. In this chapter, we consider different views on how biological communities both disagree and resolve those disagreements as part of the social process of knowledge production.

Criticism is a normal part of science (Longino 1990). Indeed, it is an essential component that has been institutionalized in processes such as peer review. Occasionally, however, scientific criticisms spark disputes that become public and lasting. These disputes, extended in time, are scientific controversies (Engelhardt & Caplan 1987; Machamer, Pera, & Baltas 2000). Unlike criticisms, controversies are not normal or essential for science, but they do offer an opportunity to understand how communities of scientists interact, engage with difficult issues, and ultimately find some form of closure. They offer an opportunity to understand science as a social process of collective reasoning.

Scientific controversies are, on the one hand, often understood as disputes among scientists and are typically thought of as resolvable by careful reasoning from some set of empirical facts. Political or social controversies, on the other hand, are usually thought of as resolvable by some process of negotiation, which is not necessarily dependent on the facts (Engelhardt & Caplan 1987, p. 1). This distinction between scientific and political controversies is often blurred in scientific controversies that also have significant relevance for science policy, the sociopolitical uses of science or its technology, or scientific controversies that are intertwined with social and political controversies. The controversy over race and IQ, for instance, was as much a controversy over genetics and the measurement of intelligence as it was over the potentially discriminatory uses of IQ scores to set educational policy (Lewontin, Rose, & Kamin 1984). These kinds of controversies that blend scientific disputes and disputes over policy, ethics, or law are called mixed controversies (McMullin 1987, p. 75–77). Some controversies, such as the disputes over climate change and the safety of vaccinations, are much more social than scientific, since the consensus within the scientific community is established and the disagreement is fomented from and exists outside of those scientific communities.

The life-spans of scientific controversies raise questions regarding how they begin, how they end, and why they persist. Typically, philosophers and science-studies scholars frame controversies in terms of changes in consensus or mutual agreement. Within this framework, controversies begin as breaches of community consensus and end when consensus is again restored. This chapter is structured around the questions that arise at different points in the life-span of a controversy, focusing primarily on the thornier issues of how controversies end and how they persist.

The issues that arise at various points in the life of a controversy are illustrated by the case of the classical-balance controversy in evolutionary genetics, a dispute over genetic variation with wide ranging implications that began in the 1950s and lasted through the 1970s. This controversy was at once about the nature of genetic variation and the ways in which natural selection acted upon it, but was wrapped up in questions regarding the genetic effects of exposure to atomic radiation. Advocates of the classical position claimed that a randomly selected individual would have very little genetic variability, measured as heterozygosity or possession of different alleles at the same genetic locus, because natural selection would have been working to eliminate harmful alleles and preserve beneficial alleles. Defenders of the balance position claimed that sometimes the most beneficial genetic combination was a heterozygote combination, and that exposure to mutation-inducing radiation might create more of these beneficial heterozygotes. For its antagonists, the fate of humanity during the Cold War was expressed in this dispute over the nature of genetic variation (Beatty 1987a, 1987b; Dietrich 1994).

Following John Beatty's analysis of controversies in evolutionary genetics, such as the classical-balance controversy, controversies in biology are rarely all-or-nothing disputes; instead periods of polarization are followed by relative depolarization (Beatty 1997). Beatty referred to these kinds of disputes as relative significance controversies, and they are very common in modern biology. In a relative significance controversy, depolarization signals the end of an active controversy but does not require perfect consensus. A taxonomy of different ways that controversies end will reveal that some kinds of resolution suggested by consensus might in fact be rare in modern biology.

12.2 Controversies and How They Begin

Controversies are born from dissent. Someone disagrees with an accepted practice, theory, or standard and in doing so potentially disrupts the consensus within the scientific community. This "modification of consensus" approach, articulated by philosophers such as Philip Kitcher (2000), assumes that science functions by community agreement and that individuals can disrupt those communities and attempt to change the beliefs that they collectively agree upon. One does not have to postulate unequivocal consensus and universal shared beliefs for this view to make sense as a starting place for scientific controversy. It is enough that controversy arises from dissent from an acknowledged position within a scientific field.

That said, most dissenting opinions never blossom into controversies. Dissenting opinions, objections, and criticisms are commonplace features of science. A criticism that is relatively easily addressed will not develop into a controversy, simply because the question raised was quickly resolved. A precondition for the community to adjudicate a continuing controversy is the recognition of some merit on both sides of the issue (McMullin 1987). A criticism that the community agrees is not relevant will not spark a controversy in that it will not foster the kind of mutual engagement that a controversy requires. The issue of when an objection is judged to have merit raises the further questions of who is included within a scientific community and whether every community member has the authority to raise objections in print or otherwise (Rudwick 1994; Panofsky 2014). For controversies among members of a particular field, these issues of community membership and authority may never arise. But, if someone from outside of a particular field wants to contribute to a new field, but they have no claim to membership because they don't have the relevant training and expertise, for instance, then their contribution, positive or negative, may be downplayed as a result.

For instance, James Watson famously lacked expertise in x-ray crystallography when he embarked on his research on DNA. Rosalyn Franklin, an established expert on the technique who produced crucial evidence of the structure of DNA, held a dismissive attitude toward Watson fueled in part by Watson's sexism and in part by his ignorance of x-ray crystallography (Sayre 2009). Alternatively, consider a scientific creationist with degrees in engineering, such as Henry Morris, who found that his lack of training and expertise in evolutionary biology made it difficult to have his taken seriously by the community of evolutionary biologists, especially in the form of publication in refereed evolutionary biology journals (Kitcher 1982). That said, other forms of publication and legislation of state and local education have been used very effectively by scientific creationists to foster controversy with evolutionary biologists. In this case, the creationists' objections were determined to have merit by legislators and some members of the public, and their actions spurred biologists to respond (Kitcher 1982; Numbers 1993).

The objects of scientific controversies themselves have been subdivided into three basic kinds: controversies of fact, theory, and principle (McMullin 1987; Machamer, Pera, & Baltas 2000). *Controversies of fact* concern disputes over the observational or experimental basis for a claim, especially when there is agreement about the assumptions used in making the observations or conducting the experiment. According to McMullin, attempts to resolve controversies of fact usually involve "attempting to reproduce the disputed results" (McMullin 1987, p. 66). That said, a number of scholars have argued that experimental replication is not decisive when it comes to settling the existence of controversial scientific phenomena (Collins 1981, p. 11; 1994; Franklin 1999;

Godin & Gingras 2002). Controversies of fact revolve around disputes over phenomena, experimentation, data, and the constitution of evidence.

Controversies of theory occur when two or more theories are put forward to account for the same question, data, or experimental results (McMullin 1987, pp. 66–67). Even though the classical-balance controversy had strong sociopolitical elements, the dispute over rival theories of the nature of genetic variation and natural selection espoused by proponents of the classical and balance positions provides a good example of a controversy of theory. The positions in this controversy were formally named in 1955 at the meeting of the Cold Spring Harbor Symposium on Quantitative Biology, when Theodosius Dobzhansky laid out two diametrically opposed positions on evolutionary genetics. The classical position, according to Dobzhansky, held that "evolutionary changes consist in the main in gradual substitution and eventual fixation of the more favorable, in place of the less favorable, gene alleles and chromosome structures." Most loci, according to the classical position, should be homozygous. Heterozygotes were rare and had four possible sources: (1) deleterious mutations that are eventually eliminated by selection; (2) adaptively neutral mutations; (3) "adaptive polymorphisms maintained by the diversity of the environments which the population inhabits"; and (4) rare beneficial mutants that are on their way toward fixation (Dobzhansky 1955). According to Dobzhansky, the main proponent of the classical position was H. J. Muller. The balance position, which Dobzhansky advocated, instead held that most loci should be heterozygous. Homozygotes would still occur, but they would not be as advantageous as heterozygous combinations with superior fitness. In terms of genetic variation, the issue at stake between the classical and balance positions was the relative number and importance of heterozygous superior loci. As we shall see, the challenge was in producing experimental data that unequivocally addressed this question of genetic variation.

Controversies of principle are in general much more difficult to resolve than controversies of fact or theory, because a controversy over methodological principles may have no clear or agreed upon means for adjudication. In McMullin's words, "the resolution of controversies of principle, in short, is slow, and oblique, and practice oriented" (McMullin 1987, p. 75). Controversies of this sort over scientific standards speak to methodological issues that are often foundational. For instance, part of the dispute between early advocates of Mendelian genetics and Biometricians revolved around the value of quantitative understanding of traits and the appropriate use of statistical methods in genetic research (Sapp 1983; Olby 1989). The contemporary replication crisis can be thought of as a similar kind of dispute. This crisis emerged as social scientists found it difficult to reproduce each other's studies. On the one hand, this could be the result of questionable research practice such as removing outliers from data in order to produce a statistically significant result. On the other hand, the crisis concerns the use of significance testing itself and what constitutes an appropriate significance level for statistical testing. Indeed, proposed solutions to the crisis include raising the p-value or significance level required or abandoning significance testing altogether (Benjamin et al. 2018).

A further defining feature of scientific controversies is that they be public. The protagonists involved in a scientific controversy "appeal to others as though to judges; the outcome of the controversy will depend, not just on the actions and arguments of the protagonists but on the response of the scientific community generally, or at least of that part of the community that concerns itself with the disputed topic" (McMullin 1987, p. 52). Scientific controversies are thus by their nature social activities; the dispute is not limited to the protagonists. Moreover, controversies require judgments to be made from individual and community perspectives. Judgments about whether or not to favor one rival theory over another and so resolve a controversy of theory, for instance, are not made only in terms of satisfying an individual scientist's goals, but also take into account community and group goals and standards.

In summary, controversies begin with a disruption of agreement with a community of scientists. What they disagree about varies widely, but can roughly be grouped as disagreements of fact, theory, and principle. Most disagreements in science are resolved quickly. Controversies, however, persist, and it is to this feature that we now turn.

12.3 Why Controversies Continue

As disputes extend in time, a controversy can remain open for both constitutive and contextual reasons. On the contextual side, commitments to social and political beliefs can create biases that cause antagonists to stray from shared standards. These social values can also alter what are considered appropriate standards of evidence. Every scientific field has a set of standards that constitute good scientific practice within that field. These standards can change and can be disputed, but they are usually held to be a good guide for judging different scientific proposals. Occasionally, though, an agreed upon standard is underdetermined when it applies equally well to the alternatives. For instance, two rival theories can generate predictions that agree equally well with the evidence at hand. Prediction then cannot determine which theory is better: Choice using the standard of predictive value is underdetermined. This kind of underdetermination means that predictive standards will not resolve the dispute, and the disagreement will continue until some form of resolution is found. In underdetermined controversies, it is possible though that when one standard is underdetermined, other standards can be appealed to as a means for resolving the dispute (Dietrich & Skipper 2007). Shared standards can also be very difficult to apply. In a controversy over theories, for instance, evidence that a community judges to be important to the controversy resolution may be difficult to obtain or ambiguous in its interpretation.

In the classical-balance controversy, the problematic nature of a key set of experiments contributed significantly to the persistence of the controversy. Bruce Wallace, a student of Dobzhansky's, had been collaborating with J. C. King to study the effects of radiation exposure in *Drosophila*. The purpose of the study was to try to come to a better understanding of the effects of radiation on populations and of the evolutionary implications of these effects. After a number of generations, samples indicated that the population that had received an acute amount of radiation (7000 r, x-ray) had the highest average frequency of wild type flies. Moreover, this acutely radiated population had a higher estimated adaptive value than the population receiving no radiation. Adaptive values were intended to be a way of measuring the effects of deleterious alleles relative to an ideal adaptive value of 1. In Wallace and King's study, the population receiving no radiation was given an adaptive value of 1.0, and the population receiving an acute amount of radiation had an adaptive value of 1.04. Populations receiving lower level, chronic amounts of radiation (5.1 r/ hr., gamma) had adaptive values of 0.92 and 0.95. From these data, Wallace and King concluded that the higher adaptive value of the acutely irradiated population "could exist not merely *in spite of* but *because of* the original treatment" (Wallace & King 1951).

Wallace and King's results were meant to invite further research, which they did. Wallace himself continued his experiments for about three more years and the acutely irradiated population still had an adaptive value of 1.03 (Wallace 1956). His thinking was that if heterozygous superior combinations were common in the genome, then radiation-induced mutations would probably produce more heterozygote loci with superior fitness. So, when Wallace observed higher adaptive values after acute radiation, he ascribed these to the creation of more heterozygote superior loci in the genome. Muller worked with a graduate student, Raphael Falk, to perform similar experiments. When Falk's replications saw no increase in fitness after irradiation, he interpreted this as evidence that radiation did not produce new superior heterozygotes. Instead, he thought, as most biologists had up until then, that new mutations were harmful and recessive, so radiation would produce new heterozygotes that would lead to harmful recessive combinations in the next generation. This supported the classical position argument that heterozygote superiority was rare and that natural selection acted to eliminate harmful new mutations leading in the long run to a genome that was mostly homozygous. These radiation experiments were not convincing in the end, in part because it was impossible to pin down the exact effects of the irradiation. One could not tell where the mutations were occurring or whether they produced new heterozygote superior loci. In fact, because the effects of radiation might change each time the experiment was conducted, the mutations produced might be very different in each experiment. Despite efforts to bring the disputants together to work out their differences, by the 1960s, after more than a decade arguing, the classical-balance controversy had stalemated (Beatty 1987b). In this case, both sides sought experimental evidence that they hoped would decisively support their theory. The ambiguity of the experiments, however, had the effect of prolonging the dispute as the experiments were repeated and refined, but with no progress.

The perceived social and political consequences of this dispute raised the stakes and the evidential standards in this dispute. In this way, the social context altered the evidential standards that constituted good practice in this field, but in doing so allowed the controversy to persist. Wallace and King's experiments firmly placed the theoretical debate in the classical-balance controversy within a larger sociopolitical debate over nuclear weapons and the effects of atomic radiation. In effect, Wallace and King claimed that exposure to acute atomic radiation created genetically superior fruit flies. Muller had a very different view of the dangers of atomic radiation. Having won a Nobel Prize in 1948 for his research on the production of mutations with X-rays, in 1950 Muller published "Our Load of Mutations," which provided a new way to assess the genetic damage created by mutation. Accepting the premise that the vast majority of mutations are harmful to some degree, Muller argued that in a population of constant size, each mutation leads to one "genetic death" – to one individual that fails to reproduce. The number of deleterious alleles possessed by an individual represented that individual's deviation from a genetic ideal – they represented that person's genetic load. Because he had pioneered much of the early work on the genetic effects of radiation, Muller was adamant about the genetic loads that exposure to radiation could produce. This concern was motivated by the use of atomic weapons in World War II and was heightened by the Cold War arms race and testing programs. The damaging effects of radiation on genetic material were a major concern, especially in the wake of the use of atomic weapons in World War II. Naturally, when Muller discussed factors that would increase genetic loads and put human populations at risk, radiation was prominent (Beatty 1987b). By linking genetic variability to radiation, the stakes in this controversy had been raised beyond those of an intellectual dispute in evolutionary genetics. Both Muller and Dobzhansky saw themselves as struggling for the future of humankind.

The impact of social values on science can be profound in cases such as the classical-balance controversy. One way that philosophers explain this impact of social values on the standards of evidence themselves is to invoke what they call inductive risk (Douglas 2009; Elliott 2017). The idea behind inductive risk is that inductive reasoning is never going to produce certainty. So how much evidence is enough to create consensus within a community is not going to be a universally fixed standard. Instead, social values enter as scientists reason about how to manage the possibility of error. We can imagine Muller asking, what are the consequences of being wrong about the harmful effects of atomic radiation? If the consequences are severe, then it might be reasonable to demand a higher standard of evidence. This escalation of standards because of perceived social consequences can extend the resolution of a scientific dispute. This certainly contributed to the persistence of the classical-balance controversy as they fought over what were arguably a vague set of experiments.

The challenge of explaining why a controversy persists comes down to explaining why reasonable people resist each other's arguments. If a community has shared standards and one position has better evidence or more robust assumptions, then continued dissent can appear irrational (Kitcher 2000, p. 25). That said, if one of the goals of science is to promote innovation, then tolerating theories and practices that are not as well supported as their rivals can foster a diversity of views that allows innovative ideas and approaches to be developed. From a community perspective, if a diverse intellectual community of scientists is more likely to foster innovation, then tolerating dissent and new, perhaps poorly supported, theories may be worth the trade-off of having some of those dissenting positions spark controversy.

The most famous case for the value of a dissenting position in evolutionary biology was made by Stephen Jay Gould and Richard Lewontin in their attack on adaptationism in "The Spandrels of San Marco and the Panglossian Paradigm: A Critique of the Adaptationist Programme" (Gould & Lewontin 1979). In this paper, Gould and Lewontin describe an adaptationist orthodoxy within evolutionary biology that explains traits only in terms of optimal adaptations. Casting themselves as champions of free thought, they offered a number of alternatives to this optimality strategy, such as recognizing that traits could be constrained by developmental and phylogenetic processes. One way to read this paper is as a call for a diversity of approaches within the evolutionary biology community. In other words, Gould and Lewontin were not arguing that adaptation had no place in evolution, but that some of the community ought to be allowed to pursue alternative explanations (See Orzack & Forber 2017 for more on adaptationism).

12.4 How Controversies End

Philosophical work on scientific controversies has placed a heavy emphasis on how they end. The 1987 Hastings Center Project on Scientific Controversy offered a taxonomy of different ways controversies end or close (Engelhardt & Caplan 1987). These different forms of closure include:

1. *Closure through loss of interest.* If interest in a controversy wanes, the dispute can be abandoned and the controversy will die a natural death (Beauchamp 1987, pp. 31–33; McMullin 1987, pp. 81–82). Put another way, if the participants lose interest, the controversy ends not on rational grounds but because the participants have ceased to appeal to rational grounds for resolution, either through sound argument or through fair principles of negotiation (Engelhardt and Caplan 1987, p. 14). An extreme version of this form of closure occurs when antagonists in a controversy die. Thomas Kuhn famously described what is known as the Planck effect. When asked when his rivals would come to accept his view on black body radiation, the physicist Max Planck is described as claiming that "a new scientific truth does not triumph by convincing its opponents and making them see the light, but rather because its opponents eventually die, and a new generation grows up that is familiar with it" (Kuhn 1970, p. 150). Waiting for opponents to die is a strategy for the young and patient, but can end a dispute without having a reasoned resolution.

2. *Closure through force.* Closure in this form is achieved on the basis of non-epistemic factors such as "the authority of the state, the loss of a research grant, or the laziness of a controversialist" (McMullin 1987, p. 78). In other words, closure through force suggests "the employment of external authority to declare a controversy closed" (McMullin 1987, p. 78). As in closure through loss of interest, closure through force terminates a controversy but does not necessarily resolve it – there are no scientific reasons to keep the dispute closed. One of the best examples of this kind of closure comes from T. D. Lysenko's attack on Mendelism in the Soviet Union under Stalin. Lysenko's scientific opponents were harassed, imprisoned, and even killed, thereby terminating any controversy. Lysenko's triumph was not permanent, and as his influence weakened, the controversy reopened, he was repudiated, and modern genetics was reintroduced to the Soviet Union (McMullin 1987, p. 80; Graham 2016).

3. *Closure through consensus.* Closure through consensus occurs "if, and only if, in a context of controversy a consensus has been reached

that some position is correct or fair and opposition views incorrect or unfair" (Beauchamp 1987, p. 30). This consensus does not have to be a result of sound argumentation; the "weight of evidence" may have no role at all in the formation of the consensus." Such a consensus need not be permanent either; it is enough that "correctness" of a position is what the community agrees on at the time. For instance, most evolutionary biologists in the early twentieth century rejected the inheritance of acquired characteristics. This view was widely shared and reflected the genetics of the time, but was reevaluated with the rise of epigenetics at the turn of the twenty-first century (Jablonka, Lamb, & Avital 1998).

4. *Closure through sound argument.* Resolution of a controversy by means of sound arguments refers to closure produced in response to arguments whose premises are considered sound or empirically verified. Put another way, the premises in this form of closure have strong empirical support and are ordered in a logically valid argument that produces a strong argument. In the strict sense, the rules of evidence and inference used in this form of closure would be "true, valid and undistorted by the history and culture of the participants" (Engelhardt & Caplan 1987, p. 14). This strict sense of sound argument closure is an ideal that can be compared to a broader sense of sound argument closure. In the broad sense, the standards of evidence and inference are contextualized; they are "historically, socially, and culturally conditioned" (Engelhardt & Caplan 1987, p. 9). Because standards of evidence can change, understanding the standards used at a particular moment and time and how they were justified can be very important. There is debate among historians, for instance, over whether Gregor Mendel's results were too good to be true. Modern statistical standards fuel these discussions, but Mendel himself could not have deployed standards that had not yet been articulated. Actual controversies are always of this broad sort, but they can be judged in terms of their distance from the ideal, strict sort. This broad sense of closure can be further subdivided into a simple and complex mode. In the simple mode, the disputing scientists, "form a community of investigators with common rules of evidence and inference, and when sufficient data can be secured to resolve the controversy" (Engelhardt and Caplan 1987, p. 15). The complex mode involves more than one community with different interpretations of rules of

evidence and inference. In this kind of situation, closure requires finding some common ground from which to resolve the issue. In effect, claiming that controversies are complex in this way amounts to claiming that they are controversies of principle. The difference between the two communities is a difference of principle, as well as a difference over other disputed issues that make up the controversy. Among the many cases of this kind of controversy closure is Martin Rudwick's description of the end of the Great Devonian Controversy where a core group of geologists accepts a standard of evidence and a body of evidence to reach consensus in stratigraphy (Rudwick 1994).

5. *Closure through negotiation.* This form of closure occurs "if, and only if, a controversy is settled through an intentionally arranged and morally unobjectionable resolution acceptable to the principals in the controversy, even if they regard the resolution as compromising their ideal solution" (Beauchamp 1987, p. 33). Indeed, compromise solutions are a prime example of closure through negotiation. Often disputes involving policy, such as radiation exposure levels, can be thought of as involving negotiated compromise between different parties. To the extent that such negotiations involve an agreed-upon procedure for resolving controversies, negotiation closure can also be called procedural closure (Beauchamp 1987, pp. 30–31). The difference between sound argument closure and negotiation closure, then, is that in negotiation closure there is no single correct position indicated by sound epistemic argumentation. Negotiation is instead a matter of arbitration and often compromise. In this way, negotiated closure is similar to the closure of relative significance controversies discussed in the next section.

When these forms of controversy closure were discussed at the Hastings Center workshop, the consensus in the group was that "in cases of 'pure' scientific controversy – that is, where the conflict is over empirical facts, the nature of evidence, scientific theory, or methodology – sound argument closure is the appropriate mode of ending controversies." In cases of political, ethical, or science policy controversies, negotiation was thought to be the appropriate mode of closure (Macklin 1987, p. 620). The form of the controversy thus seemed to dictate the appropriate form of resolution.

12.5 Relative Significance Controversies

It should not be assumed that scientific controversies are necessarily all or nothing affairs where one side wins, whether by sound argument, negotiation, or force. In biological controversies especially, disputes take the form of what John Beatty has called relative significance controversies (Beatty 1997). In these kinds of disagreements, the positions differ by a matter of degree. In the classical-balance controversy, for instance, the two positions were NOT that everyone is either homozygous at every genetic locus or heterozygous at every locus. Instead, both positions admitted that genomes had a mix of homozygous and heterozygous loci. What was in question was the proportion of heterozygous superior or over-dominant loci. Advocates of the classical position argued that such over-dominant balanced polymorphisms, as they were called, were rare. Advocates of the balance position believed that they were common. The exact percentages were notoriously vague, though (Beatty 1987b). This controversy then was about the relative significance or relative frequency of balanced polymorphisms.

Relative significance controversies will not be resolved in an all-or-nothing manner; rather, their resolution will come from agreement on a proportion of frequency. That said, these controversies tend to have a dynamic of polarization and depolarization. When the controversy begins, the positions are highly polarized and stake out strongly distinct positions. This is exactly what Dobzhansky did in 1955 when he first described the positions in the classical-balance controversy (Dobzhansky 1955). Over time, though, the positions in question change, sometimes radically, sometimes more subtly. These transformations allowed the controversies to depolarize by enabling some participants to disengage, revise their opinions, and find agreement about the empirical support for the relative proportion in question.

Consider the course of the classical-balance controversy. Radiation exposure experiments had led to stalemate by the mid-1960s. Hope of some empirical resolution depended on a way of detecting genetic differences more precisely than was possible in the radiation experiments. The tools for addressing this issue had been developing within biochemistry and molecular biology for a number of years and promised to allow for a more direct assessment of genetic variation (Lewontin 1974; Dietrich 1994).

In 1964, Richard Lewontin, a student of Dobzhansky, was looking for a technique to break the stalemate. He found his solution in electrophoresis, a technique in biochemistry as a means for separating molecules by charge and size. In the early 1960s, geneticist Jack L. Hubby began to adapt electrophoresis for use with *Drosophila*. When Lewontin moved to the University of Chicago to collaborate with him in 1964, he arrived with a list of criteria for experimentally resolving how much heterozygosity there was per locus in a population. In their words:

> Any technique that is to give the kind of clear information we need must satisfy all of the following criteria: (1) Phenotypic differences caused by allelic substitutions at single loci must be detectable in single individuals. (2) Allelic substitutions at one locus must be distinguishable from substitutions at other loci. (3) A substantial proportion of (ideally, all) allelic substitutions must be distinguishable from each other. (4) Loci studied must be an unbiased sample of the genome with respect to the physiological effects and degree of variation. (Hubby and Lewontin 1966).

Hubby and Lewontin's work tried to meet these criteria and provide a reliable measure of the amount of heterozygosity found in *D. pseudoobscura*. Their survey of 18 loci revealed what they understood to be a high degree of polymorphism; the average heterozygosity was 11.5 percent. Lewontin and Hubby proposed that this level of heterozygosity was evidence that favored the balance position (Lewontin 1974). But in 1958, Wallace had claimed based on his experiments that the balance position predicted that "on the average an individual member of the *Drosophila* population studied is heterozygous for genes at 50% or more of its loci" (Wallace 1958, p. 555). Lewontin's much lower level of heterozygosity was a limited measure, but it was much lower than the value held earlier in the controversy. Lewontin and Hubby brought new evidence into the classical-balance controversy that broke the earlier experimental stalemate and offered a new measure of the relative proportion of heterozygosity in the genome. In 1966, it looked like these empirical results would resolve the classical-balance controversy. Instead, the controversy was fundamentally altered by the introduction of evidence by Motoo Kimura that many genetic changes might be neutral and so not subject to the influence of natural selection. The heterozygosity that electrophoresis detected could be neutral, not the balanced polymorphisms that confer advantages to their possessor (Lewontin 1974; Dietrich 1994). This conceptual shift did not allow the classical-balance controversy to be resolved by sound argument or negotiation. Instead the controversy was dissolved and reformulated in different terms, which added neutral alleles and genetic drift to the theories grappling with the problem of explaining genetic variation. Neutral evolution, itself, was very controversial, which led Lewontin to claim that the classical-balance controversy had been continued under a new name. However, there are good reasons to see a more significant break between the two disputes (see Dietrich 1994).

12.6 Conclusion

Controversies offer philosophers an opportunity to understand the dynamics of scientific change by focusing on some of its most contentious moments. These episodes are not the stuff of everyday biology, but the reasoning that leads to their resolution is. Because controversies lead biologists to be more explicit about their reasoning as they try to articulate their own position and persuade others as to its correctness, controversies provide an opportunity for insight into how scientific reasoning is deployed and when it is most effective. Because scientific controversies often have a social or political dimension, they also offer an opportunity to understand the interplay between social values and scientific reasoning.

The existence of controversies in science should not be taken as evidence that an area under dispute is in danger of becoming unscientific or untrustworthy. Criticism is an essential part of everyday science. Controversies often represent extended periods of that critical process. Usually controversies point to important methodological, conceptual, or evidentiary issues that need further work and refinement. Recently, however, deniers of climate change have used the critical process of science to stall scientific work. Rather than formulate criticism that will foster useful and transformative scientific change, they make criticisms for the sake of eliciting scientific responses that have already been given or that do not constitute a new or useful line of inquiry. Their goal is to stall meaningful research and encourage doubt about scientific research in the minds of the public and policy makers. These kinds of developments have now led philosophers to try to articulate when dissent is detrimental to science and when it is useful (Biddle & Leuschner 2015). Research on the nature of dissent will certainly have ramifications for how we think about the dynamics and closure of scientific controversies in the future.

References

Beatty, J. (1987a). Natural Selection and the Null Hypothesis. In J. Dupré (ed.), *The Latest on the Best: Essays on Evolution and Optimality*. Cambridge, MA: MIT Press.

Beatty, J. (1987b). Weighing the Risks: Stalemate in the Classical/Balance Controversy. *Journal of the History of Biology* 20(3): 289–319.

Beatty, J. (1997). Why Do Biologists Argue like They Do? *Philosophy of Science* S64(4): 231–242.

Beauchamp, T. L. (1987). Ethical Theory and the Problem of Closure. In H. T. Engelhardt & A. L. Caplan (eds.), *Scientific Controversies: Case Studies in the Resolution and Closure of Disputes in Science and Technology*, pp. 27–48. Cambridge: Cambridge University Press.

Benjamin, D. J., Berger, J. O., Johannesson, M., Nosek, B. A., Wagenmakers, E. J., Berk, R., Bollen, K. A., et al. (2018). Redefine Statistical Significance. *Nature Human Behaviour* 2(1): 6–10.

Biddle, J. B. & Leuschner, A. (2015). Climate Skepticism and the Manufacture of Doubt: Can Dissent in Science Be Epistemically Detrimental? *European Journal for Philosophy of Science* 5(3): 261–278.

Collins, H. M. (1981). Son of Seven Sexes: The Social Destruction of a Physical Phenomenon *Social Studies of Science* 11(1): 33–62.

Collins, H. M. (1994). A Strong Confirmation of the Experimenters' Regress. *Studies in History and Philosophy of Science Part A* 25(3): 493–503.

Dietrich, M. R. (1994). The Origins of the Neutral Theory of Molecular Evolution. *Journal of the History of Biology* 27: 21–59.

Dietrich, M. R. (2006). From Mendel to Molecules: A Brief History of Evolutionary Genetics. In C. W. Fox & J. B. Wolf (eds.), *Evolutionary Genetics: Concepts and Case Studies*, pp. 3–13. New York: Oxford University Press.

Dietrich, M. R. & Skipper, R. (2007). Manipulating Underdetermination in Scientific Controversy: The Case of the Molecular Clock. *Perspectives on Science* 15: 295–326.

Dobzhansky, T. (1955). A Review of Some Fundamental Concepts and Problems in Population Genetics. *Cold Spring Harbor Symposium on Quantitative Biology* 20: 1–15.

Douglas, H. E. (2009). *Science, Policy, and the Value-Free ideal*. Pittsburgh: University of Pittsburgh Press.

Elliott, K. (2017). *Exploring Inductive Risk: Case Studies of Values in Science*. New York: Oxford University Press.

Engelhardt, H. T. & Caplan, A. L. (1989). *Scientific Controversies: Case Studies in the Resolution and Closure of Disputes in Science and Technology*. Cambridge: Cambridge University Press.

Franklin, A. (1999). How to Avoid the Experimenters' Regress. In *Can That Be Right? Boston Studies in the Philosophy of Science, Vol 199*, pp. 13–38. Dordrecht: Springer.

Godin, B. & Gingras, Y. (2002). The Experimenter's Regress: From Skepticism to Argumentation. *Studies in History and Philosophy of Science* 33(1): 137–152.

Gould, S. J. & Lewontin, R. C. (1979). The Spandrels of San Marco and the Panglossian Paradigm: A Critique of the Adaptationist Programme. *Proceedings of the Royal Society of London* 205(1161): 581–598.

Graham, L. R. (2016). *Lysenko's Ghost: Epigenetics and Russia*. Cambridge, MA: Harvard University Press.

Hubby, J. L. & Lewontin, R. C. (1966). A Molecular Approach to the Study of Genic Heterozygosity in Natural Populations. I. The Number of Alleles at Different Loci in *Drosophila Pseudoobscura*. *Genetics* 54(2): 577–594.

Jablonka, E., Lamb, M. J., & Avital, E. (1998). "Lamarckian" Mechanisms in Darwinian Evolution. *Trends in Ecology & Evolution* 13(5): 206–210.

Kim, K.-M. (1994). *Explaining Scientific Consensus: The Case of Mendelian Genetics*. New York: Guilford Press.

Kitcher, P. (1982). *Abusing Science: The Case against Creationism*. Cambridge, MA: MIT Press.

Kitcher, P. (2000). Patterns of Scientific Controversies. In P. K. Machamer, M. Pera, & A. Baltas (eds.), *Scientific Controversies: Philosophical and Historical Perspectives*, pp. 21–39. New York: Oxford University Press.

Kuhn, T. S. (1970). *The Structure of Scientific Revolutions*. Chicago: University of Chicago Press.

Lewontin, R. C. (1974). *The Genetic Basis of Evolutionary Change*. New York: Columbia University Press.

Lewontin, R. C. (1981). Introduction: The Scientific Work of Theodosius Dobzhansky. In R. C. Lewontin, J. A. Moore, W. B. Provine, & B. Wallace (eds.), *Dobzhansky's Genetics of Natural Populations I–XLIII*, pp. 93–115. New York: Columbia University Press.

Lewontin, R. C. (1991). Perspectives: 25 Years Ago in Genetics: Electrophoresis in the Development of Evolutionary Genetics: Milestone or Millstone? *Genetics* 128(4): 657–662.

Lewontin, R. C. & Hubby, J. L. (1966). Molecular Approach to the Study of Genic Heterozygosity in Natural Populations. II. Amount of Variation and Degree of Heterozygosity in Natural Populations of *Drosophila Pseudoobscura*. *Genetics* 54(2): 595–609.

Lewontin, R. C., Rose, S. P. R., & Kamin, L. J. (1984). *Not in Our Genes: Biology, Ideology, and Human Nature*. New York: Pantheon Books.

Longino, H. E. (1990). *Science as Social Knowledge: Values and Objectivity in Scientific Inquiry.* Princeton, NJ: Princeton University Press.

Machamer, P., Pera, M., & Baltas, A. (2000). Scientific Controversies: An Introduction. In P. K. Machamer, M. Pera, & A. Baltas (eds.), *Scientific Controversies: Philosophical and Historical Perspectives*, pp. 3–17. Oxford: Oxford University Press.

Machamer, P. K., Pera, M., & Baltas, A. (2000). *Scientific Controversies: Philosophical and Historical Perspectives.* New York: Oxford University Press.

Macklin, R. (1987). The Forms and Norms of Closure. In H. T. Engelhardt & A. L. Caplan (eds.), *Scientific Controversies: Case Studies in the Resolution and Closure of Disputes in Science and Technology*, pp. 615–624. Cambridge: Cambridge University Press.

McMullin, E. (1987). Scientific Controversy and its Termination. In H. T. Engelhardt & A. L. Caplan (eds.), *Scientific Controversies: Case Studies in the Resolution and Closure of Disputes in Science and Technology*, pp. 49–91. Cambridge: Cambridge University Press.

Muller, H. J. 1950. Our Load of Mutations. *The American Journal of Human Genetics* 2: 111–176.

Numbers, R. L. (1993). *The Creationists: The Evolution of Scientific Creationism.* Berkeley, CA: University of California Press.

Olby, R. (1989). The Dimensions of Scientific Controversy: The Biometric–Mendelian Debate. *The British Journal for the History of Science* 22(3): 299–320.

Orzack, S. H. & Forber, P. (2017). "Adaptationism." *The Stanford Encyclopedia of Philosophy.* Edward N. Zalta (ed.), https://plato.stanford.edu/archives/spr2017/entries/adaptationism/. Accessed on July 25, 2019.

Panofsky, A. (2014). *Misbehaving Science: Controversy and the Development of Behavior Genetics.* Chicago: University of Chicago Press.

Powell, J. (1994). Molecular Techniques in Population Genetics: A Brief History. In B. Schierwater et al. (eds.), *Molecular Ecology and Evolution: Approaches and Applications.* Basel: Birkhauser.

Provine, W. B. (1992). The R. A. Fisher–Sewall Wright Controversy. In *The Founders of Evolutionary Genetics*, pp. 201–229. Dordrecht: Springer.

Raynaud, D. (2017.) *Scientific Controversies.* New York: Routledge.

Rudwick, M. J. S. (1994). *The Great Devonian Controversy: The Shaping of Scientific Knowledge among Gentlemanly Specialists.* Chicago: University of Chicago Press.

Sapp, J. (1983). The Struggle for Authority in the Field of Heredity, 1900–1932: New Perspectives on the Rise of Genetics. *Journal of the History of Biology* 16(3):311–342.

Sayre, A. (2009). *Rosalind Franklin and DNA.* Bridgewater, NJ: Baker & Taylor.

Skipper, R. A. (2002). The Persistence of the R. A. Fisher–Sewall Wright Controversy. *Biology and Philosophy* 17(3): 341–367.

Smocovitis, V. B. (1996). *Unifying Biology: The Evolutionary Synthesis and Evolutionary Biology.* Princeton, NJ: Princeton University Press.

Wallace, B. (1956). Studies on Irradiated Populations of *Drosophila Melanogaster*. *Journal of Genetics* 54: 280–293.

Wallace, B. (1958). The Average Effect of Radiation-Induced Mutations on Viability in *Drosophila Melanogaster*. *Evolution* 12(4): 535–556.

Wallace, B. & King, J. C. (1951). Genetic Changes in Populations under Irradiation. *American Naturalist* 85: 209–222.

13 What Is the Relation between Facts and Values in Biological Science?

Biology *in* Society

CARRIE FRIESE AND BARBARA PRAINSACK

13.1 Facts and Values

It is well-known that the dividing line between facts and values can be blurry. This is the case for several reasons. One is that every statement about the world is expressed in human language, which always reflects values and judgments in human societies. Even if we use, for example, the seemingly neutral and factual word, "table," we cannot fully understand the meaning of the word without knowing what people normally do with tables: eating or working on it, and not putting one's dirty feet on it. Another reason that facts and values are not mutually exclusive is that statements about what happens in the world necessarily reflect how the world is organized and valued by those inhabiting it. For example, we could say that Mara is a reliable person. We could prove this "fact" by showing a record of all our interactions with Mara since we first met her to prove that she kept her promises and never let us down. At the same time, our statement that Mara is reliable is also a value statement in that it reflects societal (and possibly also personal) values in terms of what kinds of behavior are, or should be, expected from a person in our societies. Keeping a promise is valued by people because it is seen as contributing to the functioning of human interaction and social and political institutions.

As these examples show, saying that there are cases where facts cannot be clearly distinguished from values does not imply that there

We thank Giulia Cavaliere, Clemence Pinel, Kerstin Weich, and the editors of this volume for helpful comments on this manuscript. We also thank Tarquin Holmes for helping with the historical understanding of vivisection and anti-vivisection in the UK. The usual disclaimer applies.

is confusion, moral ambiguity, or relativism ("anything goes"). It means that when we talk about facts, we should ask what assumptions and processes have gone into turning a statement into a fact. Similarly, when we consider something as a value judgment, we should ask whether this perhaps is shared among so many people that it becomes "objective" in the sense that many people recognize and experience it as true.

The discussion about facts and values is of key importance for the role of biological sciences in our societies. As philosophers such as Michael Polanyi (1962) or Hilary Putnam (2002) argued, science – which is commonly associated with the creation of facts, understood as observable and measurable evidence about the world – requires agreement on values in order to function (see Kuhn 1962; Kitcher 1995; Knorr Cetina 2009; Fleck 2012). Putnam spoke of "epistemic" values when he referred to the conventions and agreements about what is valuable data, what is a robust method, and what counts as a good outcome, for example.

To demonstrate these points, we turn to four different examples that variously show how facts and values are produced and exist together, and the implications that this has for what we "know" and how people are treated within medicine and society. These examples, based on our research and teaching, are eugenics, vivisection, epigenetics, and inter-sex. We have chosen historical as well as contemporary examples in order to show that the blurring of fact and value is not a phenomenon of the past, and neither does it, as such, represent "bad science." Rather, all science contains facts *and* values. If nothing else, all research needs funding, and values inevitably shape what kinds of facts get produced. On this basis, we argue that "good science" is connected to a reflection on substantive values – such as accuracy, progress, health, or justice, depending on the field of science, as well as a consideration of procedural values such as transparency and reproducibility. We conclude this chapter by considering the extent to which the current trend towards "open science" achieves these goals.

13.2 Biological Science in Society: Historical Examples

We first turn to two well-known examples of how biology has historic-ally been enmeshed within the social values of the milieu in which its knowledge was produced: eugenics and vivisection.

13.2.1 Eugenics

Many people, when they hear the word eugenics, think of the cruelties committed by and with approval of murderous systems such as Nazi Germany. The forced sterilization or even murdering of people who were considered unworthy to procreate, as well as projects such as "Lebens-born" that encouraged the birth of children by women and couples seen as "racially pure and healthy," are outcomes of a eugenic ideology that was embedded in an inhumane ideology of racial superiority and rooted in an obsession with "purity." Such haunting examples of eugenics from the past should not, however, blind us to the long, multi-faceted, and ongoing history of eugenics in many parts of the world (Koch 2004; Camporesi & Cavaliere 2018).

First coined by the nineteenth-century English statistician Francis Galton, the term "eugenics" was inspired by the Greek word for "good stock" or "good birth" and referred to – in Galton's words – people and communities "hereditarily endowed with noble qualities" (Galton 1883; see also Allen & Turda 2015). The idea of improving the quality of the population was taken up by ideologists and politicians in many parts of the world, either by actively intervening in the reproductive choices and capacities of people – by legal prohibitions for "races" to mix, for example, or by forced interventions such as sterilizations – or by creating incentives for people to procreate in specific ways. A famous example is the American biologist Charles B. Davenport (1866–1944) who lobbied for the introduction of measures to improve the American people not only in quality but also in quantity. In this spirit, Davenport and his fellow eugenicists sought to increase the fertility of those segments of the population that were considered to carry desirable characteristics, such as health, wealth, and whiteness. Such active encouragement of specific groups of people to procreate is called positive eugenics, whereas meas-ures seeking to prevent those with less desirable characteristics from procreating is now referred to as negative eugenics.

What specific characteristics eugenicists sought to promote varied greatly between different times and places. In regions such as Latin America, for example, ideas about what kinds of people should procreate were strongly influenced by the requirements and ideologies of coloni-alism. For one, colonialism was underpinned by racist assumptions about the different abilities of different types of people to contribute to civilized and prosperous societies. In this spirit, political elites in countries such as Argentina, Mexico, Peru, and Brazil sought to increase the number of people of European descent, as these were considered to possess the most desirable characteristics. By the beginning of the twentieth century, in some Latin American countries, ideals shifted from seeking to increase the dominance of people of European origin or descent toward "the right mix" between local and indigenous populations with European immigrants to design national races that combined the best of each population's qualities (Schell 2015). As a result, in Mexico, for example the mixed-race person became the ideal of many eugenic policies (Schwartz-Marin & Silva-Zolezzi 2010).

As these examples show, one should not assume that eugenics always sought to achieve an ideal of racial purity – its goals were much more diverse. Neither has eugenics always revolved around genes and genomes, as today's discourse on eugenics might suggest. Today's discourse is strongly focused on genetic technologies such as prenatal/preimplantation genetic testing or genome editing to ensure the birth of people who fit our contemporary ideals of health and fitness. In many parts of the world, however, eugenics was driven by the assumption that acquired characteristics – such as healthy living, abstaining from alcohol, and good education – would be passed on to the next generation (e.g., Stepan 1991; Meloni 2016). Eugenicists operating under this latter assumption, besides encouraging people with such good characteristics to procreate, invested heavily in what we would, today, call "social determinants": sanitation, education, and social security.

In summary, while there are many examples of eugenic programs designed and implemented in a top-down manner, which sought or seek to increase the number of people with desirable characteristics and prevent those with undesirable characteristics from procreating – either by forced sterilizations or by "softer" means such as free access to birth control measures or tax incentives – eugenics has many faces and forms. It is also related to the practice of medicalizing or naturalizing societal issues. The understanding that alcoholism, for example, is a medical problem rather than – as it was seen in the past – a moral deficiency has had the positive effect of no longer blaming people suffering from it for making the "choice" to drink too much. Instead, the structural factors that have contributed to their addiction move into the foreground of analysis and policy. At the same time, the portrayal of alcoholism as a medical problem also means that interventions are focused on people's bodies. Once alcohol had become "medicalized" (Conrad 1992; 2005), it had become part of the person's health that she is responsible for. As a patient, the alcoholic is morally obliged to try to "heal" by seeking treatment, or by considering it when making reproductive choices. In some areas, contemporary eugenic practices are seen to overlap with population health and public health policies, and even with global health.

In each instance, however, eugenics mirrors dominant assumptions about what the good life is, and correspondingly, what characteristics ideal people and citizens should possess. Throughout almost all places and times, practices of eugenics have had only a few things in common. The first is that in virtually all instances, its proponents bolstered up their claims with science. The second is the idea of a hierarchy of people according to the traits they have and are to pass on (or not) to the next generation. These understandings, in turn, also influence how biological knowledge is made. In this manner, the way that the biology of women was interpreted has justified making them the targets of eugenic policies in different ways than men. While men were sterilized to improve their own situation (e.g., in case of mental health problems), women were often sterilized to protect society from their immoral behavior (e.g., Schell 2015). Even in more recent, "soft" measures and liberal eugenics, women are still treated as those bearing the responsibility for reproductive choices and the well-being of future societies. Finally, also the *formats* of eugenic policies reflect geographic and historic variation in what is considered acceptable state interference in people's lives. In different places and times, these eugenic policies have included forced measures such as coerced sterilizations and attempts to improve the quality of the population by improving social determinants, as well as "liberal" modes of eugenics that emphasize individual choice but "steer" this choice by prescribing what is morally and economically a "good" choice that rational and responsible people should follow (Hashiloni-Dolev 2007; Agar 2008).

Today, as philosopher Giulia Cavaliere argues, eugenics is also used as a deliberate political tool to discredit practices or technologies that people consider undesirable. The mere claim that something is "eugenic" puts it in the same corner as practices such as forced sterilizations or even the killing of people considered unworthy of living

(Cavaliere 2018). Examples range from editing human genes or allowing couples with known heritable diseases to use pre-implantation genetic diagnosis to have a healthy child. These practices have been called eugenic by those trying to restrict or outlaw their use. These examples show that the use of arguments about what human biology should and should not be like is not merely a historical practice, but it continues to be employed as a political strategy today.

13.2.2 Vivisection

While our first example has shown how facts and values have mutually shaped each other throughout the history of eugenics, the use of animals in scientific research is one area of biology where facts and values have long been in dispute, and where both have changed over time. The values of scientists as vivisectionists have long been at odds with anti-vivisectionist parts of publics (Sperling 1988; Arluke 1991; Jasper & Nelkin 1992; Lederer 1992; Herzog 1993; Rupke 1997; Bittel 2005; Elston 2006). However, this polarized view that separates publics into two camps, vivisectionists and those opposing it, is not entirely accurate. The history of regulating laboratory animals shows that there have been differences in the values of scientists as well, both at any given time as well as over time. And scientific facts regarding sentience have been important in establishing those values.

For example, in much of the early experimentation in Europe, animals were not anesthetized. There was the belief, amongst some, that animals did not suffer; and even if animals did suffer, many people did not care, because they thought that the knowledge produced through vivisection was more important than the welfare of animals. The public was growing less willing to condone the suffering of animals across the nineteenth century, particularly in Britain. And scientists were also less willing to condone unnecessary and painful experiments performed upon animals as many also believed that some animals are sentient and that not all experimental knowledge justified making animals suffer.

The United Kingdom was the first country to centrally govern the use of animals in science with the 1876 Cruelty to Animals Act (French 1975; Kean 1998). In establishing this regulation, a series of testimonies was taken from scientists who worked with animals. In his interview,

Edward Emanuel Klein reported to the committee that he preferred not to use anasthetics in animal experiments, and that he was not at all concerned about whether or not animals were suffering. His values were shocking to not only the committee and the general public, but also to other scientists. In response, T. H. Huxley wrote to Charles Darwin: "I am told that he [Klein] openly professed the most entire indifference to animal suffering, and said he only gave anæsthetics to keep animals quiet! I declare to you I did not believe a man lived who was such an unmitigated cynical brute as to profess and act upon such principles, and I would willingly agree to any law which would send him to the tread-mill." Klein's testimony did much to suggest to the public and scientists alike that the use of laboratory animals should be regulated by the state to ensure that a particular set of values regarding the sentience of animals would be respected in all experiments. The 1876 Cruelty to Animals Act established the still-existing requirement in the UK that scientists receive a license from the Home Office before conducting research involving animals.

This legislation was updated with the Animals (Scientific Procedures) Act (ASPA) in 1986. ASPA maintained the need for a Home Office license, but made adherence to the 3Rs (Reduce, Refine, and Replace animals in research) a requirement. The 3Rs is a concept developed by Russell & Burch (1959) in their *Principles of Humane Experimentation*, which aimed to make animal welfare concerns central to the conduct of science (Hobson-West 2009; Kirk 2018). Ensuring that animals were "not suffering" was no longer viewed as sufficient. In addition, scientists needed to use as few animals as possible, in well-designed experiments where there was no other option but to use animals. The 3Rs have since become the gold standard in the ethics of research involving animals in the global circulation of science (Davies et al. 2018; McLeon & Hartley 2018; Sharp 2019).

In this context, zebra fish became a way of "refining" animal experiments. This was based on the idea that fish do not experience pain, a dominant idea 20 years ago (Law & Lien 2017). However, there are now a growing number of scientists who believe that fish are sentient and do experience pain. While this remains contested by some scientists, fish are treated as sentient in European regulations (Law and Lien 2017). The ethics of animal experimentation thus changed again, such that zebra fish do not replace animals from research but rather are another site where experimentation needs to be refined to protect fish from suffering.

What we see from the history of vivisection and anti-vivisection are the ways in which facts (e.g., sentience) change values (e.g., how to experiment), and how values (e.g., humanitarianism) inform the creation of facts (e.g., fish are sentient). We also see how the values of scientists change over time, such that it is impossible to consider experimenting with animals without anesthetics in the way that had been commonplace in the nineteenth century and before.

13.3 Biological Science in Society: Contemporary Examples

Both eugenics and vivisection are often portrayed as "bad science" conducted in the past. Some might argue that these examples show how much society and biology have progressed, and that adherence to the scientific and ethical standards according to which science should be carried out will prevent such bad science from happening again. Some might even think that the best way of preventing bad science is to ensure that science does not become infused with political or other normative goals, and remains entirely objective. Using two contemporary examples, the study of how environmental factors change the regulation and expression of DNA (epigenetics) and the study of human biology leading to different results regarding the organization and relationship of biological sex and gender, we show how facts and values are always produced together (Jasanoff 2004) in contemporary biological and medical science.

13.3.1 Epigenetics

As noted, the idea that characteristics acquired by people in their lifetime could be passed on to their offspring has been around for a long time (Meloni 2016). Speculations and theories about the mechanisms by which such characteristics could be passed on, however, varied greatly throughout history. Since the second half of the twentieth century, the label of "epigenetics" has been used to describe an expanding field studying how environmental factors change the expression and regulation of DNA – which can, in some cases, be passed on to future generations. The "environment," here, denotes any factors that are outside of genes. This can refer to cells within the body. But this also includes practices such as smoking and diet and toxins in our natural environment, as well as experiences that we have due to our social environment. It has been found, for example, that trauma stemming from abuse can influence how people's genes are regulated and expressed: It can change which genes are "switched on" or "switched off." At the same time, some studies have found statistical associations between the experience of trauma in parents with psychological and physical health issues in children (for an overview, see Youssef et al. 2018), suggesting that some epigenetic effects of trauma may be heritable. Under what circumstances and how this happens is often not well understood (for an overview, see Heard & Martienssen 2014; Klosin & Lehner 2016).

Epigenetics is a rapidly expanding field worldwide. In medicine, it is heavily associated with research on how early life experience and intergenerational effects influence health and disease (see also Palsson & Lock 2016). In order to understand how epigenetics shapes medical research, it is necessary to treat it not merely as a result of scientific progress, but to see it as partly shaped by societal and political factors. For example, the increasing attention to developmental and intergenerational factors shaping health and disease in recent decades has corresponded with societal and political discourses that reject genetic determinism and blame it for justifying racial and other types of discrimination (e.g., Hedgecoe 2001). And the new scientific paradigms, in turn, also influence societal debates and policy making: If certain behaviors have the capacity to switch genes off or on, should there be policies to foster behaviors that have positive epigenetic effects and discourage, or even forbid, policies with negative effects?

This question is debated among policy makers and scholars. The answers given so far go strongly in the direction of targeting behavior at the level of individual people. For example, it has been suggested that, given the bad epigenetic effects of smoking, women of reproductive age should be prohibited from smoking (for a critical view, see Valdez 2018). This emphasis on individual-level behavioral change is part of a larger discursive and policy-shift from steering the practices of people by improving social determinants (by means of welfare, social housing, etc.) toward targeting individual "choice." At the same time, there is very strong evidence in public health and epidemiology of the impact of "social determinants" such as stable and affordable housing, financial and economic security, clean air, etc., on positive health outcomes –

which are mediated in part through epigenetic mechanisms. This evidence, however, has not translated into increasing investments into better public infrastructures and environmental policies in most countries. On the contrary, countries such as the United Kingdom or Australia have introduced "behavioral insights units" into their public administration apparatus that study individual behavior and try to "nudge" people into behaving more responsibly and more rationally. Nudging – the creation of incentives for individual behavior change – is adopted also in other fields of research and policy in various fields as diverse as global health, education, and international development. The American economist Richard Thaler, one of the inventors of nudging (2008), received the Nobel Prize for Economics in 2017.

Against this backdrop, the growth of the field of epigenetics is not merely the result of new scientific knowledge about how traits develop and are inherited; it also reflects dominant political ideologies about where the root causes of health and societal problems are located, and how they should be addressed. The interconnectedness of facts and values also becomes apparent here: While there is strong evidence to suggest that people's practices and "lifestyles" leave marks on their genomes, that we focus on this fact, and not on the fact of companies and people polluting our environments and governments neglecting to build affordable and green living spaces, is a political choice.

13.3.2 Intersex

In 1993, feminist biologist Ann Fausto-Sterling conducted a thought experiment. She argued that there are not two sexes of humans from a biological perspective, given that some people do not fit clearly within the category of male or female because their genitalia, chromosomes, and/or hormones do not all clearly align with what is taken to be the biology of one sex. The controversy surrounding middle-distance runner Caster Semenya is a more recent example of Faust-Sterling's point that there are not just two sexes from a biological perspective. With this thought experiment Fausto-Sterling provided an important intervention in the assumption that dimorphic sex (e.g., male or female) in biology is the basis for dimorphic gender (e.g., man or woman) in society. By challenging dimorphic sex through the biological fact of intersexed people, Fausto-Sterling tried to show that our ideas about gender are not only rooted in biological "facts." Rather, the way we read biological phenomena has been made to fit our categories and assumptions that separate people into either women or men. In other words, societal values around gender were shaping how and what we know about the biology of sex (Fausto-Sterling 1993, 2000).

Fausto-Sterling's work is also part of the intersex social movement that has challenged and changed the practices of gender assignment surgery in medicine. Particularly since the 1990s, these movements have provided support to people who were diagnosed as intersexed and had their body changed – almost always without their knowledge – through surgeries conducted anywhere from shortly after birth to teenage years. These movements have tried to change public opinion on what it means to be intersexed, and to change medical practices (Chase 1998; Davis 2015).

The bodies of intersexed people – diagnosed through the appearance of genitalia, genetics, and/or hormones – have been the target of medical interventions for centuries. Scientists such as the American psychologist John Money were instrumental in legitimating the practice of conducting irreversible surgeries on infants with ambiguous genitalia (Davis 2015, p. 58). Money and his colleagues believed that gender identity was not fixed until a child was 18 months and that it was created and maintained through childhood gender socialization, that is, it depended on what gender role the child was brought up. They thought that doctors were best placed to determine the future gender of an intersexed baby and use surgery to shape the genitals to match that gender. To support gender socialization, it was recommended that the child not be given information about the surgery. By the 1990s, Money's theories had come under heavy attack by those affected by imposed gender and sex assignments, as well as by feminists, medical professionals, and journalists. Protocols and procedures in many countries have changed since then. But, as sociologist Georgiann Davis (2015) has argued, being intersexed is still medicalized and stigmatized in most societies through the re-labeling of "hermaphrodite" and "intersexed" to "disorder of sex development." The genitalia of intersexed people exhibits natural biological variation and diversity. While intersex can signal an underlying metabolic problem in some instances, groups such as the Intersex Society of North America argue that it is the metabolic concern rather than the intersexed genitals that are in need of medical attention and treatment (www.isna.org/compare).

What we see from the current medical knowledge and treatment of intersexed people are the ways in which facts (e.g., not every person fits into the binary gender model) do not necessarily change values (e.g., a binary gender system that is social). The value placed upon gender and gender relations and the belief that these gendered identities have a binary biological basis in genitalia continue to determine how medicine is practiced. Here, biology continues to be remade (e.g., ambiguous sex is made male or female) to accommodate social values of gender (e.g., woman or man). Those biologies that do not fit this binary social value system are pathologized in order to make these biologies "fixable."

13.4 What Is Good Science? Values, Ethics, and Knowledge

Just as understandings about what counts as a desirable characteristic in people and populations, and about how "normal" sexes and genders have changed through time and space, notions of what good science is have also changed over time. There has been a general trend from leaving it to scientists to judge what is worthy to be researched and what is ethical or unethical to do, toward systems of external control (see also Kitcher 1995). For example, a lot of science in the nineteenth century was done by "amateur" scientists without affiliations with established scientific institutions, whose wealth allowed them to devote large amounts of their time to scientific inquiry – to study plants, stars, or animals. These amateur scientists communicated through correspondence groups, by sending each other samples and by writing letters. Many of these correspondence groups, which are precursors of modern scientific journals, were governed by informal mechanisms of social control and self-regulation of their members. In other words, the idea that the quality of science and the integrity of the scientific enterprise are best protected through embedding them into formal institutions that provide training, accreditation, and research ethics reviews, as well as formal peer review and research evaluations, is a relatively recent phenomenon. These kinds of institutionalized, procedural ethical controls have not always formed a part of what "good science" is.

Does this mean that science, before the externalization of control, was more free, and perhaps for this reason, also more innovative? (That frequent research evaluations and current pressures to obtain grants stifle innovation are common arguments against "audit cultures"

[Strathern 2000] in science and research.) Not entirely. Those who were privileged enough to devote their lives to science could do so with arguably more freedom than most people employed as scientists in research institutions today. But class and gender relations determined who could and who could not produce "facts" in this context, and who could be admitted into the informal institutions and networks of science. Reuben Message (2016), for example, has shown how in nineteenth-century aquaculture in Britain there were controversial debates over the species of "parr"; were they the young of salmon, or were they a separate species? A groundskeeper designed an experiment to address this question, which ultimately showed that the parr were young salmon. Because of the low status of the experimenter, however, his findings needed to be supported by leading naturalists of the day in order for this "fact" to be accepted in science and ultimately in law. The groundskeeper could not even present his findings at the meetings of learned societies; a person of higher status had to do that for him. If he had been a woman, it is unlikely that she would have been allowed to design and carry out an experiment in the first place.

The expansion of external controls on scientists and science – in the form of audit systems for scientific training, outputs, and scientific knowledge production in itself – could be read as a lack of trust in the capacity of scientists to self-regulate; but it could also be seen as a reflection of a greater democratization of science. When it was left to scientists alone to decide what they would research, who they would accept as their own, and what they considered ethical, they often reproduced upper- and middle-class values in their actions and institutions. Lower-class people, women, differently abled or bodied people, non-white people, or those with "wrong" religious affiliations were often excluded from the scientific enterprise. The choice of research topics reflected the interests of privileged, typically white and male scientists more than considerations of what would be valuable for publics. While most of these characteristics still matter today – in the sense that class, gender, minority status, etc., continue to affect people's opportunities in society – most of them are no longer *formal* barriers to the institutions and practices of scientific knowledge production. The introduction of mechanisms of public accountability for scientific knowledge production has contributed to a more inclusive science in terms of who gets admitted to training, who gets hired and promoted, and who gets funded and for what, as well as how public value and benefit is taken into consideration. At the same time, the introduction of these mechanisms has also increased bureaucratic burdens and audits. What makes audit systems in science so harmful in their current configuration in the United Kingdom and elsewhere (e.g., Berg et al. 2016; Olssen 2016; Waterman & Olssen 2016) is not the fact that they hold people accountable, but *how* they hold them accountable, and how they reflect the values of the same political economy that has led to a pauperization of knowledge workers (e.g., Ravetz 2016).

Besides formal training, accreditation, and peer review, another aspect that many consider important for the quality of good science is reproducibility. For a long time, the privileged status of scientific knowledge has been justified by its being based on evidence that can be reproduced by others. It was only in the twenty-first century that researchers found that many studies published in even the most prestigious scientific journals could not be replicated – so much so that there is now a discussion about the so-called "reproducibility crisis" in science (e.g., Baker 2016; see also Ioannidis 2005). Reasons for this are manifold and are mostly far removed from straightforward fraud. They range from pressure on scientists to "publish or perish" so that they publish findings prematurely, they analyze and publish findings selectively, or they conceal methodological issues (low statistical power; effects of factors that were not captured in the experiment, etc.). Another important reason is that scientists are under pressure to produce new findings, and this – rather than replicating the studies of others – is what normally gets scientists funding, journal article publications, promotions, and ultimately esteem.

Reproducibility is closely related to transparency, with the idea that enough information regarding the experiment must be provided so that it can be reproduced by another set of scientists. For example, the ARRIVE guidelines (Kilkenny et al. 2010) are part of an attempt to make scientific research involving animals more transparent. They were published as a joint statement by several biomedical journals requiring that information about not only the experimental design, but also animal husbandry practices be included with any article reporting on research involving animals. The idea was that two studies could appear to be the same in terms of experimental design but would not be comparable if husbandry differed significantly.

Transparency is indeed important for producing good science; it is crucial to reproduce and validate findings to make them into "facts," which, it is important to note, need to be open to questioning if new evidence comes to light. Doubt and questioning have long been a crucial part of scientific research, and transparency can support both in a productive way. But transparency is not an end in itself, and neither is "openness." More and more research funders and universities endorse and adopt "open science" guidelines, which require publicly funded research being transparent at various stages: Depending on the discipline and the policy, either the findings of the research ("open access publishing") or also the protocols, or even the raw data, need to be made available to the public. This comes, in part, from the understanding that the public, through taxes and other contributions, finances science and must not be excluded from access to it. At times, however, openness and transparency are used to gloss over structural problems. One such problem is the way in which higher education is increasingly seen as a service to buy, instead of a public good, and research is becoming a domain of competition rather than collaboration. Openness is transformed from a commitment to making the different steps of research reproducible into an ideology that is blind to the harms and costs of its own practices (see e.g., Mirowski 2018; Nerlich et al. 2018; Prainsack & Leonelli 2018). We need to ask what goals openness serves in this context, and also probe unintended consequences.

For example, many people inside and outside of science are familiar with open access publishing, meaning that scientific research and findings are made available to the public without cost barriers. But Open Science goes much further. It seeks to make potentially every aspect of science, from the data to the methods and tools (including protocols, software, etc.) publicly available. A precursor of the Open Science movement as we know it today was the Human Genome Project in the 1990s, a multi-national and private–public collaboration to map the first human genome. In 1996, the institutions participating in the Human Genome Project decided to release every newly mapped DNA sequence into the public domain within 24 hours (Hilgartner 2017). The idea behind this is not only that other scientists, and members of the public, need to be able to see the key elements of answering a specific question or addressing a particular problem, but also that they should see everything so that they can judge what they consider the key elements in the first place.

But there are costs to openness and transparency. As recent scholarship has pointed out, making data, processes, and outputs "open" takes considerable time and resources from researchers, and it does not guarantee that people can realistically access these data and information (e.g., Levin & Leonelli 2017). Moreover, the shift within research ethics to requiring that researchers are open and transparent about every aspect of their research also to research participants makes some research projects practically unfeasible. Take a research project looking into why people hold racist views; if people had to be told upfront that this research addresses racism, this would put many racists off participating. As this example shows, openness and transparency can have the unintended consequence of doing the opposite, namely closing things off and hindering, rather than fostering, the progress of scientific research.

13.5 Conclusion

This chapter has shown how facts and values cannot be easily disentangled and has suggested that demarcating the line between facts and values is not a fruitful way forward in evaluating what counts as good science. To make this argument we started by introducing the debates regarding facts and values amongst philosophers of science and political scientists. We then used historical examples demonstrating how biological science in the past was, from today's perspective, "unscientific" or morally wrong (e.g., eugenics, vivisection). However, we contend that this should not be interpreted as a historical problem alone. Rather, scientific norms and standards are always shaped by political, religious, and other cultural factors, and we argue that science thus is never entirely "neutral" and objective. In other words, both historical and also contemporary biological science is always part of society, and it cannot be understood in isolation from societal, political, and economic norms and imperatives. We discussed this on the occasion of two contemporary examples, the debates on "intersex" and the field of epigenetics. We then posed the question of how we know what good science is and discussed concepts such as reproducibility, transparency, openness, and ethics in connection with this. While we argue that each of them plays a role in demarcating good science from non-science (or bad science), we also warn that none of these concepts should be treated as ends in themselves; if we treat them as such then we risk turning them into ideologies. Maintaining doubt and creating the conditions in which to question facts becomes the hallmark of good science in this context.

References

Agar, N. (2008). *Liberal Eugenics: In Defence of Human Enhancement.* Hoboken, NJ: John Wiley & Sons.

Allen, G. E. & Turda, M. (2015). Eugenics as a Basis of Population Policy. In *International Encyclopedia of the Social and Behavioral Sciences*, 2nd ed., pp. 218–223. Amsterdam: Elsevier.

Arluke A. (1991). Going into the Close with Science: Information Control among Animal Experimenters. *Journal of Contemporary Ethnography* 20(3): 306–330.

Baker, M. (2016). Is There a Reproducibility Crisis? A Nature Survey Lifts the Lid on How Researchers View the "Crisis" Rocking Science and What They Think Will Help. *Nature* 533(7604): 452–455.

Berg, L. D., Huijbens, E. H., & Larsen, H. G. (2016). Producing Anxiety in the Neo-liberal University. *The Canadian Geographer* 60(2): 168–180.

Bittel, C .J. (2005). Science, Suffrage, and Experimentation: Mary Putnam Jacobi and the Controversy over Vivisection in Late Nineteenth-Century America. *Bulletin of the History of Medicine* 79: 664–694.

Camporesi, S. & Cavaliere, G. (2018). Eugenics and Enhancement in Contemporary Genomics. In S. Gibbon et al. (eds.), pp. 195–202. *Routledge Handbook of Genomics, Health and Society*.

Cavaliere, G. (2018). Looking into the Shadow: The Eugenics Argument in Debates on Reproductive Technologies and Practices. *Monash Bioethics Review* 36(1–4): 1–22.

Cetina, K. K. (2009). *Epistemic Cultures: How the Sciences Make Knowledge.* Cambridge, MA: Harvard University Press.

Chase, C. (1998). Hermaphrodites with Attitude: Mapping the Emergence of Intersex Political Activism. *GLQ: A Journal of Lesbian and Gay Studies* 4: 189–211.

Conrad, P. (1992). Medicalization and Social Control. *Annual Review of Sociology* 18(1): 209–232.

Conrad, P. (2005). The Shifting Engines of Medicalization. *Journal of Health and Social Behavior* 46(1): 3–14.

Davies, G., Greenhough, B., Hobson-West, P., et al. (2018). Science, Culture, and Care in Laboratory Animal Research: Interdisciplinary Perspectives on the History and Future of the 3Rs. *Science, Technology, & Human Values* 43: 603–621.

Davis, G. (2015). *Contesting Intersex: The Dubious Diagnosis.* New York: New York University Press.

Elston, M. A. (2006). Attacking the Foundations of Modern Medicine? In D. Keleher, J. Gabe, & G. Williams (eds.), *Challenging Medicine*, 2nd ed., pp. 162–185. London: Routledge.

Fausto-Sterling, A. (1993). The Five Sexes: Why Male and Female Are Not Enough. *The Sciences*, March–April 1993, 20–24.

Fausto-Sterling, A. (2000). *Sexing the Body: Gender Politics and the Construction of Sex*. New York: Basic Books.

Fleck, L. (2012 [1935]). *Genesis and Development of a Scientific Fact*. Chicago: University of Chicago Press.

French, R. D. (1975). *Antivivisection and Medical Science in Victorian Society*. Princeton. NJ: Princeton University Press.

Galton, F. (1883). *Inquiries into Human Faculty and Its Development*. London: Macmillan.

Hashiloni-Dolev, Y. (2007). *A Life (Un) Worthy of Living: Reproductive Genetics in Israel and Germany*. Dordrecht: Springer Science & Business Media.

Heard, E. & Martienssen, R. A. (2014). Transgenerational Epigenetic Inheritance: Myths and Mechanisms. *Cell* 157(1): 95–109.

Hedgecoe, A. (2001). Schizophrenia and the Narrative of Enlightened Geneticization. *Social Studies of Science* 31(6): 875–911.

Herzog, H. A. (1993). The Movement Is My Life: The Psychology of Animal Rights Activism. *Journal of Social Issues* 49: 103–119.

Hilgartner, S. (2017). *Reordering Life: Knowledge and Control in the Genomics Revolution*. Cambridge, MA: MIT Press.

Hobson-West, P. (2009). What Kind of Animal Is the "Three Rs"? *ATLA* 37: 95–99.

Ioannidis, J. P. A. (2005) Why Most Published Research Findings Are False. *PLOS Medicine* 2(8): e124.

Jasanoff, S. (ed.) (2004). *States of Knowledge: The Co-production of Science and the Social Order*. London: Routledge.

Jasper, J. M. & Nelkin, D. (1992). *The Animal Rights Crusade: The Growth of Moral Protest*. New York: Free Press.

Jurmain, R., Nelson, H., & Turnbaugh, W. A. (1990). Understanding Physical Anthropology and Archeology. 4th ed. St Paul, MN: West Publishing.

Kean, H. (1998). *Animal Rights: Political and Social Change in Britain since 1800*. London: Reaktion Books.

Kilkenny, C., Browne, W. J., Cuthill, I. C., et al. (2010). The ARRIVE Guidelines. Animal Research: Reporting in Vivo Experiments. *PLOS Biology* 8: e1000412.

Kirk, R. G. W. (2018). Recovering the Principles of Humane Experimental Technique: The 3Rs and the Human Essence of Animal Research. *Science, Technology, & Human Values* 43(4): 622–648.

Kitcher, P. (1995). *The Advancement of Science: Science without Legend, Objectivity without Illusions*. Oxford: Oxford University Press.

Klosin, A. & Lehner, B. (2016). Mechanisms, Timescales and Principles of Trans-Generational Epigenetic Inheritance in Animals. *Current Opinion in Genetics & Development* 36: 41–49.

Koch, L. (2004). The Meaning of Eugenics: Reflections on the Government of Genetic Knowledge in the Past and the Present. *Science in Context* 17(3): 315–331.

Kuhn, T. S. (1962). *The Structure of Scientific Revolutions*. Orig. ed. Cambridge, MA: Harvard University Press.

Laqueur, T. (1992). *Making Sex: Body and Gender from the Greeks to Freud*. Cambridge, MA: Harvard University Press.

Law, J. & Lien, M. E. (2017). The Practices of Fishy Sentience. In K. Asdal, T. Druglitro, & S. Hinchliffe (eds.), *Humans, Animals and Biopolitics: The More than Human Condition*. London: Routledge.

Lederer, S. E. (1992). Political Animals: The Shaping of Biomedical Research Literature in Twentieth Century America. *Isis* 83(1): 61–79.

Levin, N. & Leonelli, S. (2017). How Does One "Open" Science? Questions of Value in Biological Research. *Science, Technology, & Human Values* 42(2): 280–305.

Lock, M. & Palsson, G. (2016). *Can Science Resolve the Nature/Nurture Debate?* Hoboken, NJ: John Wiley & Sons.

McLeon, C. & Hartley, S. (2018). Responsibility and Laboratory Animal Research Governance. *Science, Technology, & Human Values* 43(4): 723–741.

Meloni, M. (2016). *Political Biology: Science and Social Values in Human Heredity from Eugenics to Epigenetics*. Dordrecht: Springer.

Message, R. (2016). To Assist, and Control, and Improve, the Operations of Nature: Fish Culture, Reproductive Technology, and Social Order in Victorian Britain. Doctoral Dissertation, London School of Economics and Political Science (LSE).

Mirowski, P. (2018). The Future(s) of Open Science. *Social Studies of Science* 48(2): 171–203.

Nerlich, B., Hartley, S., Raman, S., & Smith, A. (eds.) (2018). *Science and the Politics of Openness*. Manchester, UK: Manchester University Press.

Olssen, M. (2016). Neoliberal Competition in Higher Education Today: Research, Accountability and Impact. *British Journal of Sociology of Education* 37(1): 129–148.

Polanyi, M. (1962). The Republic of Science: Its Political and Economic Theory. *Minerva* 38(1): 54–74.

Prainsack, B. & Leonelli, S. (2018). Responsibility. In B. Nerlich et al. (eds.), *Science and the Politics of Openness: Here Be Monsters*, pp. 97–106. Manchester, UK: University of Manchester Press.

Putnam, H. (2002). *The Collapse of the Fact/Value Dichotomy and Other Essays*. Cambridge, MA: Harvard University Press.

Ravetz, J. (2016). How Should We Treat Science's Growing Pains? *The Guardian* (June 8, 2016). www.theguardian.com/science/political-science/2016/jun/08/how-should-we-treat-sciences-growing-pains [accessed December 28, 2019].

Rupke, N. A. (1997). *Vivisection in Historical Perspective*. London: Croom Helm.

Russell, W. M. S. & Burch, R. L. (1959). *The Principles of Human Experimental Technique*. London: Methuen.

Schell, P. (2015). Eugenics in the Americas. In *International Encyclopedia of the Social and Behavioral Sciences*, 2nd ed., pp. 246–252. Amsterdam: Elsevier.

Schwartz-Marín, E. & Silva-Zolezzi, I. (2010). "The Map of the Mexican's Genome": Overlapping National Identity, and Population Genomics. *Identity in the Information Society* 3(3): 489–514.

Sharp, L. A. (2019). *Animal Ethos: The Morality of Human–Animal Encounters in Experimental Lab Science*. Berkeley, CA: University of California Press.

Sperling, S. (1988). *Animal Liberators: Research and Mortality*, Berkeley: University of California Press.

Stepan, N. (1991). *"The Hour of Eugenics": Race, Gender, and Nation in Latin America*. Ithaca, NY: Cornell University Press.

Strathern, M., (ed.) (2000). *Audit Cultures: Anthropological Studies in Accountability, Ethics, and the Academy*. London: Routledge.

Thaler, R. H. (2008). *Nudge: Improving Decisions about Health, Wealth, and Happiness*. New Haven, CT: Yale University Press.

Trimble, S. W. (1997). Streambank Fish-Shelter Structures Help Stabilize Tributary Streams in Wisconsin. *Environmental Geology* 32(3): 230–234.

Valdez, N. (2018). The Redistribution of Reproductive Responsibility: On the Epigenetics of "Environment" in Prenatal Interventions. *Medical Anthropology Quarterly* [online first: https://doi.org/10.1111/maq.12424]

Walport, M. (2017). *Animal Research: Then and Now*. The 80th Stephen Paget Memorial Lecture at Understanding Animal Research's Openness Awards. www.gov.uk/government/speeches/animal-research-then-and-now, accessed January 27, 2020.

Watermeyer, R. & Olssen, M. (2016). "Excellence" and Exclusion: The Individual Costs of Institutional Competitiveness. *Minerva* 54(2): 201–218.

Youssef, N., Lockwood, L., Su, S., Hao, G., & Rutten, B. (2018). The Effects of Trauma, with or without PTSD, on the Transgenerational DNA Methylation Alterations in Human Offsprings. *Brain Sciences* 8(5): 83.

14 A Philosopher in the Age of Creationism

What Have I Learned after Fifty Years Doing Philosophy of Biology That I Want to Pass On to Biologists

MICHAEL RUSE

14.1 Creationism

Modern-day Creationism, taking the Bible literally, lays claim to being the truly authentic Christianity dating back to the Gospels. But while it is true that there have always been those inclined to read scripture more literally than others, from the first there have been interpretations and more, taking one away from the actual words of the text. St. Augustine, the most influential figure in Western Christianity, was clear on this. The Bible is inspired, the Word of God. God, however, knew that He could not always talk literally. The ancient Jews were not sophisticated, fourth-century Romans and would have had little understanding if, say, rainbows were described in scientific terms. Metaphor or allegory was essential (Augustine 1982).

With various phases, this has been the stance down to the present. It is true that the Reformers were more literalistic than the Catholics, but they too were in the business of interpretation. No one ever thought the Whore of Babylon was an actual woman – it was always the Pope or the Catholic Church or (more ecumenically) Saladin or others outside the faith. Down through to the nineteenth century, thanks especially first to science – the universality of Noah's Flood crumbled before modern geology – and then to so-called Higher Criticism – looking at scriptural passages as human-produced documents of their time – increasingly literalistic readings fell out of favor.

The exception, as so often is the case when it comes to religion, is the USA. The Deism of the founders proved an inadequate ideology for the farmers and artisans of the new country, and by the 1830s a homegrown evangelical religion was taking root (Porterfield 2012). Historian Ronald Numbers (2006) has shown that the Seventh-day Adventists had a major role in structuring the ways in which the Bible was approached and read – they were literalists, grounded in their need to take the 6 days of Creation absolutely as 6 periods of 24 hours. After the Civil War, in the South particularly, literalism found much favor as a belief system to use as a barrier against the modernism of the hated North. Although then, as now, in all big cities, those feeling left out – drawn from the lower middle-class and respectable working-class – feeling threatened by the Catholics and others (especially Jews) flooding America and taking jobs and with alien habits, often turned to some form of evangelical literalism.

This was enshrined in a series of pamphlets, the *Fundamentals*, published at the beginning of the twentieth century. After the First World War, all came to a head when a schoolteacher in Tennessee, John Thomas Scopes, was prosecuted for teaching evolution – anathema to Creationists (Larson 1997). Although his conviction was overturned on appeal, the trial had a chilling effect on American science education, and mention of Darwin and his theory was always omitted. Until Sputnik, in 1957, and similar events and phenomena, the success of which was taken as a sad commentary on American science education. New texts were commissioned and distributed and the evangelicals found that their children were being taught the hated evolution – as gospel, to use a phrase. There was immediate reaction. A biblical scholar and a hydraulic engineer, John C. Whitcomb, and Henry M. Morris published *Genesis Flood* (1961), arguing against evolution and for the literal truth of the whole Holy Bible – six-day creation, universal deluge, parting of the Red Sea, and so on down to the present. It was hugely successful – the emphasis on the Flood, rather than Creation, reflected the fear of atomic conflict and the feeling that it was a foreteller of worse to come, Armageddon. Through the sixties and seventies, Creationism (as it was now called, rather than the more traditional Fundamentalism) gained strength, and things came to a climax in 1981, when the State of Arkansas passed a law insisting that in publicly funded schools of the state, Creationism (or Creation Science) be given "balanced treatment" along with the teaching of evolution (Ruse 1988).

This is the background to this chapter. What had any of this to do with the philosophy of science in general, with the philosophy of biology in particular, and with this writer in person? What does this have to say to biologists? Let us start with the general and move to the more particular and personal.

14.2 Founding the Philosophy of Biology

Although it had roots going back to Newton or earlier, Anglo-American philosophy of science in the post-war period was based on the nature of physics and much influenced by the beginning-of-the-century work of people such as Bertrand Russell and Alfred N. Whitehead on formalizing mathematics. The mature scientific theory was seen as a "hypothetico-deductive" system, that is, an axiom system made of empirical generalizations, laws, bound together deductively, with the upper levels making reference to unseen theoretical entities, and the latter to observable empirical consequences. Gas theory was a paradigmatic example, with the upper-level laws about unseen particles in motion, governed by Newton's laws, and the lower-level laws, such as Boyle's Law, making claims about empirical phenomena that could be observed and measured. The twin mirror aims were explanation and prediction – one past, one future – so you explained what the gas did and why and predicted what would happen if you changed circumstances and parameters (Braithwaite 1953; Nagel 1961; Hempel 1965, 1966).

This was the dominant position of the mid-1960s, when David Hull and I and one or two others entered the field and turned our attention to biology (Honenberger 2018; Ruse 2018b). Or, rather, we turned our attention to evolutionary biology. I think we both sensed that it would be there that the rich vein of philosophical problems would lie. As Theodosius Dobzhansky used to say: "Nothing in biology makes sense except in the light of evolution." I don't think there was anything particularly idealistic about this turn. We were young academics seeking fields to conquer – and the most promising are those with little literature and that is almost uniformly bad.

Hull and I felt this strongly about philosophical work on the life sciences. It had a noble history. Aristotle in the *Parts of Animals* for a start. Kant in the *Third Critique* for a second. But the work of the twentieth century left much to be desired. J. H. Woodger (1952), for instance, had put much effort into formalizing biology in the fashion of Russell and Whitehead, with ever-diminishing results for ever-increasing effort. Someone such as the Canadian philosopher Thomas Goudge (1961) based his whole analysis on secondary works for the layperson rather than serious publications in the journals. And more sophisticated thinkers, such as Jack Smart (1963), simply looked for quick examples in biology in order to show the dominant physics-based philosophy of science of the day was inadequate – no laws because biology dealt with unique instances, and its style was narrative and not deductive.

We, pioneers, labored in the vineyard for 10 years. Toward the end, we both produced, not so much texts, as introductions to the field. Mine, published in England, was *The Philosophy of Biology* (1973). Hull's, published in the USA, was *The Philosophy of Biological Science* (1974). There were differences. Hull had been a student of Norwood Russell Hanson, author of *Patterns of Discovery* (1958), and always was a little more wary of the dominant model (known as "logical empiricism") than I. But the similarities were far greater. We both homed in on the significance of population genetics for evolutionists – something ignored by earlier philosophical writers – and we both concluded that although, because of the magnitudes of the problems, sometimes modern biological thinking has to be a little looser than physico-chemical thinking, in theory and in practice, physicists and biologists think alike. They have the same aims and the same standards.

In looking back, I feel well satisfied with what we had done. The modern-day version of the philosophy of biology was off to a good start. I think, without being too immodest, this is shown by its vigorous nature today. I myself was to turn from doing this kind of analytic philosophy of science. I did however, in 1985, start a new journal devoted to the field. I edited *Biology and Philosophy* for 15 years, until 2000, when I handed over the reins to Kim Sterelny. Although I wrote a short "Booknotes" for every issue, by the end of my time, the journal had reached a level of sophistication that it would never have accepted anything by me!

14.3 History of Biology

Tremendously important for my own intellectual development and life was the publication of Thomas Kuhn's *The Structure of Scientific Revolutions* in 1962. I was just then beginning graduate work and it was not until around 1966 or so, when I was teaching and also beginning work on the philosophy of biology, that I became fully conscious of the book and its contents. I never was a Kuhnian and am not one now. Yet, he influenced me tremendously in two ways. First, by stressing the cultural nature of science – that it is not just facts – and, although it came out more in his later writings, the importance of metaphor (see Chapter 6). Scientists construct theories as much as they discover them. Second, by stressing the importance of history of science. That in order to understand the nature of today's science, one has to understand the nature of yesterday's science.

My nonexistent God is a Calvinist. He does not believe too much in the virtues of human free will. It was preordained that someone working on evolutionary theory would turn to Darwin and his *The Origin of Species*. Hugely excited by Kuhn's message, as soon as I had completed my introduction to the philosophy of biology, I turned to the history of biology. I took my first sabbatical (1972–1973) in Cambridge and – under the tutelage of the Marxist historian Robert Young and the historian of geology Martin Rudwick, not to mention the philosopher and booster of metaphor, Mary Hesse – I retooled as a serious historian of science.

I joke that, whereas I am a rather logical empiricist in the philosophy of science, I am an out-and-out constructivist in the history of science. By this I mean in philosophy I take Newtonian physics as the model – axioms, deductions, empirical laws confirmed by experience, unseen or theoretical entities explaining all – whereas in history I see social and other cultural factors all important. Philosophers, in the tradition of Karl Popper, regard science as "knowledge without a knower" – meaning it is value-free and culturally free. Historians think values and culture define the beast. As I shall show later, to say that I drew this hard distinction is not quite true. But it is not that far off the mark. To complement my introduction to the philosophy of biology, I then wrote an introduction to one of the greatest events in the history of biology, the coming of Darwinian evolutionary theory in the nineteenth century. Thanks especially to the influence of Bob Young, in *The Darwinian Revolution: Science Red in Tooth and Claw* (1979), I made much of the extent to which Darwin's theorizing was influenced by outside factors. Logical empiricists might agree that this was so, but invoking the hallowed distinction between the context of discovery and the context of justification, they would argue that once the theory is on the table, as it were, outside factors are irrelevant. Constructivists would disagree and I was right with them on this. Culture remains significant, and in Darwin's case this is especially so in religion. Intending in his youth to become an Anglican clergyman, Darwin imbibed all sorts of Christian concepts such as the tree of life, which he used as a metaphor for the history of organic life, and about the nature of organisms, that they are – as natural theologians such as Archdeacon William Paley (1802) were insisting – design-like. Paley thought this was because God designed them. Darwin argued that it was because his mechanism of natural selection fashioned them to function – those that did, survived and reproduced, and those that didn't, didn't. I did not then argue that this means that Darwin's theory is a kind of secular religion, but I did argue that the relationship between Darwin's theory and religion (Christianity in particular) is probably going to be much more complex than one of straight opposition. In some respects, Darwin's theory is the bastard offspring of Christianity and, as with humans, there is an ongoing love–hate relationship.

14.4 Arkansas

So, Creationism comes into my story. I was born in England in 1940, emigrated to Canada in 1962, and then came south to Florida in 2000. Although English schools were not very big on American history – I knew all about the Armada and Trafalgar but nothing about the War of Independence – I had certainly heard of Fundamentalism, one more mark of US daftness! (After the war, it became clear to the British that they were no longer *Numero Uno* – that had passed to the USA. Naturally, there was resentment and, although best friends, a certain amount of snarkiness at US culture.) Toward the end of the 1970s, I became personally aware of the modern-day form, because I was asked to debate against two of the leading Creationists – Henry M. Morris and Duane T. Gish (author of *Evolution: The Fossils Say No!* – more than 150,000 copies sold). I discovered I had a talent for that sort of thing. Not only was I preadapted – I had spent 10 years as a philosopher of biology and was now writing the primer on the Darwinian Revolution – but I had the kind of slick mind that is needed for these encounters. 15 years of teaching undergraduates had taught me that a good joke will often get you further than a serious argument. I did not get upset by the outrageous statements made by Morris and Gish – I just responded in kind!

When the law was passed in Arkansas, the American Civil Liberties Union (ACLU) sprang into action, mounting an attack. Obviously, they needed theologians as expert witnesses – they got Langdon Gilkey, the

leading Protestant theologian of the day – and they needed scientists –
they got Francisco Ayala, the geneticist, and Stephen Jay Gould, the
paleontologist and just then starting his meteoric rise as the best-known,
popular science writer in the United States. Did they need a philosopher? I was the obvious choice, but there were worries about this, right
up to the last moment. Not so much about me personally but about
whether they were leading with their chin to offer up someone who
worked in such a hazy area, where disagreement is the norm and not the
exception. However, the way that the Creationists (and their scientific
opponents) bandied around the names of Karl Popper – was Creationism
falsifiable? – and Thomas Kuhn – was Creationism a new paradigm? –
made my appearance on the witness stand imperative.

My testimony was the heart of the case. I do not mean this in a
boastful way. The theologians and the scientists made the essential
arguments. The most moving on the stand were the science teachers
of Arkansas, who pleaded not to be forced to teach this bogus nonsense.
I, however, offered the argument that proved crucial – that Creationism,
or Creation Science, is not genuine science but religion, and hence may
not be taught in science classes in state-funded schools. The judge made
my points the crucial part of his argument in discrediting the law.
Having looked at more sociological characterizations of science, he said
flatly:

> More precisely, the essential characteristics of science are:
>
> (1) It is guided by natural law;
> (2) It has to be explanatory by reference to natural law;
> (3) It is testable against the empirical world;
> (4) Its conclusions are tentative, i.e., are not necessarily the final
>   word; and
> (5) It is falsifiable. (Ruse and other science witnesses).
>   Creation Science ... fails to meet these essential characteristics.
> (Ruse 1988)

The law was overturned. There were three major consequences. First,
the Creationists quit trying to do things through legislation. They
worked rather (and very successfully) at nagging away at individual
school boards and the like, getting textbooks banned from classes and
so forth. They also developed a smoother version of their position,

Creationism-lite, which they called "Intelligent Design Theory" (Johnson 1991). Now the emphasis was less on the Bible taken literally and more on the need of an intelligence – more precisely, Intelligence – to get all started and going. Literalism was not discarded but put off for a while. Second, the philosophers went berserk over my testimony and the judge's ruling. People far more eminent in the field than I – Larry Laudan, for instance – argued that my criteria of demarcation were worthless and that the aim should have been simply to show that Creation Science is bad science. Fortunately, I have a hide like a politician – say what you like about me so long as you spell my name right – and I responded forcefully, arguing especially the point that teaching bad science is not unconstitutional. Teaching religion is.

Third – particularly as a result of jeers by Gish (with whom, incidentally, I always had a very warm personal relationship, as did Ron Numbers with Morris) – I realized that it was not just enough to criticize Creationism. As a committed Darwinian, I needed to articulate my own world system. If not the God of Genesis, then what? This led me into my one and only excursion into real philosophy – epistemology and ethics – as I argued that Darwinism brings about a kind of neo-Kantian human mind, structured by certain rules or imperatives, that have proven their worth and been naturally selected (Ruse 1986). Expectedly, if I got scorn for my views on scientific demarcation, I got super-scorn for my excursion into the real world of my discipline. I should say however that my arguments about ethics – a form of moral-nonrealism now known as the "debunking argument" – have found supporters in recent years. My revenge – I did this first with my critics about my Arkansas testimony – was to publish a collection that heaps coals of fire on the heads of my detractors, treating them as worthy opponents that it is a privilege to comment on. They love the attention and I get to make my totally devastating points all over again (Ruse & Richards 2017).

14.5 Accommodation

You might think that this is the end of the story, and in a way it is. To be honest, Creation Science and its successor, Intelligent Design Theory, are political tools rather than important intellectual achievements. We are not looking here at the Platonic theory of Forms or the Kantian critical philosophy. I was prepared to fight Creationism politically, but

I was not interested in pursuing Creationism or IDT as scholarly activity. With one of the leading ID supporters I did publish a collection, but as previously for me it was more to draw people's attention to my position than much else (Dembski & Ruse 2004). You might push things further, and others did, but I was not interested in doing so.

At least, not directly. One interesting consequence of my Creationism activities was that I joined up with a group of liberal Christians interested in science – the Institute for Religion in an Age of Science (IRAS) – and I annually used to go to their conference. I saw that there was interesting work here for me. Traditional Christianity, both Protestant and Catholic, may want to have little to do with extreme literalism, but, when it comes to science, it still has issues that need to be ironed out. If, for instance, Darwinism does point to a world of moral nonrealism – that there are no ultimate foundations for ethical claims – then how does one reconcile this with a world created by an all-loving and all-powerful God, who made humans in His own image? Again, if Darwinism be right and we descended from monkeys, then what price Adam and Eve? And if no Adam and Eve, then what sense do we make of substitutionary atonement, Christ's sacrifice on the Cross to counter the inherited sin of Adam?

I thought about these issues for a number of years. Influenced by my logical-empiricist background, I did not want to buy answers too cheaply, as I think was the inclination of many of my liberal Christian friends. I did not, for example, want to go against the way of modern Darwinism and cast off powerful explanatory notions such as selfish genes in favor of a more group-oriented perspective (as favored by people such as my good friends in IRAS, not to mention Marxists and others such as Stephen Jay Gould, who had their own particular group-friendly axes to grind). I did not want to be trapped into saying things like the Christian religion is holistic inasmuch as it includes all people and this corresponds to evolutionary biology, which is likewise holistic. Apart from anything else, I am rather wary of holism. It can be a good thing, but not necessarily. The biggest holists of the twentieth century were the National Socialists. *Ein Volk, ein Reich, ein Führer.* I felt – I still feel – that if science and religion are to be harmonized – I mean able to live together amicably, not that they agree – then it must be the best science and the best religion. Not science (or religion) tailored to the occasion.

In my *Can a Darwinian Be a Christian? The Relationship between Science and Religion* I tried to show that such harmony is possible. Not easy, but then, as I concluded cutely, whoever said that the important things in life are easy? I published these ideas at the beginning of the new century, just as the so-called New Atheists were gathering steam. They – Richard Dawkins and company – didn't much like Christians. Even more, they disliked harmonizers like me – giving us the intendedly demeaning name of "accommodationists." (Like the Quakers of old, I wear the name with pride.)

Yet, all was not lost and negative. One important point that these critics did raise about the arguments of my book focused in on the notion of biological progress. It is clearly essential for the Christian story that humans appear. You may not much care for the nigh-exclusive concern with humans, but there it is. Now, it is all very well to claim – correctly I think – that it matters not theologically whether we are the end point of evolution or the end point of six days of creation, but appear we must. The trouble is that Darwinism seems not to guarantee this. There is no built-in progress to the process fueled by natural selection – no better (aka more human-like) and worse (aka less human-like). The paleontologist Jack Sepkoski made this point colorfully. "I see intelligence as just one of a variety of adaptations among tetrapods for survival. Running fast in a herd while being as dumb as shit, I think, is a very good adaptation for survival" (Ruse 1996, p. 486).

At the time of writing the book, I rather glossed over this problem, following St. Augustine in saying that, if God wanted humans to appear, He could do it. The critics pointed out – properly – that this gets me close to a guided form of evolution, IDT by another name or no name at all (Coyne 2002). A number of ways of getting around this problem have been proposed since I wrote my book. One, favored by Richard Dawkins (1986), argues that an important aspect of evolutionary change is that of an "arms race." As in human arms races, groups or lines compete against each other, with ongoing improvement. Yesterday, guns shooting shells at enemy ships. Today, guided missiles. Yesterday, prey running from predators at 10 miles an hour in a straight line. Today, prey running at 20 miles an hour, zigzagging all the way. Dawkins notes that one common feature of human arms races is that as the weaponry gets more sophisticated, it tends toward artificial intelligence and the machines or contraptions that support this. Dawkins argues that in evolution we find something similar, with better and better onboard computers. Until, we reach the expected apotheosis, humans!

Another way of getting humans, favored by the paleontologist Simon Conway Morris (2003), argues that evolution makes its way to preexisting ecological niches. The saber-toothed tiger niche, for example, was occupied independently by both placental mammals and marsupials. Since obviously an intelligence niche did preexist, even if we had not occupied it, in time some human-like creature (humanoid) would have done so. A third way, promoted by the Duke paleontologist Daniel McShea and Duke philosopher Robert Brandon (2010), simply argues that complexity is a natural consequence of any evolutionary process. And so again human-like beings would have appeared.

I confess I am not convinced that any of these suggestions yield quite that result needed – that humans (or humanoids) *must* have evolved. Sometimes, even with weaponry, Keep It Simple Stupid is the best policy. Arms races might go that way in some circumstances, especially if the needed materials are not necessarily available. Even if ecological niches do preexist, and some biologists question this and argue instead that organisms create their own niches, it is not guaranteed that we will find all that do so preexist. Suppose the dinosaurs had not been wiped out. Mammals could not have thriven as they have and perhaps dinosaurs would not have been up to occupying the humanoid niche. Cold-bloodedness cannot support the brain power needed. And as for an innate drive to complexity, even if such exists, it is quite another step to say that such complexity must be consciousness producing and supporting.

Are there other options? Somewhat hesitantly, I would appeal to multiverses. If they exist, then presumably, given an infinite number and infinite time, life will reappear, and that will kick off the evolutionary process. Although arms races might not usually work, keep at it over and over again, and something human-like somewhere sometime will appear. Likewise with niches. After all, it happened with us so it must be possible, and an infinite number of attempts will turn the possible into the certain. Monkeys will type Shakespeare and Darwinian evolution must produce humans. (I doubt the Christian story would be much changed if we were all bright green rather than the colors of us down here on Planet Earth.) Of course, it does mean that these multiverses are filled with non-producible plays and would-be humans with the IQs of cabbages. No necessary progress with a vengeance!

14.6 Teleology

When I published *Can a Darwinian Be a Christian*, in 2001, I had been a philosopher and historian of biology for rather more than 30 years, and for the past 15 years of that span – as I noted earlier – I had moved away from analytic philosophy. For a while, enthusiastic about history, I thought I might give up philosophy entirely, but this could never really be the case. I have always had the sneaking feeling that, although history is fun, it is not really a full-time occupation of grown-ups. So, my move proved to be to the history of ideas – rather than analyzing philosophical problems as they are to us today, using history to solve these philosophical problems. My work on teleology, final cause, illustrates this perfectly. Even as I was writing my little book on the philosophy of biology, I was starting to worry about this problem. Aristotle (and Darwin) spoke of and referred to "final causes" – explanations in terms of the future, as in the plates on the backs of stegosaurus are for heat control. As Kant noted in the *Third Critique*, they are essential in biology. Brought on by natural selection, they focus on adaptations, the key causal component of evolutionary thinking. Yet, they are illicit in physics – the moon does not exist in order to light the way home for drunken philosophers (even if it does). In the physical sciences, "efficient causes" in terms of the past are all that are needed and all that are allowed. Why should biology be different? I didn't solve the problem then, but noted it as important. Somehow it didn't seem to fit within the logical empiricist mold, and attempts to force it in – Nagel suggested that teleology flags that you are talking about goal-directed systems – simply don't work.

The problem continued to nag. As the years went by and I became more historically competent and confident – the history of philosophy as much as the history of science – I came to realize that teleology has a very long history. Plato had one position, seeing an actual designer behind it all. It was not, as with the Christian God, a creator. But, thanks to St. Augustine's Neo-Platonism, there are striking similarities. Aristotle had another position seeing a force acting in nature, the forerunner of the vital forces postulated by people such as the early twentieth-century French philosopher Henri Bergson. Kant had yet a third position, thinking teleology purely heuristic but necessary nevertheless. I saw Darwin as cutting the Gordian knot, keeping final cause – as in the purpose of the eye – but, thanks to natural selection, making unnecessary a designer or vital forces, also showing the Kantian why final cause is necessary. The thinking of boosters of metaphor such as Thomas Kuhn (1993) and Mary Hesse (1966) – arguing that metaphor introduces an essential and irreplaceable element in scientific understanding – is crucial here. Design is the metaphor through which we view organisms, and to reject this is to reject a vital tool of understanding – something not needed in the physical sciences because objects such as the sun and the moon are not design-like. They just are.

At last, thanks to history, I had my philosophical solution. I should say that I regard this as deeply threatening to the logical empiricist program. While I would deny with indignation the idea that biology is second rate, I would strongly agree that it is not like physics and never could be. Hopes of "reduction" – the deduction of biology to physics, the holy grail of the logical empiricists – are doomed to failure. Biology is forever different. Not because it is looser or more narrative-like – I don't see much looseness or narration in the double helix – but because it is teleological in some profound sense. Unlike inorganic substances – the moon for instance – organisms – trilobites for instance – are design-like and call for a different kind of understanding. Not second-rate. Different. Women are not second-rate men. They are different. Same for the biological and physical sciences. Biology is not second-rate physics – as Kant rather thought ("there will never be a Newton of the blade of grass") – it is first-rate biology. It is (or can be) excellent science. Unlike the Dutchman, Braithwaite, Nagel, and Hempel will forever be at sea. Organisms are molecules in motion, but they have a distinctive organization missing in the physical sciences. Biology can never be a deduction from physics.

For all that I had worries when I wrote my first book, *The Philosophy of Biology*, I was confident that tinkering would lead to a solution. Not a wholesale revision of my thinking about science. This is why my quip about being conservative in philosophy but radical in history is not quite true. My faith in the conservative philosophy of logical empiricism crumbled rather in the face of my radical views about history. Conversely, the work on teleology shows I was never an out-and-out constructivist, thinking science is all made up by us through our culture. The eye really is like a telescope in a way that the moon is not like a street lamp.

14.7 Progress

The work on teleology was spaced out, from my early days as an analytic philosopher to the very recent past, by which time I had become a full-time and explicit historian of ideas. In fact, I dropped the topic of teleology for 30 years until the beginning of this century, when I took it up again, although it was not until my book of a year past that I really think I did the topic justice (Ruse 2018b). My earlier book on the topic, *Darwin and Design: Does Evolution Have a Purpose?* (2003) was about teleology, but I would not pretend that I wrote it simply because the topic continued to nag. It was really not so much about final cause in its own right, but a corollary to work that I had been doing for 15 years, from the end of my more direct involvement in Creationism – work that was part cause and part effect of my move from analytic philosophy.

Showing again that I really did not separate my logical empiricism from my social constructivism, I was worrying increasingly about the question of values in science. If the philosophers were right, then values get expelled from mature science. It is all a matter of the context of discovery against the context of justification. You can of course have relative values – the diesel engine is more efficient than the gas (petrol) engine–but you cannot have absolute values. Science cannot tell you if the diesel engine is a good thing. So final causes don't really give you much trouble. No one is saying, for instance, that the fang of the viper is a good thing in itself. It is a good thing for the viper. And a bad thing for us!

If the constructivists are right, all of this is pie in the sky. Culture is drenched in absolute values – democracy is better than fascism – and so consequently is science. You cannot get away from this. Popper's (1972) celebrated saying that science is knowledge without a knower – meaning the nature of the knower is irrelevant – is bunkum. Much science is racist and anti-feminist. Absolute values – white better than black, male better than female. And if you change it the other way around, you still have absolute values. Black better than white, female better than male. Even those in the middle are stuck with such values. Black and white equal, male and female equal.

I am a philosophical naturalist. I believe that the best knowledge we have is science and hence others like philosophers should use the results of science as far as possible and emulate the methods of science as far as possible. This does not mean that philosophy should vanish into the sciences. It has its own problems, like the nature of values in science. There are problems of philosophy that I am not sure are susceptible to scientific answers (Ruse 2010). The problem of consciousness may be one. The very nature of existence – Why is there something rather than nothing? – is surely another. But science is the great tool and example. Naturalism! In the case of moral non-realism, this was exactly my strategy, using the nature and findings of evolutionary biology – its non-directionality – to destroy hopes of real moral foundations.

In approaching the question of values in science, 1 likewise wanted a naturalist strategy, this time using the methods of science to attack the problem. I needed a case study against which I could test rival hypotheses. I have just mentioned the non-directionality of evolutionary change and this was the obvious example. Does evolutionary change show a pattern? Does it really show a lot of direction, from the simple to the complex, from the blob to the human, from (as they used to say in the nineteenth century) the monad to the man? Having written a book on the Darwinian Revolution, I knew that a lot of people back in the nineteenth century were keen on the idea, for all that apparently Darwin had brought it tumbling down. His mechanism of change is the epitome of non-directionality. First, natural selection is relativistic – being dumb in the middle of the herd can facilitate survival better than being big brained – and, second, the building blocks of evolution, new variations, are random – not in the sense of being uncaused but in the sense of not occurring with respect to a particular need. Darwin himself insisted on that, even though he had little idea of the real truth about new variations, mutations as we now call them.

And yet, for all that I today was using the non-directionality of Darwin's theory to make a case for moral non-realism, and for all that I am right in line with modern evolutionary biology – you search in vain the pages of journals such as *Evolution* and *American Naturalist* for mentions of biological progress – Darwin himself was an ardent biological progressionist! These are the closing lines of the *Origin*.

> Thus, from the war of nature, from famine and death, the most exalted object which we are capable of conceiving, namely, the production of the higher animals, directly follows. There is grandeur in this view of life, with its several powers, having been originally breathed into a few

forms or into one; and that, whilst this planet has gone cycling on according to the fixed law of gravity, from so simple a beginning endless forms most beautiful and most wonderful have been, and are being, evolved. (Darwin 1859, p. 490)

He was not an outlier. One might expect progress before Darwin. Indeed, it was the reason why so many were attracted to evolution, despite the lack of solid evidence. But after Darwin, if anything progress was even more popular. Today's most eminent evolutionist, Harvard ant specialist and sociobiologist Edward O. Wilson (1992), is blunt on the subject. "Progress" he tells us "is a property of the evolution of life as a whole by almost any conceivable intuitive standard, including the acquisition of goals and intentions in the behavior of animals" (p. 187).

Biological progress was an ideal case study through which to understand the place of (absolute) value in science. It is hard to think of anything more loaded with absolute value than the claim that evolution leads to humankind. So, what can it tell us about values in science? With respect to the history of evolutionary biology, the problem to be solved is why early evolutionists (including Charles Darwin) were all completely committed to biological progress, whereas if you look at today's journals such as *Evolution*, despite Wilson's enthusiasm, there is not a mention of it.

There are two obvious hypotheses. The first is the logical empiricist hypothesis. This hypothesis was that as science matures, it shows (in the jargon of philosophers) more "epistemic excellence" – meaning, in normal language, that it shows the merits of science we value (it is good at prediction and so forth). Simultaneously, cultural values like progress – evolution tends to excellence (*Homo sapiens*) – get kicked out. Natural selection and mutation theory lead to prediction and unification and simplicity, to the satisfaction of values that we take to be characteristic of the best science. So progress must and does go. The second is the constructivist hypothesis. Culture is still there, it is just that the values have changed. Who today in an age of global warming and atomic weapons and world poverty can still believe in progress culture and hence progress in biology? Absolute values are there as before. It is just that they now anti-progress.

As a naturalist, I did my own empirical research, going to the archives and interviewing evolutionists to find the facts (Ruse 1996). It turned out that both hypotheses were wrong! Evolutionists still believe in bio-logical progress but they keep it out of their professional science because it is antithetical to their being considered quality scientists! There are still values in the best kind of science. However, these are not the values of external culture – four legs good, two legs better – but more the values of the internal culture of science itself – do what is going to get the respect of your fellow scientists, especially physical scientists. Exactly why today's evolutionists continue as progressionists is another ques-tion, but my guess is that it is bound up with the nature of science where you do see genuine progress: Copernicus is better than Ptolemy, Darwin is better than Genesis, Einstein is better than Newton. You almost have to believe in cultural progress if you are to be a successful scientist, and the rest flows. From culture to biology in one easy step, except most of them would not think it a step at all.

There is a difference between my work on teleology and my work on progress. I thought teleology was there and was functioning in biology. It was just that a lot of people, including biologists, thought it should not be there and were rather ashamed of it. My solution showed that this was a mistaken sentiment. In the case of progress, ostensibly it is no longer there and is non-functioning. In reality, many biologists have a hankering after it, and it keeps popping up in more popular venues such as presidential speeches and museums. It functions perhaps a little too well. There is nevertheless a connection between the two, in that they both seem future-directed – the course of evolution is toward humans and the eye exists in order to enable us to see – and this was the reason I wrote my first book on teleology, as a kind of follow-up to the work on progress. I see now that the connection is less than I supposed. Progress involves absolute value, which is illicit in science, whereas final causes involve relative values, which are allowable in science. More work on teleology had to wait, because the problem of progress continued press-ing. My troubles with the New Atheists showed me that notions of progress are, let us say, insidious and pervasive. A bit more personal than I anticipated.

14.8 Progress versus Providence

What I did spot increasingly was that progress came up in discussions about religion, in opposition to, or as a substitute for, deep Christian commitments. In my discussion of the necessity of human beings, I was (illicitly I now believe) implicitly implying something like progress. I was putting progress (or Progress if you want to include all kinds, cultural and biological) against the Christian doctrine of God doing the job directly – Providence. Casting thought back to the work on biologists and progress, it was clear that most had no religious beliefs – there were major exceptions such as R. A. Fisher and Theodosius Dobzhansky, sincere Christians both – and to a certain extent their beliefs in Progress were substituting. This is certainly the case of someone like Julian Huxley in the past. *Religion without Revelation* (1927) tells all. The same is true of Richard Dawkins (2006) in the present. He claims not to be religious, but truly a progressivist evolutionary humanism underlies his thinking. Ed Wilson (1978) is open in his desire to replace Christianity with a Progress-based secular world picture.

This all brought me back once again to Creationism. I always found slightly puzzling their extreme animosity to Darwinism. It could not be simple literalism. As I explained, no one who claims to be a biblical literalist takes the whore of Babylon seriously, and the most important theologians have always played fast and loose with the text. Luther did not much like James with its message about the importance of action over belief. "A right strawy epistle," he called it. Moreover, no one – meaning no evangelical Christian – today rejects Copernicus on biblical lines. The sun may have stopped for Joshua implying that the sun usually moves. So what? No one – including all those teetotal Baptists – rejects the Miracle at Cana (water into wine) even though it goes against modern chemistry.

In any case, today's Creationists accept a fair amount of evolution! Creationists now distinguish between "macroevolution," which they do not accept, and "microevolution," which they do accept. To explain why there are all those finches on the Galapagos, instead of saying that Noah took every species into the Ark, the claim is that he took "kinds" into the Ark, and then when released they evolved into today's different species. I took a crowd of graduate students to the museums of the midwest and we all agreed that the display on natural selection in the Creationism Museum, just south of Cincinnati, was far superior to the display on natural selection in the Field Museum, the natural history museum of Chicago.

What's going on here? Increasingly, I saw the issue as one of Progress (understood in the inclusive sense) versus Providence. In the case of

Providence, God does it all and we can best go along with this and prepare for the future, the Second Coming, both by living in faith and in converting others. In the case of Progress, it is up to us. Roll up your sleeves and get cracking on improving things here on Earth. In theological terms, the former is known as premillennialism. You think that Jesus will return and then there will be a fight with Satan (Armageddon) after which Jesus will reign for a thousand years and then the Day of Judgment. The latter is known as postmillennialism. You make things better yourself so that the thousand years become a reality before the Day of Judgment. Jesus will not return until the end of the millennium. (Amillennialism, believed by Catholics and others, thinks that we are already living in the millennium.)

In *The Evolution–Creation Struggle* (2005), I argued that Creationists are premillennialists and Evolutionists tend to a secular form of postmillennialism. Creationists are preparing the way for the coming of the Lord, trying to get themselves on board with what God wants them to believe. Evolutionists are secular postmillennialists, thinking that things are improving and it is their job to keep the upwards momentum. The point is that in respects they are both in the same business and hence the bitterness of the fight. Hindus and Muslims after the partition of India, Catholics, and Protestants in Northern Ireland, Palestinians and Jews in the Near East.

A lot of people on both sides scoffed at my somewhat cynical thesis, although shortly before he died Henry Morris wrote a sympathetic (if at times critical) review. I did not then think too explicitly about rival religions. However, over the years my thinking has crystalized firmly in that direction, and broadened in that I think a lot of Christians – not just the evangelical extremists (although much including these) – are uncomfortable with Darwinism because it is promoted as a rival (secular) religion. Recently I wrote a book, based on literature – fiction and poetry – arguing that from the beginning evolution was used and seen in that way and it only intensified after Darwin published *The Origin of Species* in 1859 (Ruse 2017a). Even more recently, I seized on the topic of war as a case study to make my point (Ruse 2018a). I argued that, because we are tainted with original sin, Christians consider war a bad thing but inevitable. Only Providence can save us. Darwinians have to think that war was a good thing, because it led progressively to us. Now, thanks to our evolved intelligence, it is possible to move on beyond war.

I want to stress, in arguing Darwinism functions as a secular religion, I have not at all given up on my original thesis, that Darwinian evolutionary theory is a genuine science, that measures to the standards of the physical sciences. I wrote a book stressing just that (Ruse 2006). My point is that Darwinism as religion is fortified by the belief in – the backbone of – Progress. Progress is not a scientific concept – taken as science it is false – and, judged by the criteria I gave the judge in Arkansas, it is a religious concept. And this sets up the extreme antipathy of the evangelicals who argue for a form of biblical literalism. Even though they believe in Noah's Flood, they don't lie awake at night worrying about those who don't believe in the Flood. They do lie awake worrying about feminism and abortion and LBGT and Muslims and the like. And they see that the secular religion of Darwinism promotes all of these in ways they find offensive. So the fight is on. This is, incidentally, as historians have shown, exactly the fight that was going on a century ago in the Scopes Monkey Trial of 1925 (Larson 1997). People like William Jennings Bryan, the politician who took on the task of prosecuting Scopes, were not worried about flooding. They were worried about the modernism coming down from the Hades of the North, contaminating the school children of the South and of rural America in the West.

14.9 Summing Up

After 50 years as a historian and philosopher of science, where do I come out at the end? My career obviously has not been exclusively one of fighting Creationism, but as obviously my interactions with Creationism have been personally and intellectually important, and in significant respects shaped or guided the moves I have made. I was raised intensely as a Quaker, and I am sure this shows in every thought I have and every move I make – starting with being a lifelong teacher. I am not a Christian. I don't accept the Christian God or Jesus as His son. I reject Providence. Neither am I a secular humanist, subscriber to Darwinism as religion. I don't accept Progress – certainly, I don't accept biological progress and I blow hot and cold on cultural progress. The growth of the number of women at universities in the past half century is a good thing. The even-greater growth of the number of administrators at universities is very much not a good thing.

I have recently written or edited several books on atheism (Bullivant & Ruse 2013; Ruse 2015). Not only am I an atheist about Christianity, this skepticism extends to other religions such as Buddhism (Ruse 2019). I am sympathetic to most of them – I can see much good in both Christianity and Buddhism, for all the failings. After writing a book on Gaia, the idea of the world as an organism, I would even extend this sympathy to Paganism (Ruse 2013). My researches led me into their world and I left, not convinced, but nothing like as cynical as before. I recently revised a little dialogue on evolution and religion, adding a new participant – Peaseblossom, a transsexual Pagan (Ruse 2016). At least the Pagans care about the environment. Against this, the sight of a group of middle-aged people putting together their magazine – typing and printing and stapling – all dead serious and totally naked ("skyclad") is one of the costs of doing research seriously.

However, friendly as I am to religions, I don't believe in any of them. I describe myself as a "Darwinian existentialist" (Ruse 2019). The existence of God is not of first importance to me. What does matter is using the opportunities and talents given to you in this life. Quakers are cavalier about the truth-status of the Old Testament. They can outdo any Creationist when it comes to literal readings of the words of Jesus. The Parable of the Talents has ruled my life. To whom much is given, from whom much is expected. We never live up to our self-set standards; but, in this chapter, I have tried to show you that this has been my guiding aim during my life as a student of the philosophy and history of evolutionary biology. For the rest, with the great population geneticist J. B. S. Haldane (1927), I am simply awed by the very fact of existence. "My own suspicion is that the Universe is not only queerer than we suppose, but queerer than we can suppose" (p. 286).

14.10 Envoi

What has any of this to do with biologists and their science? Why should they care about my life's work and its results? At one level, I should say that I don't at all care if they care or not. I did my work as a philosopher and historian, not as a therapist to the life sciences. That said, I see a huge amount of importance. Most obviously, fighting *Creationism*. As I say, I was not necessarily the most important or impressive witness in Arkansas. I was the most crucial. No bragging. That is true. Given the ongoing pressure from Creationists, I beg you to read the judge's ruling (given in Ruse 1988). More theoretically, I take my work on *final causes* to be very significant. Even biologists – especially biologists – feel insecure about using final causes, and often – too often – one gets editorials urging the dropping of final causes. Or, like Kant, feeling second-rate because one cannot drop them. Nonsense! Organisms are different, they call for understanding in terms of design (even if it is natural selection and not the Great Designer in the Sky); final causes cannot be dropped, they should be paraded and celebrated. A final point for this chapter: I have helped to disentangle a lot of the worries about *values* in biology. When it is legitimate to use them – as with final cause – and when it is not legitimate to use them – as with progress. This is not to say that you cannot have value-laden notions like progress. You should not put them into your science. You should not extract them from your science.

"I have answered three questions, and that is enough,"
Said his father. "Don't give yourself airs!
Do you think I can listen all day to such stuff?
Be off, or I'll kick you downstairs!"
                                          (Lewis Carroll)

References

Augustine, St. (1982). *The Literal Meaning of Genesis*. J. H. Taylor, trans. New York: Newman.

Braithwaite, R. (1953). *Scientific Explanation*. Cambridge: Cambridge University Press.

Bullivant, S. B. & Ruse, M. (eds.) (2013.) *Oxford Handbook to Atheism*. Oxford: Oxford University Press.

Conway Morris, S. (2003). *Life's Solution: Inevitable Humans in a Lonely Universe*. Cambridge: Cambridge University Press.

Coyne, J. (2002). Intergalactic Jesus. *London Review of Books* 24(9).

Darwin, C. (1859). *On the Origin of Species by Means of Natural Selection, or the Preservation of Favoured Races in the Struggle for Life*. London: John Murray.

Dawkins, R. (1986). *The Blind Watchmaker*. New York: Norton.

Dawkins, R. (2006). *The God Delusion*. New York: Houghton, Mifflin, Harcourt.

Dembski, W. A. & Ruse, M. (eds.) (2004). *Debating Design: Darwin to DNA*. Cambridge: Cambridge University Press.

Gish, D. (1973). *Evolution: The Fossils Say No!* San Diego: Creation-Life.

Goudge, T. A. (1961). *The Ascent of Life*. Toronto: University of Toronto Press.

Haldane, J. B. S. (1927). *Possible Worlds and Other Essays*. London: Chatto and Windus.

Hanson, N. R. (1958). *Patterns of Discovery*. Cambridge: Cambridge University Press.

Hempel, C. G. (1965). *Aspects of Scientific Explanation*. New York: Free Press.

Hempel, C. G. (1966). *Philosophy of Natural Science*. Englewood Cliffs, NJ: Prentice-Hall.

Hesse, M. (1966). *Models and Analogies in Science*. Notre Dame, IN: University of Notre Dame Press.

Honenberger, P. (2018). Darwin among the Philosophers: Hull and Ruse on Darwin, Herschel, and Whewell. *HOPOS* 8: 278–309.

Hull, D. L. (1974). *The Philosophy of Biological Science*. Englewood Cliffs, NJ: Prentice-Hall.

Huxley, J. S. (1927). *Religion without Revelation*. London: Ernest Benn.

Johnson, P. E. (1991). *Darwin on Trial*. Washington, DC: Regnery Gateway.

Kuhn, T. (1962). *The Structure of Scientific Revolutions*. Chicago: University of Chicago Press.

Kuhn, T. (1993). Metaphor in Science. In Andrew Ortony (ed.), *Metaphor and Thought*, 2nd ed. , p. 533–542. Cambridge: Cambridge University Press.

Larson, E. J. (1997). *Summer for the Gods: The Scopes Trial and America's Continuing Debate over Science and Religion*. New York: Basic Books.

McShea, D. W. & Brandon, R. (2010). *Biology's First Law: The Tendency for Diversity and Complexity to Increase in Evolutionary Systems*. Chicago: University of Chicago Press.

Nagel, E. (1961). *The Structure of Science: Problems in the Logic of Scientific Explanation*. New York: Harcourt, Brace and World.

Numbers, R. L. 2006. *The Creationists: From Scientific Creationism to Intelligent Design*. Standard ed. Cambridge, MA: Harvard University Press.

Paley, W. ([1802] 1819). *Natural Theology (Collected Works: IV)*. London: Rivington.

Popper, K. R. 1972. *Objective Knowledge*. Oxford: Oxford University Press.

Porterfield, A. 2012. *Conceived in Doubt: Religion and Politics in the New American Nation (American Beginnings, 1500–1900)*. Chicago: University of Chicago Press.

Ruse, M. (1973). *The Philosophy of Biology*. London: Hutchinson.

Ruse, M. (1979). *The Darwinian Revolution: Science Red in Tooth and Claw*. Chicago: University of Chicago Press.

Ruse, M. (1986). *Taking Darwin Seriously: A Naturalistic Approach to Philosophy*. Oxford: Blackwell.

Ruse, M (ed.) (1988). *But Is It Science? The Philosophical Question in the Creation/Evolution Controversy*. Buffalo, NY: Prometheus.

Ruse, M. (1996). *Monad to Man: The Concept of Progress in Evolutionary Biology*. Cambridge, MA: Harvard University Press.

Ruse, M. (2001). *Can a Darwinian Be a Christian? The Relationship between Science and Religion*. Cambridge: Cambridge University Press.

Ruse, M. (2003). *Darwin and Design: Does Evolution Have a Purpose?* Cambridge, MA: Harvard University Press.

Ruse, M. (2005). *The Evolution–Creation Struggle.* Cambridge, MA: Harvard University Press.

Ruse, M. (2006). *Darwinism and Its Discontents.* Cambridge: Cambridge University Press.

Ruse, M. (2010). *Science and Spirituality: Making Room for Faith in the Age of Science.* Cambridge: Cambridge University Press.

Ruse, M. (2013). *The Gaia Hypothesis: Science on a Pagan Planet.* Chicago: University of Chicago Press.

Ruse, M. (2015). *Atheism: What Everyone Needs to Know.* Oxford: Oxford University Press.

Ruse, M. (2016.). *Evolution and Religion: A Dialogue.* 2nd ed.. Lanham, MD.: Rowman and Littlefield.

Ruse, M. (2017a). *Darwinism as Religion: What Literature Tells Us About Evolution.* Oxford: Oxford University Press.

Ruse, M. (2017b). *On Purpose.* Princeton, NJ: Princeton University Press.

Ruse, M. (2018a). *The Problem of War: Darwinism, Christianity, and Their Battle to Understand Human Conflict.* Oxford: Oxford University Press.

Ruse, M. (2018b). *The Darwinian Revolution: What Philosophers Should Know.* Cambridge: Cambridge University Press.

Ruse, M. (2019). *A Meaning to Life.* Oxford: Oxford University Press.

Ruse, M. & Richards, R. J. (eds.) (2017). *The Cambridge Handbook of Evolutionary Ethics.* Cambridge: Cambridge University Press.

Smart, J. J. C. (1963). *Philosophy and Scientific Realism.* London: Routledge and Kegan Paul.

Whitcomb, J. C. & Morris, H. M. (1961). *The Genesis Flood: The Biblical Record and Its Scientific Implications.* Philadelphia: Presbyterian and Reformed Publishing Company.

Wilson, E. O. (1978). *On Human Nature.* Cambridge, MA: Harvard University Press.

Wilson, E. O. (1992). *The Diversity of Life.* Cambridge, MA: Harvard University Press.

Woodger, J. H. (1952). *Biology and Language.* Cambridge: Cambridge University Press.

15 How Can We Teach Philosophy of Science to Biologists?

KOSTAS KAMPOURAKIS AND
TOBIAS ULLER

15.1 Introduction

So, we now have a book on philosophy of science written for biologists, which contains topics that its editors (both of them biologists) thought would be useful. Whereas we hope that many biologists will read it and reflect on philosophy of science, this is not the only role we envisaged for this book. We also believe that the book could be used in courses for biology undergraduates and graduates. It could be used in a course on philosophy of science, or as a supplement to a biology-focused course that would at the same time aim to promote philosophical reflection among participants. In either case, what is important for anyone who would use this book for teaching purposes is that the contributors to the present book have highlighted some major themes to be discussed.

In particular, someone who would decide to teach a philosophy of science course to biologists might wish to structure a whole semester course around the topics of this book, following the recommendations of this chapter, and base their lessons on several or all of the chapters. The outcome would be a philosophy of science course that would touch upon central issues in philosophy of science, properly contextualized with numerous examples from the life sciences. The objectives of each class could be summarized in the form of answers to the questions that constitute the title of each chapter. The question that each chapter asks in the title is addressed by its author(s) and the contents of the chapters could thus be the basis for developing an interactive discussion class for each topic. If someone did not have the opportunity (we might say the luxury) to teach a philosophy of science course to biologists, one might at least contextualize the topics of a biology course by using the present book as a supplementary reading, ideally in a study group directly related to that course. We are sure that many teachers can find creative ways to do this.

In the following sections, we provide some recommendations about teaching philosophy of science to biologists in any of the aforementioned (or any other) ways, based on our own experience. We begin with a few suggestions of what not to do when teaching philosophy of science to biologists, followed by suggestions of what one had better do. We must note that there is limited empirical research on teaching philosophy of science to biologists. However, there is a related area of teaching history and philosophy of science in science education, especially to pre-service and in-service teachers, upon which one can draw for ideas (e.g., Monk & Osborne 1997; Abd-El-Khalick & Lederman 2000; Galili & Hazan 2001; Abd-El-Khalick 2005; Höttecke & Silva 2011; Matthews 2014; Kampourakis 2020). To facilitate readers, we present the main points as lists of "not to do's" and "to do's."

15.2 What NOT TO DO When Teaching Philosophy of Science to Biologists

Philosophers of science know their topic very well – and certainly a lot better than philosophically minded biologists such as ourselves. Nevertheless, our experience is that those very same people can easily put off non-philosophers from listening further. There are several reasons for this, ranging from lack of a common language, expectations of students and teachers, attitudes toward what constitutes intellectual skills and knowledge, and what topics one considers interesting or useful. Here, then, is what we think anyone teaching philosophy of science to biologists should refrain from doing.

15.2.1 Do Not Present Philosophical Topics in Extreme Detail, But Focus on the Aspect That Matters

Philosophy is about asking questions that usually do not have a single answer. That philosophers have been preoccupied with particular questions for at least 2,500 years is not a sign of intellectual laziness. It is just that some questions are hard to answer, and the more one explores them, the more difficult it becomes to provide a definitive answer. Think for instance about causes and causation, which are central in the process of explanation. If we accept that a main aim of science is to identify the causes of particular events or of general patterns, then we need a solid account of what a cause is.

In philosophy of science, there are at least four different types of theories of causation: (i) Regularity theories; (ii) Probabilistic theories; (iii) Counterfactual theories; and (iv) Manipulability theories. These theories perceive causes in different ways that can be summarized as follows (Hitchcock 2008):

- According to regularity theories of causation, the presence or absence of the cause C makes a difference for whether the effect E regularly follows (from the conjunction of additional factors ABD).
- According to probabilistic theories of causation, the presence or absence of the cause C makes a difference to the probability of the effect E (from the conjunction of additional factors ABD).
- According to counterfactual theories of causation, the presence or absence of the cause C in nearby possible worlds makes a difference to whether the effect E occurs in those worlds (from the conjunction of additional factors ABD).
- According to manipulability theories of causation, interventions that make C occur or fail to occur make a difference to whether or not effect E occurs (from the conjunction of additional factors ABD).

Instead of presenting all these theories in detail, one had better focus on the common idea they all share: Causes are difference-makers for their effects, or that a cause makes a difference to whether or not the effect occurs. The various approaches differ over how precisely the notion of making a difference is to be understood. But the point made here is that it would be more useful for biologists to understand the notion of difference-maker, rather learn all theories of causation in detail. Genes for instance are difference-makers for their effects. Whereas many genes contribute to the development of a particular trait, in some cases a change in one of these genes is sufficient to bring about a difference in that particular trait (see Waters 2007). Armed with this understanding of causation, it will be easier to grab the biologist's attention, for example, by demonstrating how a conceptual analysis of genetic causation can reveal that some features of genes that biologists often take for granted in fact are highly dubious (Kampourakis 2017).

15.2.2 Do Not Use Complex Definitions of Concepts, but Provide Simple and Comprehensible Ones

The gene concept mentioned in the previous section is an important concept in contemporary biology research. However, the notion of genes is characterized by many discrepancies, as scientists dealing with different problems or working in different domains use distinct gene concepts. These different gene concepts stem from different scientific aims, conceptualizations about the boundaries of genes, their precise localization, and the biological functions that can be assigned to genes. A suggestion to address this problem has been to teach students, especially those who are not planning to work on genetics, about the ways in which biologists handle fundamental questions about genetics before discussing the various gene concepts (Burian & Kampourakis 2013). After a detailed analysis of the various gene concepts, the authors arrived at the following concept of "genetic material," which might replace the concept of gene in genetics education (pp. 623–624):

> Genetic material: any nucleic acid [composition] in the cell [localization] that interacts with other cellular components and transmits a specific message determining the sequence of other molecules [structure] and thus results in particular, but often quite variable, outcomes inside or outside the cell [function]. These nucleic acids are (usually) reliably copied and maintained from generation to generation, preserving their structure and resulting in the same functions in similar environments (robustness), though with a range of variation in functions and consequences that depends on cellular and environmental conditions (plasticity). The functions of particular portions of the genetic material may affect or be implicated in cellular processes with local (cellular) or extended (organismal and even environmental) impact; this allows the assignment of fitnesses to particular differences in the genetic material.

This is a concept that is in many ways more accurate and more inclusive than the typical gene concepts. One important advantage is that it allows for a clear distinction between genetic and epigenetic inheritance, as well as for a consideration of both "structural" and "regulatory" DNA sequences.

However, we assume that you are already wondering how students might learn this definition, assuming they would be able to understand it. Being aware of this, the authors provided a simpler form of the concept (Burian and Kampourakis 2013, p. 624):

> Genetic material: any nucleic acid with the propensity to be inherited and to interact with other cellular components as a source of sequence information, eventually affecting or being implicated in cellular processes with local or extended impact.

It is obvious that this second definition of the concept is much shorter and simpler, while it contains all the essential information contained in the detailed definition above. The point made here is that whereas the first definition is richer and more detailed, it might be better to use the second definition that contains the essential elements and has a lower risk of putting off biologists.

15.2.3 Make Use of History but Do Not Teach Chronologically

It is tempting to teach philosophy of science chronologically, perhaps starting with Popper and moving forward to the best thinkers of today. However, this is usually not a good idea for getting the attention of biologists since they are largely indifferent to the history of philosophy of science. In fact, biology students may also be indifferent to the history of biology; they care little for what happened in, say, the 1950s since the cutting-edge science they are interested in and the problems they wish to solve are taking place here and now. Yet, we are convinced that the history of science and the history of philosophy of science have important roles to play in science education. Perhaps most notably, examples of what happened in a particular field, such as biology, help to illustrate that concepts, theories, and knowledge are not "out there" to be discovered but need to be constructed. Another important point to note is that scientific theories, and everything that they encompass, such as models, principles, and concepts, are historical entities. As such, they originate in particular contexts, and thus have specific features directly related to those contexts. The influence of scientific exemplars in physics on the development of biology is illustrated very well by theoretical population genetics (Provine 1971). For example, Fisher successfully applied methods from statistical physics to put evolutionary theory on an equally sure footing as, for example, the theory of ideal gases. To do so, he had to apply a particular lens of seeing organisms and populations, one that has had an immense effect on biology and continues to shape contemporary evolutionary thought (e.g., Stoltzfus 2019). Examples such as these thus illustrate the relationship between science and philosophy of science. Used effectively, important historical landmarks in philosophy of science can help biology students make sense of the role that philosophical issues play in contemporary biology, and how making those issues transparent can make it easier to grasp complex biological concepts and theories.

However, as amply illustrated by the chapters in this book, scientific models, principles, and concepts also evolve across time. For instance, contemporary evolutionary theory by and large shares the fundamental principles of the theory proposed by Charles Darwin 160 years ago. However, there also exist differences because the theory has evolved since then. Another example is the gene concept: The definition that Wilhelm Johannsen provided in 1909 is entirely insufficient to cover the various gene concepts used in science in 2019. Tracing these changes can often be a good way to ask students, which of our current concepts will stand the test of time, and which will not?

15.3 What TO DO When Teaching Philosophy of Science to Biologists

We have argued that 1) focusing on the aspects of a topic that matters, 2) providing simple definitions and 3) using history but refraining from teaching it chronologically, might be more effective in teaching philosophy of science to biologists. But these will not be enough, so we now turn to what else we should do when teaching philosophy of science to biologists.

15.3.1 Present the Philosophical Topic of Interest Beginning with Simple Decontextualized Examples

Whereas we would probably all agree that the teaching of philosophy of science to biologists should be carefully contextualized with concrete biology examples, we argue that there is a crucial step one could take before doing that: use simple contextualized examples to introduce the main idea.

Let us consider the idea of difference-makers, discussed in Section 15.2.1. Whereas genes are the typical example of difference-makers in biology, we argue that one had better start introducing the idea at a simpler level. As we already explained, the main idea is that whereas several causes of a particular effect can exist, it is usually the case that one of them is responsible for bringing about a particular effect. The classic example is that of a forest fire, which requires several causal factors such as a lighted match, oxygen, and flammable materials such as dried grass. These are all important; there will be no forest fire if there isn't anything such as a lighted match to start it. Flammable material such as dried grass is also a causal factor because humid material is difficult to burn, and oxygen is of course necessary for the process of combustion. However, in the case of an actual destruction of a forest because of a fire, it is a lighted match that is usually considered to have caused the fire, and none of the other factors. The reason for this is that oxygen and dried grass can exist in a forest for a long time without any fire occurring. When a fire occurs, it is not due to a change in their condition, but in the condition of a match that makes the difference and brings about the fire. In this sense, the lighted match is the difference-maker. This is an example that is easy to understand and to convey the main idea, before one turns to, for example, genes as difference-makers. Students can actually be guided to infer on their own that the match is the difference-maker by being asked questions about the causal influence of these factors.

Here is another example. Many of the topics taught in biology are related to models. For instance, in molecular biology we teach students about the double helix structure of the DNA molecule. Imagine that we would like to take this opportunity to explain to our students the properties of this model. We could thus explain that structural models of DNA are not true, that they only exhibit similarities with the actual molecule of DNA and that they do not portray how the molecule of DNA actually is. In the end of the class we might be satisfied that we did not only teach our students about the specific model, but that we also taught them some general features of such models. However, as the particular model is something they are not well familiar with, and as abstraction and idealization are the main features of models, we should not be certain that our students really understood. To facilitate understanding, it is better to begin, again, with a de-contextualized example that our students are probably well familiar with. Geographical maps are such an example. If we asked our students to prepare a list of the properties of a map, based on their own experience they would likely note that maps are means of representation; that they exhibit similarities with aspects of real systems; that they do not represent whole systems and they are not true of them; that they are neither entirely precise, nor entirely accurate; and that they are used by scientists for a particular purpose in a specific context.

15.3.2 Present the Philosophical Topic of Interest in Detail with Many Contextualized Examples

To teach philosophical topics to biology students, it is necessary for the students to see that the topics are relevant to them. This can be done either by using contemporary research in biology, or by referring to the history of science. Let us provide a concrete example. Philosopher Ernan McMullin has provided a useful list of the virtues of a good scientific theory: (1) empirical fit (support by data); (2) internal consistency (no contradictions); (3) internal coherence (no additional assumptions); (4) simplicity (testability and applicability); (5) external consistency (consonance with other theories); (6) optimality (comparative success over other theories); (7) fertility (novel predictions, anomalies, change); (8) consilience (unification); (9) durability (survival over tests); and (10) explanatory power, which is a consequence of all the aforementioned virtues. This looks nice, but it would likely come across as meaningless to students, and might simply seem as a list to memorize. In order for this list to become meaningful and useful, we argue that careful contextualization is required. As an example, consider some of the virtues through the lens of evolutionary theory (see Kampourakis 2014, pp. 208–212 for a more complete description).

Empirical fit or support by data is the first consideration for a scientific theory. This does not mean that all theories with empirical basis are scientific; however, scientific theories need to be supported by the available data. In particular, all the available data from such diverse disciplines as paleontology, biogeography, developmental biology, cellular biology, and molecular biology support the two main propositions of evolutionary theory: that all organisms on Earth share a common ancestry and that they all have evolved from preexisting ones through natural processes. This theory can be easily tested, for example, by making predictions about the distributions of species and comparing it to biogeographical data. The theory could for instance be applied to predict where fossils of particular organisms and of particular age could have been found, and in several cases they have indeed been found. Evolutionary theory also exhibits internal consistency and no contradictions among its propositions. As we saw in Chapter 1, evolutionary explanations typically rely on representations that make many additional assumptions about living beings. However, none of these evolutionary explanations require violation of the two main features of the theory, common ancestry and descent with modification. Populations living close to each other are more closely related to each other – genetically speaking – than others living in more remote areas, even if their environments are not similar. Similar structures may evolve in otherwise very different organisms because of natural selection. And so on and on.

An especially interesting aspect of evolutionary theory is its consilience: the enormous potential for unification, as it brings together and explains different kinds of data (fossils, biogeography, morphology, genomics). Another virtue of evolutionary theory is its fertility – the core features of evolutionary theory continue to motivate research that reveals ever more about the history of life. Often this involves the revision of auxiliary hypotheses while the core of the theory is retained. Another relevant virtue is the durability of evolutionary theory: its ability to surpass all tests. So far everything that we have found is in agreement with common descent and descent with modification. This perhaps is where another virtue comes from: optimality, or the comparative success over other theories. Indeed, evolutionary theory offers the best explanations available for the characters of organisms by relying exclusively on natural processes. It is so successful that there is currently no other scientific theory for the origin and diversification of life that scientists seriously consider.

Perhaps the greatest virtue of evolutionary theory, which is actually a consequence of all the other virtues discussed above, is its enormous explanatory power. Evolutionary theory can explain a wide variety of phenomena based on a small number of propositions and models. It can explain the peculiar features and properties of organisms; the vast variety of life forms on Earth; the similarities of the DNA sequences of organisms that are morphologically very different; and much more. By discussing what makes evolutionary theory a good scientific theory, the students can discover the criteria that we listed above for themselves and then be encouraged to apply those to other biological theories.

15.3.3 Try to Discuss Philosophy of Science Topics in Standard Biology Classes, Too

Biologists may think that philosophy of science is a topic distinct from science itself and that having any philosophical discussion during a science class would be irrelevant, or at best an add-on. However, this is far from true. Philosophical topics are actually inherent in the science that we teach, and so we always address them but in an implicit manner. Therefore, what one could do is to make the philosophical dimensions of a topic explicit and have students reflect upon them.

The concept of adaptation provides a nice example. If you open a biology textbook such as the one by Raven and colleagues (2008), you will find two different definitions of adaptation. In the main text, adaptation is defined as "The ability of an individual to alter its physiology, morphology, or behavior is itself an evolutionary adaptation, the result of natural selection" (p. 1147). According to this definition, an adaptation – in this case the ability to alter some phenotypic traits – is the outcome of natural selection. This means that this ability has conferred an advantage to the ancestors of the current population and was selected because of this advantage as its bearers survived and reproduced better than the others, and eventually this ability became prevalent in the population. According to this definition, a feature is an adaptation if its prevalence in a population is the outcome of natural selection. This can be described as a historical definition of adaptation. However, if you look at the glossary of that very same textbook, you will find a quite different definition for adaptation: "A peculiarity of structure, physiology, or behaviour that promotes the likelihood of an organism's survival and reproduction in a particular environment" (p. G-1). According to this definition, an adaptation is something that currently makes a positive contribution to its bearers' survival and reproduction. Whether or not it has become prevalent in the population due to natural selection does not seem to matter. This can be described as an ahistorical definition of adaptation.

The philosophical issue that emerges here is, under which conditions a trait can be considered as an adaptation. What is more important, past selection or current selection? Some scholars have argued that past selection is what matters, because a trait can only be considered as an adaptation if it is the outcome of natural selection. If something contributes now but emerged accidentally, it cannot be considered an adaptation. Other scholars have argued that history does not matter in order for a trait to be considered as an adaptation; it is only the current contribution that matters. According to them, there is no point in calling a trait an adaptation if it does not currently make any positive contribution to the survival and reproduction of its bearers. In other words, according to the historical definition of adaptation, past selection is a necessary and sufficient condition for a trait to be considered as an adaptation; current contribution is not that important. Whereas according to the ahistorical definition of adaptation current contribution to fitness is a necessary and sufficient condition for a trait to be considered as an adaptation, past selection is not that important (see Kampourakis 2013; Kampourakis & Stern 2018).

What this example shows is that two different definitions in the same textbook can form the basis for a philosophical discussion about how to define a concept. The aim here is not to get into all the details, even though this could be an interesting exercise for students. Rather, the aim is to show to students that even defining a concept is not as simple or straightforward as they might think. Once this issue has been raised, one can use it to make sense of the different research agendas that may arise around different notions of adaptation, whether or not natural selection on genetic variation is a "good enough" explanation for adaptation, and so on (some of these issues were discussed more fully in Chapter 1). Ultimately, this may make students better prepared to position and connect the different kinds of biological research they hear about in their education.

Conclusion

In our experience, science teaching at the university level does not make students aware of the importance and usefulness of philosophy of science for doing and understanding science. This is related to a broader question about the kind of education that science departments should offer to their students, the future scientists: science courses only, or some humanities courses as well? Training to work in the lab or in the field, or also training to reflect upon important issues? To give an example, we want healthcare professionals to know how to deal with patients in the clinic; but do we want them to also have a good understanding of ethics? We think, yes. Not only because ethics is directly relevant to their work but also because we believe that people with a university education should have a broad scope of intellectual experiences and reflections. University science education should not only provide future scientists with what they will need to know for their research and teaching, but also with a broader culture that will make them better researchers and better teachers. This, of course, depends upon the members of the science faculties to decide and implement. There is a long way to go, but we can try. Teaching philosophy of science to biologists could be a modest first step in this direction.

References

Abd-El-Khalick, F. (2005). Developing Deeper Understandings of Nature of Science: The Impact of a Philosophy of Science Course on Preservice Science Teachers' Views and Instructional Planning. *International Journal of Science Education* 27(1): 15–42.

Abd-El-Khalick, F. & Lederman, N. G. (2000). The Influence of History of Science Courses on Students' Views of Nature of Science. *Journal of Research in Science Teaching: The Official Journal of the National Association for Research in Science Teaching* 37(10): 1057–1095.

Burian, M. & Kampourakis, K. (2013). Against "Genes For": Could an Inclusive Concept of Genetic Material Effectively Replace Gene Concepts? In K. Kampourakis (ed.), *The Philosophy of Biology: A Companion for Educators*, pp. 597–628. Dordrecht: Springer.

Galili, I. & Hazan, A. (2001). Experts' Views on Using History and Philosophy of Science in the Practice of Physics Instruction. *Science & Education* 10(4): 345–367.

Hitchcock C. (2008). Causation. In S. Psillos & M. Curd (eds.), *The Routledge Companion to Philosophy of Science*, pp. 317–326. New York: Routledge.

Höttecke, D. & Silva, C. C. (2011). Why Implementing History and Philosophy in School Science Education Is a Challenge: An Analysis of Obstacles. *Science & Education* 20(3–4): 293–316.

Kampourakis, K. (2013). Teaching about Adaptation: Why Evolutionary History Matters. *Science & Education* 22(2): 173–188.

Kampourakis, K. (2014). *Understanding Evolution.* Cambridge: Cambridge University Press.

Kampourakis, K. (2017). *Making Sense of Genes.* Cambridge: Cambridge University Press.

Kampourakis, K. (2020). Supporting Science Teachers' Nature of Science Understandings through a Specially Developed Philosophy of Science Course. In W. F. McComas (ed.), *The Nature of Science in Science Instruction: Rationales and Strategies.* Dordrecht: Springer.

Kampourakis, K. & Stern, F. (2018). Reconsidering the Meaning of Concepts in Biology: Why Distinctions Are So Important. *BioEssays* 40: 1800148.

Matthews, M. R. (2014). *Science Teaching: The Contribution of History and Philosophy of Science.* New York: Routledge.

McMullin, E. (2008). The Virtues of a Good Theory. In S. Psillos & M. Curd (eds.), *The Routledge Companion to Philosophy of Science*, pp. 498–508. New York: Routledge.

Monk, M. & Osborne, J. (1997). Placing the History and Philosophy of Science on the Curriculum: A Model for the Development of Pedagogy. *Science Education* 81(4): 405–424.

Provine, W. B. (1971). *The Origins of Theoretical Population Genetics.* Chicago: University of Chicago Press.

Raven, P. H., Johnson, G. B., Losos, J. B., Mason, K. A., & Singer, S. R. (2008.) *Biology.* 8th ed. New York: McGraw-Hill.

Stoltzfus, A. (2019). Understanding Bias in the Introduction of Variation as an Evolutionary Cause. In T. Uller & K. N. Laland, *Evolutionary Causation: Biological and Philosophical Reflections*, 8th ed., pp. 29–62. Cambridge, MA: MIT Press.

Waters, C. K. (2007). Causes That Make a Difference. *Journal of Philosophy* 104: 551–579.

Further Reading

This section provides some suggestions for where to find out more about the philosophy of science, and its relationship to biology. The list that follows is short relative to the literature that actually is out there, and many more books and articles can be found by browsing the literature cited in each chapter. For simplicity, only the main titles are given in the text; for a full reference, please see the list at the end of this section.

It can be useful to begin with a general introduction to the philosophy of science, especially if some of the chapters in this book felt challenging. *The Meaning of Science* by Tim Lewens, *Philosophy of Science: A Very Short Introduction* by Samir Okasha, and *Philosophy of Science: A New Introduction* by Gillian Barker and Philip Kitcher are particularly accessible due to their style, conciseness, and extensive use of biological examples. Other highly accessible introductions are *Theory and Reality: An Introduction to the Philosophy of Science* by Peter Godfrey Smith, and *Recipes for Science: An Introduction to Scientific Methods and Reasoning* by Angela Potochnik, Cory Wright, and Matteo Colombo.

Several of the classics in philosophy of science will still be highly readable by biologists today, including Thomas Kuhn's *the Structure of Scientific Revolutions* and Imre Lakatos's *The Methodology of Scientific Research Programmes*.

For more on the nature of science, we recommend *Ignorance* by Stuart Firenstein and *Uncertainty* by Kostas Kampourakis and Kevin McCain. For how to discriminate between science and pseudoscience, see Massimo Pigliucci's *Nonsense on Stilts* and Pigliucci and Boudry's *Philosophy of Pseudoscience*.

An alternative route into the philosophy of science are introductions to the philosophy of biology. These are more directly concerned with the aims, methods, and concepts of the biological sciences, which means they are more specialized. However, the topics they cover will of course be directly relevant to many biologists. Good recent introductions include *Philosophy of Biology* by Peter Godfrey-Smith and *Philosophy of Biology: A Very Short Introduction* by Samir Okasha. Two somewhat older, but still very useful, books are *Philosophy of Biology: A Contemporary Introduction* by Alex Rosenberg and Daniel McShea and *Sex and Death* by Kim Sterelny and Paul Griffiths. The series *Cambridge Introductions to Philosophy and Biology* includes many useful titles, such as *Genetics and Philosophy: An*

Introduction by Paul Griffiths and Karola Stotz. Concise and biologist-friendly introductions to many of the key themes of philosophy of biology are found in *The Philosophy of Biology: A Companion for Educators* edited by Kostas Kampourakis. A more popular but highly recommended book that weaves together philosophy of science and biology is *Other Minds: The Octopus and the Evolution of Intelligent Life* by Peter Godfrey-Smith.

The specialized literature on explanation, understanding, and knowledge can be difficult to explore for biologists, and it may be hard to know where to start. *Idealization and the Aims of Science* by Angela Potochnik is biologist-friendly not only because of its scope but also because she frequently relies on examples from physiology, ecology, and evolutionary biology. The arguments in James Woodward's *Making Things Happen* will likely appeal to many biologists, and it is an important book that has shaped much of the recent discussion about causal explanation in biology. Philosophical theories of causation meet scientific practice in *Causality* by Phyllis Illari and Federica Russo. A recent award-winning account of the relationship between explanation and understanding is *Understanding Scientific Understanding* by Henk De Regt. The *Nature of Scientific Knowledge* by Kevin McCain covers many more ideas than what we could fit within the pages of the book you are reading now.

The chapters in this book give many good pointers to papers that explore the role of conceptual analysis in biology. For a book-length overview of the role of metaphors in science, we particularly recommend *Metaphors We Live by* by George Lakoff and Mark Johnson. Nancy Nersessian explores how concepts arise and evolve in *Creating Scientific Concepts*. One of the trickiest concepts in biology is undoubtedly that of a mechanism. *In Search of Mechanisms* by Carl Craver and Lindley Darden offers an argument for why mechanism is central to biology and how biologists go on to discover and characterize them. If mechanisms lie at one end of the explanatory agenda, then we find historical events like the extinction of dinosaurs at the other. Adrian Currie's *Bone, Rock and Ruin* is a passionate argument for the scientific value of historical biology.

Many of the topics covered by the authors in the present book are examined in detail from the perspective of cancer research by Anya Plutuynski in *Understanding Cancer*. *Making Sense of Genes* by Kostas Kampourakis explains genetics by making extensive use of philosophical analysis. Biology today is very data intense, and this raises important philosophical issues that are explored in *Data-Centric Biology: A Philosophical Study* by Sabina Leonelli. Heather Douglas discusses the role of values in science in *Science, Policy, and the Value-Free Ideal*. Bernard Rollin's *Science and Ethics* is a very informative and accessible book on bioethics.

Looking to the history of a science can tell us much about how it works in practice. *The Century of the Gene* by Evelyn Fox Keller is an engaging account of classical and molecular genetics, rich in philosophical insight. Ron Amundson's history of evolutionary developmental biology – *The Changing Role of the Embryo in Evolutionary Thought* – illustrates how concepts and idealizations shape entire scientific disciplines. Daniel Sepkoski's *Rereading the Fossil Record: The Growth of Paleobiology as an Evolutionary Discipline* is an excellent illustration of the social and scientific interplay during the establishment of a research program. At the heart of many biological controversies are different views on the organism itself, which is the theme of Jessica Riskin's *The Restless Clock* and Erik Peterson's *The Life Organic*. While it is not about biology, Hasok Chang's *Is Water H₂O?* is a captivating case study of conceptual change and a thought-provoking argument for scientific pluralism.

In addition to these suggestions and the references in individual chapters, many entries to topics in philosophy of science can be found in the Stanford Encyclopedia of Philosophy. Some professional journals devoted to the philosophy of biology are *Biology and Philosophy* and *Philosophy, Theory and Practice in Biology*.

References

Amundsen, R. (2005). *The Changing Role of the Embryo in Evolutionary Thought*. New York: Cambridge University Press.

Barker, G. & Kitcher, P. (2013). *Philosophy of Science: A New Introduction*. New York: Oxford University Press.

Chang, H. (2013). *Is Water H₂O? Evidence, Realism and Pluralism*. Dordrecht: Springer.

Craver, C. F. & Darden, L. (2013). *In Search of Mechanisms: Discoveries across the Life Sciences*. Chicago: University of Chicago Press.

Currie, A. (2018). *Rock, Bone and Ruin: An Optimist's Guide to the Historical Sciences*. Cambridge, MA: MIT Press.

De Regt, H. (2017). *Understanding Scientific Understanding*. New York: Oxford University Press.

Douglas, H. E. (2009). *Science, Policy, and the Value-Free Ideal*. Pittsburgh, PA: University of Pittsburgh Press.

Firenstein, S. (2012). *Ignorance: How It Drives Science*. New York: Oxford University Press.

Godfrey-Smith, P. (2003). *Theory and Reality: An Introduction to the Philosophy of Science*. Chicago: University of Chicago Press.

Godfrey-Smith, P. (2014). *Philosophy of Biology*. Princeton, NJ: Princeton University Press.

Godfrey-Smith, P. (2017). *Other Minds: The Octopus and the Evolution of Intelligent Life*. London: William Collins.

Griffiths, P. & Stotz, K. (2013). *Genetics and Philosophy: An Introduction*. New York: Cambridge University Press.

Illari, P. & Russo, F. (2014). *Causality: Philosophical Theory Meets Scientific Practice*. New York: Oxford University Press.

Kampourakis, K. (2013). *The Philosophy of Biology: A Companion for Educators*. Dordrecht: Springer.

Kampourakis, K. (2017). *Making Sense of Genes*. Cambridge: Cambridge University Press.

Kampourakis, K. & McCain, K. (2019). *Uncertainty: How It Makes Science Advance*. New York: Oxford University Press.

Keller, E. F. (2000). *The Century of the Gene*. Cambridge, MA: Harvard University Press.

Kuhn, T. S. (2012). *The Structure of Scientific Revolutions: 50th Anniversary Edition*. Chicago: University of Chicago Press.

Lakatos, I. (1978). *The Methodology of Scientific Research Programmes: Philosophical Papers Volume 1* (eds. J. Worrall & G. Currie). Cambridge: Cambridge University Press.

Lakoff, G. & Johnson, M. (1980). *Metaphors We Live By*. Chicago: University of Chicago Press.

Leonelli, S. (2016). *Data-Centric Biology: A Philosophical Study*. Chicago: University of Chicago Press.

Lewens, T. (2015). *The Meaning of Science*. Milton Keynes, UK: Penguin Random House.

Nersessian, N. J. (2010). *Creating Scientific Concepts*. Cambridge, MA: MIT Press.

Okasha, S. (2016). *Philosophy of Science: A Very Short Introduction*. New York: Oxford University Press.

Peterson, E. L. (2016). *The Life Organic: The Theoretical Biology Club and the Roots of Epigenetics*. Pittsburgh, PA: University of Pittsburgh Press.

Pigliucci, M. 2010. *Nonsense on Stilts: How to Tell Science from Bunk*. Chicago: University of Chicago Press.

Pigliucci, M. & Boudry, M. (2013). *Philosophy of Pseudoscience*. Chicago: University of Chicago Press.

Plutynski, A. (2018). *Explaining Cancer: Finding Order in Disorder*. New York: Oxford University Press.

Potochnik, A. (2017). *Idealization and the Aims of Science*. Chicago: University of Chicago Press.

Potochnik, A., Colombo, M., & Wright, C. (2019). *Recipes for Science: An Introduction to Scientific Methods and Reasoning*. New York: Routledge.

Riskin, J. (2018). *The Restless Clock: A History of the Centuries-Long Argument over What Makes Living Things Tick*. Chicago: University of Chicago Press.

Rollin, B. (2006). *Science and Ethics*. Cambridge: Cambridge University Press.

Rosenberg, A. & McShea, D. W. (2007). *Philosophy of Biology: A Contemporary Introduction*. New York: Routledge.

Sepkoski, D. (2017). *Rereading the Fossil Record: The Growth of Paleobiology As an Evolutionary Discipline*. Chicago: University of Chicago Press.

Sterelny, K. & Griffiths, P. (1999). *Sex and Death: An Introduction to the Philosophy of Biology*. Chicago: University of Chicago Press.

Woodward, J. (2003). *Making Things Happen: A Theory of Causal Explanation*. New York: Oxford University Press.

Index

abstraction
  in models, 60
  in scientific explanation, 7–8
acquaintance knowledge, 37
acquired traits, 131
adaptationism, 244
adaptive evolution
  certainty and resistance to, 50
  in Lenski's long-term evolution experiment, 66–67
  models for, 55
  plasticity, non-genetic inheritance and niche construction in, 10, 17
  scientific explanations of, 9–11, 31–32
Against Method (Feyerabend), 171–172
Agassiz, Louis, 224–225
aims, scientific, 6–11
allele surfing, 13
alleles
  BRCA, 50
  enzymes and, 88
  gene frequency in, 147
  genes and, 87
  phenotypic traits and, 87–88
alternative splicing, 90
altruism, 65, 74–75
Alzheimer's, 169–170
amyloid-ß, 169–170, 188–189
animals, scientific use of, 260–262
Animals (Scientific Procedures) Act (ASPA), 261
Animals Cruelty Act of 1876, 260–261
antisense strand, 90–91
aristogenesis, 132
Aristotle, 128–129, 174–189
  classification by, 221–222, 224–225
Arkansas, law on teaching Creationism in, 276, 280–282
arms race, evolutionary, 284–285
ARRIVE guidelines, 268–269
artificial selection, 113–116, 125, 128–130
artificial systems of classification, 218–221
asocial learning, 158
ASPA. *See* Animals (Scientific Procedures) Act
asymmetry of overdetermination, 194, 198–199
atheism, 295

audit systems, in good science, 266–268
Augustine (Saint), 275
Avida, 69

Bacon, Francis, 124, 174–189
Baconian empiricism, 124
Baconian method, 175
  Darwin and, 181–183
  dead organisms *in vitro*, 177–178
  intervention and machine analogies, 176
  Mill and, 181
  natural historians and, 175–176
  Whewell and, 180–181
bad science, facts in values in, 256, 262
bias, in science, 2–4
Bible, interpretation of, 275–276
biodiversity, 216–218
biological generalizations, 58–59
biological knowledge, 51. *See also* theoretical knowledge
  certainty of, 49–51
  construction of, 48–49
  creation of, 235
  genus of, 36–39
  IBE in generation of theoretical, 40–48
  kinds of propositional, 39–40
  misunderstandings about, 48–51
biological progress, 288–294
  human evolution and, 284–285
biological science in society
  contemporary examples of, 262–266
  facts and values relation in, 255–256, 270–271
  historical examples of, 256–262
biological species concept, 94–95
biological systems, complexity of, 7–11
biology
  history of, 278–280
  scientific explanations in, 32–33
    causes, patterns, and causal patterns in, 22–25
    from mechanisms to large-scale causes, 25–28
    multiple compatible, 28–32
    nature of, 21–22
    teleology in, 286–287

biology concepts, 97–99
  change and transformation of, 86–91
  content of, 80
  as dynamic entities, 80
  formation of, 124–125
  forward-looking nature of, 81–86
  metaphorical nature of, 102–105, 119
  open-endedness of, 86–87, 98–99
  pluralism and conceptual diversity, 92–97
  scientific knowledge compared with, 79
  terms and, 79–80
biometricians, 131
birds, coloration of, 23–24, 29–31
Bonner, José, 184–185
book of life, genome as, 102–103
Brandon, Robert, 285
BRCA, 50
breast cancer, 50
Buffon, Georges-Louis, 175–176

cancer
  evolutionary models of, 69
  genetic testing and, 50
catachresis, 105–106
causal patterns
  as center of scientific explanations, 22–25
  in mechanisms compared with large-scale
    causes, 25–28
  multiple compatible, 28–32
causal theory of explanation, 6–11
  in biology, 22–25
causation theories, 300–302
causes
  mechanisms compared with large-scale,
    25–28
  in scientific explanation, 22–25
Cech, Tom, 209–210
cell theory, 179–180
Central Dogma of Molecular Biology, 209–210
certainty, of biological knowledge, 49–51
channels, membrane, concepts of, 15–16
Chicxulub meteorite, 201–202, 204–206,
    212–213
cholera, 47–48
Christianity. See also Creationism
  in Darwin's theory of evolution, 279–280
  Progress versus, 291–294
  science accommodation with, 282–285
chromosome theory of inheritance, 87
cichlid fish, 10–11
citrate metabolism, 68
classical gene concept, 88–89, 92–93

classical-balance controversy, 236–237,
    241–244, 248–250
classification
  as aim of science, 6
  by Aristotle, 221–222, 224–225
  biology concepts for, 81, 83
  by Darwin, 225–228
  difficulties with, 231–232
  of diversity, 216–218
  evolutionary theory and, 230–231
  general knowledge and, 220
  of genes, 228–230
  by Linnaeus, 222–225
  natural and artificial systems of, 218–221
  of species, 226–228
climate change
  controversies over, 235–236, 250
  models of, 64
  uncertainty and, 50–51
closeness, of models, 63–64
closure of controversy
  through consensus, 245–246
  through force, 245
  through loss of interest, 245
  through negotiation, 247
  through sound argument, 246–247
collaboration
  in biological knowledge, 48–49
  in philosophy of science, 4
coloration, of birds, 23–24, 29–31
communication, in philosophy of science, 4
comparison view of metaphor, 105–107
competition, 150
completeness, 44
complex systems, models for, 64
complexity
  management of, 7–11
  as natural consequence of evolution,
    285
computation cell lattice models, 62
computational models
  of altruism, 65
  comparisons with, 61–63
  examples of, 57
concepts. See also scientific concepts
  terminology compared with, 14–15
conceptual analysis
  in advancement of science, 14–17
  alternative approach to, 146–147
conceptual clarification
  empirical discoveries with models,
    134–137

statistical turn and Darwinism, 130–134
conceptual diversity
in gene concept, 92–93
in species concept, 93–97
conceptual ecology, 92
conceptual frameworks, 16–17
competing, 137–141
concrete models, 57, 61–62
conservation biology, 96–97
construction of biological knowledge, 48–49
convergent evolution, scientific explanations of, 10–11
cooperation, 65, 74–75
corpuscular philosophy, 178–179
counterfactual theories of causation, 300–302
Creationism, 275–276
Arkansas law on teaching of, 276, 280–282
Darwinism as rival religion to, 291–294
CRISPR-Cas9, 139–140, 188
criticism, 235, 250
crossing-over of chromosomes, 87
cultural accumulation, 149
cultural evolution
emergence of, 151
epigenetic inheritance and, 147–148
gene frequency and, 147
key concepts for, 146–147
philosophy and, 148–149
cultural evolutionary theories
cognitive representations and, 151–152
culture and, 155–156
goals of, 148–149
lactase persistence, 152–155
learning and, 148
local environments and, 159–160
motivations for, 150–152
natural selection and, 149–150
culture, 149, 163–164
cultural evolutionary theories and, 155–156
in Darwin's theory of evolution, 279–280
genes and, 152–155
local environments and, 159–160
localised phenomena and, 160
scepticism of, 160
social learning and, 155–157
cumulative adaptation, 150, 162–163
cumulative cultural change, 163–164
definition of, 161
observational learning and, 160–161
social learning and, 161–163
cumulative cultural evolution (CCE), 161

dairying, 152–155
Darwin, Charles
biological progressionist views of, 289–290
classification by, 225–228
cultural influences on, 279–280
gradualist concept of natural selection, 125–130
IBE and, 42
life science revolution and, 174
LUCA and, 206
natural selection and, 110, 113–117
scientific method and, 180–183
on species, 227–228
statistical population dynamics and, 130–134
Darwinism
conceptual issues in, 130–134
Creationist fight against, 275–276, 280–282
religion accommodation with, 282–285
as rival religion to Creationism, 291–294
Social, 128
Davenport, Charles B., 257
Dawkins, Richard, 117, 138–139, 284–285
Deccan Traps eruptions, 202–203, 205–206
declarative knowledge, 37
deduction, 174–189
deductive-nomological explanations, 22–23
definition, biology concept and, 80, 86–87
DePalma, Robert, 204–205
dependence, in scientific explanation, 22–25
derivations, explanations as, 22–23
Descartes, René, 110, 174–189
description, as aim of science, 6
developmental bias, 79–80
developmental constraint, 79–80, 84
developmental processes, in evolution, 31–32
developmental resources, 140–141
Dewey, John, 170
digital evolving systems, 69
dinosaur extinction
end-Cretaceous mass extinction, 201–206, 211–213
inference of, 44–45
scientific explanation of, 6–7
direct/observational knowledge, 39–40
discovery, psychology of, 124
discovery of biological knowledge, 48–49
dissent, 237–238, 250–251
diversity, classification of, 216–218
DNA, as metaphor, 102–103
Dobzhansky, Theodosius, 59, 239, 241–244, 248–250

Dobzhansky-Muller model, 59
dual-inheritance theory, 152–155

ecological inheritance, 160–163
ecological niches, evolution into pre-existing, 285
ecological species concept, 95–96
education. *See also* teaching
  science, history of American, 276
empirical extrapolation, 70–71
Encyclopedia of DNA Elements (ENCODE) project, 102–103
end-Cretaceous mass extinction, 201–206, 211–213
environment, epigenetic effects of, 262–263
epigenetics
  concept of, 14–15
  facts and values role in, 262–264
  inheritance, 147–148
epistemic aims, 81
epistemic values, 256
epistemological and practical classification, 218–219
Escherichia coli
  Avida comparison with, 69
  in Lenski's long-term evolution experiment, 66–67
  lessons about, 67–69
ethics, in good science, 266–270
eugenics, facts and values role in, 257–260
evolution
  biological progress and, 288–291
  Creationist fight against, 275–276, 280–282
  cultural influences in Darwin's theory of, 279–280
  history of teaching of, 276
  human appearance in, 284–285
  religion, accommodation with, 282–285
  as rival religion to Creationism, 291–294
evolutionary adaptation
  certainty and resistance to, 50
  in Lenski's long-term evolution experiment, 66–67
  models for, 55
  plasticity, non-genetic inheritance and niche construction in, 10, 17
  scientific explanations of, 9–11, 31–32
evolutionary biology, 82, 136
evolutionary developmental biology, novelty and, 83–84
evolutionary novelty, 82–85
evolutionary species concept, 95–96

evolvability, 79–80, 84
experimental science, 13–14
  historical science and, 193–194, 197–199
  robustness and, 210
  scientific method and, 195–197
experiments
  models compared with, 72–75
  types of, 195
explanations. *See also* inference to the best explanation; scientific explanations
  actual sequence of, 25
  as aim of science, 6
  biology concepts for, 81
  causal theory of, 6–11
  completeness of, 44
  description of, 40–41
  goodness of, 43–44
  potential, 42
  theoretical knowledge from, 41–44
explanatory agenda
  of evolvability, 84
  of novelty, 83–84
explanatory considerations, 47
explanatory power, 42–43
explanatory standards, 11
explanatory theories, 83–84
explanatory virtues
  description of, 42–43
  scientific hypotheses and, 47–48
  truth and, 46–48
extinction, 198
  of dinosaurs, 6–7
  end-Cretaceous mass, 201–206, 211–213
extra-genetic inheritance, in adaptive evolution, 10, 17

facts
  in biological science in society
    contemporary examples of, 262–266
    historical examples of, 256–262
  in good science, 266–270
  knowledge compared with, 36
  scientific controversies of, 238–239
  values relation to, 255–256, 270–271
Falk, Raphael, 241–244
false negative, 196–197, 211–213
false positive, 196–197, 211–212
falsification, of scientific hypothesis, 4–5, 7
  methods of, 12–14
falsificationism, 171, 196–197
Fausto-Sterling, Ann, 264–265
Feyerabend, Paul, 171–172, 188–189

final cause, 286–287
Fisher, R. A., 131–133
fitness, natural selection and, 133
focus of metaphor, 106
frame of metaphor, 106
frameworks, conceptual, 16–17
Franklin, Rosalind, 238
function, 79–80
function-of-metaphor (FoM) scheme,
    107–109
  organism-is-machine metaphor and, 111
Fundamentalism, 275–276
  Arkansas law on teaching of Creationism,
    276, 280–282

Galton, Francis, 132–133, 183, 257
gemmules, 182–183
gender assignment, facts and values role in,
    264–266
gene concept, 14–15, 230
  changes in, 79, 86–89
  chromosome theory and, 87
  complications of, 90–91
  conceptual diversity in, 92–93
  conceptual ecology and, 92
  context-dependent, 91–97
  development of, 87
  diverse conceptions of, 89–90
  molecular gene hypothesis, 88
  phenotypic traits and, 88
  RNA splicing and, 90
gene frequency, 147
gene-culture coevolution, 152–155
general knowledge, classification and, 220
generality, in models, 65–66
genes
  classification of, 228–230
  as functionally defined entities, 229–230
  metaphors of, 104–105
  natural systems of classification for, 230
Genesis Flood (Whitcomb and Morris, H.),
    276
genetic drift, 64–65, 133
genetic equilibrium, 132–133
genetic programs, concepts of, 15–16
genetic species concept, 94–95
genetic testing, cancer and, 50
genetics
  concept of, 14–15
  eugenics and, 258–260
  natural selection and, 131
genocentrism, 138–140, 147

genome
  as book of life, 102–105
  genes and, 229–230
glycolysis, scientific explanation of, 6–7, 25
good genes hypothesis, 12–13
good science
  facts in values in, 256, 262
  values, ethics, and knowledge role in,
    266–270
Gray, Asa, 130
Great Devonian Controversy, 246–247
Gulliver's Travels (Swift), 124

Haeckel, Ernst, 132
Hastings Center Project on Scientific
    Controversy, 244–247
Hempel, Carl, 22–23
Herschel, John, 125
heuristic function of metaphors, 107–109
  of natural selection metaphor, 117–118
  of organism-is-machine metaphor, 111
high-grade knowledge, 43
historical science, 13–14, 210–213
  end-Cretaceous mass extinction, 201–206,
    211–213
  examples of, 193–194
  experimental science and, 194
  hypotheses investigated in, 197–199
  information destroying processes and,
    199–200
  LUCA and phylogenetic tree of life, 206–208
  overdetermination and, 198–199
  proxy experiments, 211
  underdetermination and, 198, 200–201
homology concept, 92
Hox gene, 82–83
Hull, David, in development of philosophy of
    biology, 277–278
Human Genome Project, 102–103, 269–270
humans, evolutionary appearance of, 284–285
Huxley, T. H., 127, 260–261
hybrid incompatibility, 59
hybridization, 94–95
hypothetico-deductive method, 170–171,
    179–181
  new, 183–185

IBE. See inference to the best explanation
idealization
  in models, 60, 65–66
  in scientific explanation, 7–10
idealized systems, 59

imitation, 151
in silico observations, 186
in vitro observations, 177–178, 186
in vivo observations, 186
inclusive fitness, 138–139
individual learning, social learning and, 157–158
induction, 126–127, 174–189
　Descartes on, 178–179
　in vitro observations, 177–178
　microscopy and, 177–178
　problem of, 195–196
　vivisection and, 176
　Whewell and Mill on, 180–181
inductive risk, 243–244
inference to the best explanation (IBE)
　clarifications of, 42–44
　Darwin and, 42
　everyday use of, 47–48
　explanatoriness and truth, 46–48
　formalization of, 41–42
　multiple plausible rivals and, 44–46
　theoretical knowledge and, 40–41
　trustworthiness of, 44
　ubiquity of, 41
information
　culture as, 155
　in DNA, 102–103
inheritance, 150
Institute for Religion in an Age of Science (IRAS), 283
intelligent design, 125, 134–135
Intelligent Design Network, 193–194
Intelligent Design Theory, 281–282
interaction view of metaphor, 106
　comparison of, 106–107
　components of, 107–108
interference, 186–187
intersex, facts and values role in, 264–266
intracellular signaling, 103–104
IQ scores, 150–151, 235–236
IRAS. *See* Institute for Religion in an Age of Science

Johannsen, Wilhelm, 229
junk DNA, 138–139
justification, knowledge and, 39
justified true belief account of knowledge, 38–39

kin selection, 138–139
King, J. C., 241–244

Kleiber's law, 58
Klein, Edward Emanuel, 260–261
knock-out experiments, 62–63
knowledge. *See also* biological knowledge; theoretical knowledge
　facts compared with, 36
　in good science, 266–270
　grades of, 43
　justified true belief of, 38–39
　kinds of, 36–39
　methods of generating, 4–5
knowledge how, 37–39
Kuhn, Thomas, 4–5, 123–124, 245, 278–279

laboratory animals, facts and values role in treatment of, 260–262
lactase persistence, 152–155
language, facts and values expressed in, 255
large-scale causes, in scientific explanation, 25–28
last universal common ancestor (LUCA), 206–208
Latin America, eugenics in, 257–258
law of segregation, 58–59
laws, scientific theories and, 58–59
Lawson, Anton, 184–185
learning. *See also* social learning
　cultural evolution and, 148
　inheritable variation in, 151
　observational, 160–161
Lenski, Richard, 66–67
Lenski's long-term evolution experiment
　lessons from, 67–69
　overview of, 66–67
Lewontin, Richard, 248–250
limbs of tetrapods, 82–83
Linnaean System, 219
Linnaeus, Carl, 175–176
　classification by, 222–225
literalism, in Bible interpretation, 275–276
living fossils, 85–86
logical empiricism, teleological nature of biology and, 286–287
Lotka-Volterra equation, 57, 63
low-grade knowledge, 43
LUCA. *See* last universal common ancestor
Lyell, Charles, 126, 128
Lysenko, T. D., 245

machine analogies, 176
malaria, 94
manipulability theories of causation, 300–302

maps, models as, 61
material models
  comparisons with, 61–63
  examples of, 57
  model organisms as, 70
mathematical models
  comparisons with, 61–63
  examples of, 57
  track record of, 55–56
Mayr, Ernst, 94, 134
McShea, Daniel, 285
mechanisms, in scientific explanation, 25–28
mechanistic determinism, teleology and, 134
mechanist-vitalist debates, 176
Mendel's laws, 58–59
metaphors
  Aristotle on, 128–129
  biology concepts as, 102–105, 119
  comparison view of, 105–107
  description of, 103
  functions of, 107–109
  of genes, 104–105
  interaction view of, 105–110
  natural selection, 113–119, 128–129
  organism-is-machine, 110–111
  problems with, 104
  scientific concepts as, 15–16, 102–103, 119
  scientific theories and, 125–130
  signal and receptor proteins, 103–104
  substitution view of, 105–106
metaphysical classification, 218–219
meteorite hypothesis, 44–45
methodological naturalism, 224
methods. See scientific methods
Mill, John Stuart, 125, 180–181
mixed controversies, 235–236
model organisms, 57, 75
  comparisons of, 69
  E. coli as, 67–69
  experiments compared with, 72–75
  as models, 69–71
  uses of, 62–63
models, 75
  as constructs, 137
  experiments compared with, 72–75
  as fictions, 61
  functions of, 55
  as maps, 61
  model organisms as, 69–71
  science meets world through, 63–66, 137
  scientific theories and, 56–59
  similarity of, 63–64

  target system of, 60, 74
  types of, 57, 61–63
  views of, 55–56
Modern Evolutionary Synthesis, 134–138
Moir, Robert, 169–170, 188–189
molecular gene concept, 88–90, 92–93, 137–138
Money, John, 265
Morgan, Thomas Hunt, 184
morphological species concept, 94–95
Morris, Henry, 238, 276
Morris, Simon Conway, 285
mRNA, 90
Muller, H. J., 59, 239, 241–244
Muller's ratchet, 57, 59
multicellularity, 55, 63–64
multiple plausible rivals, 44–46
multiple scientific explanations, compatibility
  of, 28–32
multiverses, 285
mutational accumulation, 59
mutations
  natural selection and, 131, 137–138
  rates of, 67–68, 73–74
mutator hitchhiking, 67–68

natural factors, 220–221
natural kinds, 220–221
natural selection
  in adaptive evolution, 9–11, 17, 31–32
  altruism and, 65
  artificial selection and, 115–116, 128–130
  cultural evolutionary theories and, 149–150
  Darwin on, 113–115
  fitness and, 133
  genetic drift and, 133
  genetics and, 131
  Herschel on, 125–126
  as metaphor, 110, 113–119
  Mill on, 125
  regression and, 132–133
  species concept and, 134
  Spencer on, 127–128
  survival of the fittest and, 115–117, 127
  system dynamics and, 131–133
  Whewell on, 126–127
natural systems of classification, 218–221
  Aristotle and, 222
  assumption of, 224–225, 231–232
  Darwin on, 225–226
  for genes, 230
  Linnaeus, 222–224
  search for, 232

natural world
  biology concepts and, 81
  model organisms and, 74
  model relationship with, 63–66
  scientific theories and, 56
negative eugenics, 257
neo-Darwinian Modern Synthesis, 84
Newton, 129–130, 178–179
Newton's laws, 136–137
niche construction, in adaptive evolution, 10, 17
non-genetic inheritance, in adaptive evolution, 10, 17
novel trait emergence, 68
novelty, 82–85
nucleic acids, LUCA and, 206–207, 209–210
nudging, epigenetics and, 263–264

observational knowledge, 39–40
observational learning, 160–161
open science, 256, 269–270
organism-is-machine metaphor, 111–113
  FoM scheme and, 111
  instantiations of, 112
  natural selection and, 110
  similarities and differences in, 110–111
  SoM scheme and, 112–113
  unwanted consequences of, 112
The Origin of the Species (Darwin), 42, 113–117, 181–183
  biological progressionist views in, 289–290
  on classification, 225–226
  drama in, 125
  Herschel on, 125–126
  metaphor in, 128
  Mill on, 125
  tree-like structure in, 226
  Whewell on, 126–127
orthogenesis, 132
overdetermination, 198–199, 208
overlapping genes, 90–91

Paley, William, 279–280
paradigm shifts, 4–5, 123
patterns, in scientific explanation, 22–25
Periodic System, 219
phenomena, 131–132, 136
phenotypic function, 87
phenotypic traits
  alleles and, 87–88
  chromosome theory and, 87
  genes and, 88–90

genocentrism and, 139–140
The Philosophy of Biological Science (Hull), 278
philosophy of biology
  definition of, 3
  history of, 277–278
The Philosophy of Biology (Ruse), 278
philosophy of science
  communication and collaboration in, 4
  history of, 277–278
  questions asked in, 2–4
  reasons for engaging in, 1–2, 17–18
    to appreciate diversity of scientific standards, 4–5
    to study science, 2–4
    to understand scientific aims, 6–11
    to understand scientific concepts, 14–17
    to understand scientific methods, 12–14
  recommendations for teaching of, 299–300, 304, 309–310
    begin with simple decontextualed examples, 304–306
    discuss philosophy of science topics in standard biology classes, 308–309
    focus on overarching ideas rather than extreme detail, 300–302
    present topics in detail with many contextualized examples, 306–308
    use history but not chronologically, 303–304
    use simple rather than complex definitions of concepts, 302–303
phylogenetic species concept, 94–96
phylogenetic systematics, 231
phylogenetic tree of life, 206–208
physical models, 57, 61–62
physiologists, 176
Planck effect, 245
plasticity
  in adaptive evolution, 10–11, 17
  early development and, 127–128
  IQ scores and, 150–151
pluralism
  of conceptual analysis, 146–147
  of species concept, 96–97
  political controversies and, 235–236
Popper, Karl, 4–5, 171, 196–197
population genetic models, 131–132, 134–135, 137–138
population genetic theory, 13, 60
  gene frequency and, 147
  genetic changes in, 147–148

positive eugenics, 257
postgenomic era, 91–97
pragmatic views of scientific theories, 56–57
precision, in models, 65–66
predator-prey dynamics, 60
prediction, as aim of science, 6
predictive power, 42–43
pre-mRNA, 90
Price equation, 57, 62
principles
    falsification of, 13
    scientific controversies of, 239–240
Principles of Biology (Spencer), 127–128
probabilistic theories of causation, 300–302
problem of induction, 195–196
procedural knowledge, 37–39
progress, biological, 288–294
    human evolution and, 284–285
Progress, Providence versus, 291–294
propositional knowledge, 37–40
protein hypothesis, 44–45
protein synthesis, 102–103
proteins, LUCA and, 206–207
Providence, Progress versus, 291–294
proximate explanations, 9, 29–30
    in adaptive evolution, 9–11, 31–32
proxy experiments, 211
pseudohitchhiking model, 64–65
public policy, epigenetics in, 263–264
punctuated evolution, 68, 73–74
Putnam, Hilary, 256
p-values, 12

race, IQ and, 235–236
radiation, 241–244, 248–250
radical theory change, 4–5
realism, in models, 65–66
regression to the mean, 25
regularity theories of causation, 300–302
relative significance controversies,
    248–250
relevant points in common, 63–64
religion. *See also* Creationism
    in Darwin's theory of evolution, 279–280
    Progress versus, 291–294
    science accommodation with, 282–285
reproducibility, in good science, 268
reproductive advantage, 133
reproductive isolation, 94
research agenda
    multiple scientific explanations and, 31–32
    scientific concept role in, 15

rhetorical function of metaphors, 107–109
    of natural selection metaphor, 118–119
    of organism-is-machine metaphor, 111
ribozyme discovery, 208–210
RNA editing, 90–91, 102–103
RNA splicing, 90, 209–210
RNA synthesis, 102–103
RNA world hypothesis, 44–47
robust process explanations, 25
robustness testing, 210
Ruse, Michael
    Arkansas testimony of, 280–282
    in development of philosophy of biology,
        277–278
    The Philosophy of Biology by, 278

saltation, 127
scaling relationships, 58
science. *See also* biological science in society
    conceptual analysis in advancement of,
        14–17
    facts in values in good compared with bad,
        256, 262
    falsification and, 171
    history of, 278–280
    history of American education in, 276
    nature of values in, 288–291
    new pedagogy of, 141–142
    philosophy of science as study of, 2–4
    practice of, 97–99
    religion accommodation with, 282–285
    values, ethics, and knowledge role in good,
        266–270
scientific aims, 6–11
scientific concepts, 14–17. *See also* biology
        concepts
    expression of, 79–80
    as metaphors, 15–16, 102–103, 119
    in research agendas, 15
scientific controversies, 250–251
    adaptationism, 244
    beginning of, 237–240
    classical-balance controversy, 236–237,
        241–244, 248–250
    closure of, 244–247
    community and, 240
    continuation of, 240–244
    development of, 235
    of fact, 238–239
    life spans of, 236
    political and social controversies and,
        235–236

scientific controversies (cont.)
of principle, 239–240
relative significance, 248–250
shared standards and, 240–241
of theory, 239
types of, 235
scientific creationists, 193–194
scientific explanations
in biology, 32–33
causes, patterns, and causal patterns in,
22–25
from mechanisms to large-scale causes,
25–28
multiple compatible, 28–32
nature of, 21–22
reasons for applying philosophy of science
to, 6–11
scientific hypothesis
Baconian method and, 175–176
experimentation and, 195–197
explanatory virtues and, 46–48
falsification of, 4–5, 7
methods of, 12–14
in historical science, 197–199
scientific knowledge
biology concepts compared with, 79
methods of generating, 4–5
scientific methods, 12–14. *See also* Baconian
method
alternatives, 172–173
categorization of, 185–187
Darwin and, 180–183
experimental science and, 195–197
Feyerabend's anarchy, 171–172
five-step, 170–171
new hypothetico-deductive methods and,
183–185
new theoretical entities and, 183–185
number of, 168, 173
for theoretical entities, 178–180
tool use and, 173–174
scientific problem agenda, of evolutionary
novelty, 83–85
scientific purpose, of biology concepts, 81
scientific standards, diversity of, 4–5
scientific theories
concepts in, 79
laws and, 58–59
metaphors and, 125–130
models and, 56–59
natural world and, 56
representation of, 57
views of, 56–57

scope, in scientific explanation, 22–25
Scopes, John Thomas, 276
selfish gene hypothesis, 138–140
semantic knowledge, 37–39
semantic views of scientific theories, 56–57
Semenya, Caster, 264–265
Semmelweis, Ignaz, 47–48
sex assignment, facts and values role in,
264–266
similarity, of models, 63–64
simplicity, 42–43, 47
Snow, John, 47–48
Sober, Elliott, 131–132
social aspects, of science, 2–4
social constructionism, 124
social controversies and, 235–236
Social Darwinism, 128
social determinants
epigenetics and, 263–264
eugenics and, 258
social learning, 163–164
asocial and, 158
culture and, 155–156
cumulative culture and, 161–163
definition of, 158
individual learning and, 157–158
localised phenomena and, 160
terminology for, 156–157
society, biological science in
contemporary examples of, 262–266
facts and values relation in, 255–256, 270–271
historical examples of, 256–262
soft inheritance, 131
SoM scheme. *See* structure-of-metaphor
scheme
speciation
biology concept and, 86–87
idealized system of, 59
models of, 60
species
classification of, 218
Darwin on, 227–228
groups of, 226–228
natural selection and, 134
species concept, 79, 228
biological, 94–95
conceptual diversity of, 93–94, 97
ecological, 95–96
evolutionary, 95–96
genetic, 94–95
morphological, 94–95
phylogenetic, 94–96
pluralism of, 96–99

Spencer, Herbert, 116, 127–128
Standard Model, 219
standards
  explanatory, 11
  scientific, 4–5
statistical patterns
  phenomena and, 136
  in scientific explanation, 25
statistical population dynamics, 130–134
sterilisation, 257–260
structural causes, in scientific explanation, 25–28
The Structure of Scientific Revolutions (Kuhn), 123–124, 278–279
structure-of-metaphor (SoM) scheme, 107–108
  natural selection metaphor and, 117–119
  organism-is-machine metaphor and, 112–113
substitution view of metaphor, 105–106
survival of the fittest, 115–117, 127–128
Swift, Jonathan, 124
synapomorphy, 95–96
syntactic views of scientific theories, 56–57

Tanzi, Rudolph, 169–170, 188–189
target system, of models, 60, 74
taxa, classification of, 218
taxonomic splitting, 97
teaching, 151
  recommendations for philosophy of science, 299–300, 304, 309–310
    begin with simple decontextualized examples, 304–306
    discuss philosophy of science topics in standard biology classes, 308–309
    focus on overarching ideas rather than extreme detail, 300–302
    present topics in detail with many contextualized examples, 306–308
    use history but not chronologically, 303–304
    use simple rather than complex definitions of concepts, 302–303
teleology
  in biology, 286–287
  mechanistic determinism and, 134
terminology, concepts compared with, 14–15
tetrapod limbs, 82–83
Thaler, Richard, 263–264
theoretical entities
  genes as, 229
  new, 183–185
  scientific methods for, 178–180

theoretical function of metaphors, 107–109
  of natural selection metaphor, 118
  of organism-is-machine metaphor, 111
theoretical knowledge
  development of, 40
  from explanations, 41–44
  IBE and, 40–41
theoretical models, model organisms as, 70–71
theory, controversies of, 239
time asymmetry of causation, 194, 198–199
traditional account of knowledge, 38–39
traits
  acquired, 131
  function of, 79–80
  rates of change of, 85–86
  scientific explanation of, 25, 29–31
transcription, 102–103
translation, 102–103
transparency, in good science, 268–270
trans-splicing, 90
trauma, epigenetic effects of, 262–263
trustworthiness, 44
truth
  explanatory virtues and, 46–48
  knowledge and, 39

ultimate explanations, 9, 29–30
uncertainties, in biological knowledge, 50–51
unconscious selection, 115
underdetermination, 198, 200–201, 208–209
understanding, as aim of science, 6
unification, as function of scientific explanation, 22–24
"units of selection" debates, 42–43
universal generalizations, 58–59

vaccination
  controversies over, 235–236
  uncertainty and, 50–51
values
  in biological science in society
    contemporary examples of, 262–266
    historical examples of, 256–262
  facts relation to, 255–256, 270–271
  in good science, 266–270
  nature of, in science, 288–291
variation, 150
Virchow, Rudolf, 179–180
virtues, explanatory, 42–43

vivisection, 176, 260–262
Voltaire's Objection, 46–47

Wallace, A. R., 127
Wallace, Bruce, 241–244
Watson, James, 238

Weismann, August, 131
Whewell, William, 125, 180–181
Whitcomb, John C., 276
Wright, Sewall, 133, 147

zebra fish, 261

Printed in the United States
by Baker & Taylor Publisher Services

Printed in the United States
by Baker & Taylor Publisher Services